THE SEA IN NORDIC LITERATURE

The first collection to address the role of the oceanic in literary texts and traditions from across the Nordic region, this book explores applications of ecocriticism, postcolonial theory, new materialism, and gender studies at a time when interest in marine ecosystems and their vulnerabilities to anthropogenic climate change is generating new theoretical insights and interest across the humanities.

The Nordic countries with their long coastlines and archipelagic geographies have long histories of seafaring and mariculture. This book simultaneously introduces Nordic literatures about the sea to international scholars with an interest in the oceanic and blue humanities to scholars working within Scandinavian/Nordic Studies. Its historical sweep incorporates the work of Nordic writers of particular significance for European and world literature, such as Henrik Ibsen and Tove Jansson; it also explores texts known chiefly within the Northern countries, including contemporary literature engaging with the sea; and historical texts, such as the Icelandic sagas and Early Modern seafaring guides. The book takes an innovative geographical orientation, grouping chapters to correspond to the various pelagic ecosystems that surround the Scandinavian Peninsula and connect it to the rest of the globe.

The combination of canonical and lesser-known texts with key theoretical approaches makes the volume ideal for teaching. With its broad historical and theoretical scope, *The Sea in Nordic Literature* is a valuable resource for scholars and students in blue humanities, Scandinavian and Nordic studies, and comparative literature.

Frederike Felcht is Professor of Scandinavian Literature and Culture at Goethe University Frankfurt. Her current research focuses on literature and the environment and on poverty in Scandinavian literature.

Søren Frank is Professor of Nordic Literature at the Department of Nordic Studies and Linguistics at the University of Copenhagen. His research focuses on ocean and coastal literatures, blue humanities, and the Anthropocene.

Katie Ritson is Assistant Professor in the Scandinavian Department at the University of California, Berkeley. Her research focuses on literature and environment in the context of northern Europe, in particular on the North Sea and its coastlines.

Routledge Environmental Humanities

Series editors: Scott Slovic (*University of Idaho, USA*), Joni Adamson (*Arizona State University, USA*) and Yuki Masami (*Aoyama Gakuin University, Japan*)

The *Routledge Environmental Humanities* series is an original and inspiring venture recognising that today's world agricultural and water crises, ocean pollution and resource depletion, global warming from greenhouse gases, urban sprawl, overpopulation, food insecurity and environmental justice are all *crises of culture*.

The reality of understanding and finding adaptive solutions to our present and future environmental challenges has shifted the epicenter of environmental studies away from an exclusively scientific and technological framework to one that depends on the human-focused disciplines and ideas of the humanities and allied social sciences.

We thus welcome book proposals from all humanities and social sciences disciplines for an inclusive and interdisciplinary series. We favour manuscripts aimed at an international readership and written in a lively and accessible style. The readership comprises scholars and students from the humanities and social sciences and thoughtful readers concerned about the human dimensions of environmental change.

For more information about this series, please visit: www.routledge.com/Routledge-Environmental-Humanities/book-series/REH

THE SEA IN NORDIC LITERATURE

Oceanic Criticism and Blue Humanities

Edited by Frederike Felcht, Søren Frank and Katie Ritson

Routledge
Taylor & Francis Group
LONDON AND NEW YORK

earthscan
from Routledge

Designed cover image: Anna Boberg: Among the Fishing Boats in Svolvaer. Study from Lofoten, 1934,
Nationalmuseum (Photographer: Linn Ahlgren)

First published 2027
by Routledge
4 Park Square, Milton Park, Abingdon, Oxon OX14 4RN

and by Routledge
605 Third Avenue, New York, NY 10158

Routledge is an imprint of the Taylor & Francis Group, an informa business

British Library Cataloguing-in-Publication Data
A catalogue record for this book is available from the British Library

Library of Congress Cataloging-in-Publication Data
Names: Felcht, Frederike, 1982– editor | Frank, Søren editor | Ritson,Katie editor
Title: The sea in Nordic literature : oceanic criticism and blue humanities / edited by Frederike Felcht, Søren Frank, and Katie Ritson.
Description: London ; New York, NY : Routledge, 2026. | Series: Routledge environmental humanities | Includes bibliographical references and index.
Identifiers: LCCN 2026017463 (print) | LCCN 2026017464 (ebook) | ISBN 9781041089414 hardback | ISBN 9781041089391 paperback | ISBN 9781003647638 ebook
Subjects: LCSH: Scandinavian literature—History and criticism | Sea inliterature | Ecocriticism | LCGFT: Literary criticism
Classification: LCC PT7048 .S43 2026 (print) | LCC PT7048 (ebook) | DDC839/.5—dc23/eng/20260407
LC record available at https://lccn.loc.gov/2026017463
LC ebook record available at https://lccn.loc.gov/2026017464

ISBN: 978-1-041-08941-4 (hbk)
ISBN: 978-1-041-08939-1 (pbk)
ISBN: 978-1-003-64763-8 (ebk)

DOI: 10.4324/9781003647638

Typeset in Times New Roman
by Apex CoVantage, LLC

The authors are grateful to the Carlsberg Foundation, the German Research Foundation, and the Centre for Modern European Studies (CEMES) at the University of Copenhagen for their support for this project.

CONTENTS

List of Contributors *xii*

Introduction: Blue humanities into the North, Nordic sea literature into the world 1
Frederike Felcht, Søren Frank, and Katie Ritson

PART I
The Baltic Sea **17**

1 The sea as an empty space in early modern literature: Johan Månsson's *Sea-book about Seafaring on the Baltic Sea* 19
Joachim Grage

2 Gulf of Finland temporalities: Petroculture, geopolitics, and Tove Jansson's resonant island 33
Pia Ahlbäck and Mia Österlund

3 Obstructed passages: Archipelagic thinking and planetary reach in Tomas Tranströmer's *Östersjöar* 45
Dan Ringgaard

4 Writing the deluge: Drachmann and the 1872 Baltic Sea flood 57
Søren Frank

PART II
The Wadden Sea, the North Sea, and the Norwegian Sea **71**

5 Coastal writing: The Wadden Sea in the works of Steen Steensen Blicher, Peter Seeberg, and Dorthe Nors 73
Anders Ehlers Dam

6 The industrialised North Sea: Introducing tough oil fiction 86
Karl Emil Rosenbæk Reetz

7 Rimbereid's waterscapes: The North Sea viewed from Stavanger 97
Brandon Kaaz

8 Longing for what's lost: Evolution, ecology, and multispecies entanglements in Henrik Ibsen's *The Lady from the Sea* 113
Frederike Felcht

9 Mapping seascapes with Olav Duun's novel *Floodtide of Fate* 126
Sissel Furuseth

PART III
The Iceland Sea, the Greenland Sea, and the Arctic Sea **137**

10 "Narrating the sea": The journey to Iceland in travel narratives of the modern period 139
Alessia Bauer

11 Plenty and peril: The sea in the sagas of Icelanders 152
Annette Lassen

12 Investigating the Arctic ecotone: The Arctic Ocean in Peter Høeg's *Miss Smilla's Feeling for Snow* 165
Hanna Eglinger

PART IV
The Atlantic Ocean **177**

13 The "troll whale" demystified in the North Atlantic in the early modern period: From Olaus Magnus' map to the witchcraft trials 179
Ronny Spaans and Liv Helene Willumsen

14 Sexuality at sea: The challenge of entanglement in Amalie Skram's *Betrayed* (1892) 192
Christine Hamm

15 The Atlantification of Faroese literature: An archipelagic reading of Jørgen-Frantz Jacobsen's *Barbara* 203
Bergur Rønne Moberg

16 Atlantic crossings, departures, and returns in Marjam Idriss' *Halvt* 216
Lill-Ann Körber and Hanna Rinderle

PART V
The Global Sea and the Undersea **229**

17 "All the world was blue": The sea in Isak Dinesen's stories 231
Annegret Heitmann

18 How to read a wave: August Strindberg in the surf 243
Sylvain Briens

19 "A sea without direction": The deep in contemporary Icelandic literature 257
Auður Aðalsteinsdóttir

20 Risen from the foam: Hans Christian Andersen's fairy tales as aquatic texts 269
Klaus Müller-Wille

Index *282*

CONTRIBUTORS

Pia Ahlbäck is University Lecturer and Adjunct Professor in Comparative Literature, particularly Literature and the Environment at Åbo Akademi University. Her research interests include the environmental humanities, petrocultures, ecocriticism, utopias and dystopias, and the theory of history.

Auður Aðalsteinsdóttir is Research Assistant Professor and Director of the Research Centre of Environmental Humanities of the University of Iceland in Þingeyjarsveit. She has a research focus on literary history, literary criticism, and aesthetics, as well as modern literature and environmental humanities.

Alessia Bauer is Professor of Scandinavian Studies and Chair of Scandinavian and Runological Studies at the École Pratique des Hautes Études in Paris. She is a leading runologist and specialises in Old Norse literature and edition philology.

Sylvain Briens is Professor of Nordic Literature at Sorbonne University in Paris. His research focuses on modern and contemporary Scandinavian literature, with a particular emphasis on the blue humanities in the Nordic and Arctic cultural spheres.

Hanna Eglinger is Professor of Comparative Literature with a focus on Northern European Literature & Scandinavian Studies at Friedrich-Alexander University of Erlangen–Nuremberg. Her research interests include Scandinavian literature and culture of the nineteenth, twentieth, and twenty-first centuries, body theory, intermedial relations, and Arctic literature.

Anders Ehlers Dam is Professor of Danish Literature and Culture at the Department of Danish at Europa-Universität Flensburg. His research focuses on literary history, literary theory, interart studies, and cultural studies.

Frederike Felcht is Professor of Scandinavian Literature and Culture at Goethe University Frankfurt. Her current research focuses on literature and the environment and on poverty in Scandinavian literature.

Søren Frank is Professor of Nordic Literature at the Department of Nordic Studies and Linguistics at the University of Copenhagen. His research focuses on ocean and coastal literatures, blue humanities, and the Anthropocene.

Sissel Furuseth is Professor of Nordic Literature at the Department of Linguistics and Scandinavian Studies at the University of Oslo. Her research focus lies within literary criticism and periodical studies, ecocriticism and environmental literature, as well as versification and poetry studies in literature from the eighteenth century to the present.

Joachim Grage is Professor of North Germanic Philology at the Scandinavian Seminar at the University of Freiburg. His research focus is Scandinavian literature from the Baroque to the present, as well as German-Scandinavian literary relations.

Christine Hamm is Professor of Scandinavian Literature at the Department of Linguistic, Literary and Aesthetic Studies at the University of Bergen. Her current research interests are sexuality, motherhood, and the role of class in Scandinavian literature, ranging from the nineteenth century to contemporary works.

Annegret Heitmann is Professor Emerita of Scandinavian Studies at the Institute for Nordic Philology at Ludwig-Maximilian University Munich where she held the chair until 2018. Her research interests include Scandinavian literature from the nineteenth and twentieth centuries, with a focus on gender studies, female authorship, intermediality, and popular culture.

Lill-Ann Körber is Professor of Modern Nordic Literatures and Cultures and Head of the Department of Scandinavian Studies at the University of Göttingen. Her current research focuses on Nordic colonialism, its continued impact, and its literary and cultural representations.

Brandon Kaaz is Assistant Researcher at the Institute of Scandinavian Studies at Goethe University Frankfurt. His research interest is on twentieth- and twenty-first-century Scandinavian poetry with a focus on the blue humanities, environmental humanities, Anthropocene studies, and psychoanalysis.

Annette Lassen is Professor of Icelandic Literature and Language at the Department of Nordic Studies and Linguistics at the University of Copenhagen. Her research focuses on Old Norse/Icelandic literature (sagas and Eddic poems) and culture, as well as Norse mythology, and the reception of Old Norse/Icelandic literature in Danish culture.

Bergur Rønne Moberg is Associate Professor at the Department of Nordic Studies and Linguistics at the University of Copenhagen. His research interests are in (ultraminor) world literatures, Island Studies, literary anthropology, literary geographies, narrative medicine, and culturally based archival studies.

Klaus Müller-Wille is Professor of Nordic Philology, Department of Scandinavian Studies at the University of Zurich. His research focuses include Scandinavian Romanticism, avant-gardes and neo-avant-gardes, as well as theory of material culture, edition philology, and book history.

Hanna Rinderle is Assistant Researcher in the Department of Modern Scandinavian Literatures at Humboldt University of Berlin. Her research focus is on intercultural and postcolonial literary studies, as well as migration, and post-migration literature in Swedish and German literature from the nineteenth century onwards.

Dan Ringgaard is Professor of Scandinavian Studies at the School of Communication and Culture at Aarhus University. His research interests include poetry and poetics, Nordic literary history, creative writing, and place and environment.

Katie Ritson is Assistant Professor in the Scandinavian Department at the University of California, Berkeley. Her research focuses on literature and environment in the context of northern Europe, in particular on the North Sea and its coastlines.

Karl Emil Rosenbæk Reetz is Postdoctoral Researcher at the University of Copenhagen. His work is generously funded by the Carlsberg Foundation (CF25–0498). His research interests include petroculture and petrofiction, energy humanities and energy fiction, blue humanities, coastal literature, Nordic literature, ecocriticism, and environmental humanities.

Ronny Spaans is Professor of Nordic Literature at the Department of Linguistics and Scandinavian Studies at the University of Oslo. His research focuses on literary history, literary historiography, and early modern literature, as well as poetry, new Norse literary history, and Norwegian-Dutch relations.

Liv Helene Willumsen is Professor Emerita of History at the Department of Archaeology, History, Religious Studies, and Theology at the University of Tromsø. She has a background in both history and literary studies with a focus on women's history, particularly witchcraft trials in Northern Norway and Scotland.

Mia Österlund is Professor of Comparative Literature at the Department of Culture, History, and Philosophy at Åbo Akademi University. Her research focuses on literature in relation to body and environment studies, queer and girlhood studies.

INTRODUCTION

Blue humanities into the North, Nordic sea literature into the world

Frederike Felcht, Søren Frank, and Katie Ritson

Sällsamma fiskar glida i djupen, okända blommor lysa på stranden; jag har sett rött och gult och alla andra färger,— men det granna, granna havet är farligast att se, det gör en törstig och vaken för väntande äventyr: vad som har hänt i sagan, skall hända även mig! *—Edith Södergran, "Det underliga havet", 1916*	Strange fishes glide in the depths, unfamiliar flowers glow on the shore; I have seen red and yellow and all the other colours,— but the gaudy gay sea is the most dangerous to look upon, it makes one thirsty and wide-awake for waiting adventures: what happened in the fairy-tale will happen also to me! *—Edith Södergran, "Strange Sea", 1916*

Edith Södergran's poem "Strange Sea" opens with a vision of strangeness and wonder. Peculiar fish glide through the depths; unknown flowers glow along the shore. The speaker has seen red and yellow and "every other colour", but it is "the dazzling, dazzling sea" that proves most arresting. Beautiful and dangerous, the sea arouses thirst, awakens the senses, and stirs a longing for adventure: "what happened in the fairy-tale [*sagan*] will happen also to me!" Sea and saga become inseparable—literature and oceanic life intertwined.

Södergran, a leading figure in Finland–Swedish modernism, spent most of her short life in Raivola on the Karelian Isthmus, with the Gulf of Finland as her nearest sea. It is from this liminal borderland between East and West, forest and coast, that her poetry draws its vivid, often fantastical imagery—a sea both real and imagined, close and mythic. Her work thus becomes a fitting point of departure for this volume: a Nordic imagination of the sea that is rooted in geography but oriented towards poetic transformation.

DOI: 10.4324/9781003647638-1

Södergran's poem, too, hints at the sea's shifting palette: red, yellow, and beyond. Water is never simply blue. As we explore later in this introduction, blue is a starting point, not an end point—a chromatic entry into planetary thinking and the sea's elemental primacy, but also a surface from which other hues emerge. To attend to the sea's colours is to attend to its material and metaphorical variability.

Our book sets sail under the influence of Södergran's poetic vision. Like the speaker in "Strange Sea", we are drawn to the sea's vivid enchantment—its strange creatures, luminous shores, and chromatic seductions. The sea dazzles not only with its physical richness but with its narrative promise: what has happened in the tale, *sagan*, the poem suggests, shall happen again. This merging of sea and story sets the tone for our inquiry. Yet we proceed with caution. As the American literary scholar Elizabeth DeLoughrey reminds us, the ocean's romanticisation as a site of adventure has historically served to justify extractive violence (2022, 150). Rather than approach "the sea" as a singular, abstract expanse, we focus on the literary representation of specific bodies of water, each with its own histories, ecologies, and cultural meanings.

These multiple registers—mythic, aesthetic enchantment, and historical and material critique—call for a relational framework that neither romanticises nor abstracts the sea. This book, accordingly, charts a mode of *sympoiesis*, a "making-with", in American feminist theorist Donna Haraway's terms (2016, 58): a mutual shaping of sea and story. Literature not only reflects the ocean's force and form but helps constitute how we know it and dwell with it. In tracing these entanglements, our volume explores how Nordic literature has responded to and been transformed by the mutable surfaces and depths of its surrounding seas.

Sympoiesis, in fact, is not only a theme of this volume—it is integral to its very formation. A workshop in April 2025 at the University of Copenhagen, supported by the Carlsberg Foundation and the Centre for Modern European Studies, brought together the contributors in a format designed to foster genuine collaboration: instead of presentations, we circulated draft chapters in advance, and across three intensive days, each text was discussed collectively, allowing for detailed, generative feedback at a formative stage of the writing process. These conversations also fostered shared reflection on themes such as methodology, genre, periodisation, and colour, forging the sense of collective inquiry and common purpose that runs through the book.[1]

The Sea in Nordic Literature has a dual ambition: firstly, to introduce the conceptual and methodological frameworks of blue humanities and critical ocean studies to scholars and students of Nordic literature; and secondly, to present the depth and variety of Nordic sea literature to an international audience engaged with oceanic paradigms.

Much of the pioneering work in blue humanities has focused on the Pacific, the Atlantic, or the Mediterranean, emphasising their histories of colonialism, migration, and cultural exchange. By contrast, this volume brings into view the lesser-known literary seascapes of the Nordic region. A specifically Nordic blue humanities, we suggest, adds new dimensions to the field: it centres colder and more liminal waters; explores the poetics of "nordicité" (Hamelin 1968); and traces the distinctive interweaving of genres such as the sagas of Icelanders and Scandinavian social realism with the *longue durée* of environmental change. The Nordic countries—with their long coastlines, histories of seafaring and coastal subsistence, and cultural imaginaries shaped by the proximity of the sea—have long "told the sea" in literature. From medieval sagas of seafaring to early nautical writings by Jens Munk, Johan Månsson, and Erik Valkendorf, and onward to the maritime romanticism, realism, and modernism of writers such as Hans Christian Andersen, Amalie Skram, August Strindberg, and Olav Duun, as well as the mythopoetic narratives of Isak Dinesen, the Atlantic poetics of Jørgen-Frantz Jacobsen, and the island-centred storytelling of Tove Jansson, Nordic literature has long turned to the ocean as both setting and agent, danger and sustenance, and metaphor and material. In recent decades, writers such as Øyvind Rimbereid, Tomas Tranströmer, Magnus Dahlstrøm Johnsgaard, and Gerður Kristný have continued this tradition—rendering coastal and oceanic environments as sites of memory, loss, extraction, and ecological entanglement, attuned to both cultural heritage and contemporary crises. This book seeks to bring this maritime tradition to the surface and into global conversation: to situate Nordic sea literature within transnational oceanic thought, and to invite blue humanities to think further north.

The sea and the Nordic region

It is worth pausing to clarify what we mean by "Nordic" in this volume. The Nordic region refers to the northern part of Europe, grouped around three major bodies of water: the Norwegian Sea, the North Sea, and the Baltic Sea. The term typically encompasses the sovereign states of Denmark, Sweden, Norway, Finland, and Iceland, along with autonomous territories such as Greenland, the Faroe Islands, and Åland.

These are not simply nations adjacent to water; they are cultures fundamentally shaped by the sea. Norway has the longest coastline in Europe—and the second longest in the world. Denmark is a nation of islands; Sweden's and Finland's *skärgård* comprises thousands of small islands along its coasts. Iceland, the Faroes, Greenland, Svalbard, and Åland are geographically and culturally defined by their maritime location and histories. In short, the sea is not a backdrop to the Nordic region—it is constitutive of its geographies, economies, ways of life, and aesthetic sensibilities. While the welfare-state imaginaries that have dominated twentieth-century views of Nordic societies tend to foreground terrestrial spaces, it is the ocean that historically enabled mobility, sustenance, and exchange. Fishing, shipping, naval warfare, colonisation, trade, and, later, oil and gas extraction have all tied the Nordic countries intimately to the sea.

Swedish maritime archaeologist Christer Westerdahl (1992) has suggested the concept of *sjöbruk* (literally "sea use", analogous to *jordbruk*, or agriculture)—a term introduced by Swedish maritime ethnologist Olof Hasslöf in the 1950s—to describe the traditional Nordic mode of subsistence mariculture that combines farming with sea-based livelihoods such as fishing, seaweed gathering, and small-scale transport. Beyond its designation of specific maritime practices, *sjöbruk* names a maritime lifeway, deeply embedded in the cultural fabric of coastal and island communities throughout the region. This compound relation between land and sea, domestic economy and oceanic engagement, informs not only material practices but also the region's stories, poetics, and aesthetic traditions. This book follows Westerdahl's cue in framing the Nordic as both a linguistic and geographical category. We focus primarily on literary works written in the North Germanic languages—Danish, Swedish, Norwegian, Icelandic, and Faroese—which share historical continuities and mutual intelligibility. This scope also highlights the shared cultural archive through which many oceanic imaginaries have been shaped in the Nordic world.

At the same time, we recognise that this definition is necessarily partial. The seas surrounding the Nordic region are also traversed by texts in Sámi, Finnish, and Greenlandic—literary traditions that represent vital cultural engagements with maritime environments, often from Indigenous or non-Germanic perspectives. These voices are largely absent from this volume, not due to a lack of relevance or scholarly interest. Their absence reminds us that the Nordic is a contested and heterogeneous space, and that blue humanities in the North must continue expanding its linguistic, cultural, and political compass. Rather than presenting this volume as comprehensive, we see it as a contribution to an expanding field, and we hope it can serve as a foundation for future work that further explores the rich, polyvocal seascape of Nordic literature.

Navigating regional seas

In this volume, we have embraced a broad spectrum of literary genres and historical periods, which approach the sea in different ways. However, one of our key editorial decisions was to organise the book geographically—by specific maritime regions—rather than chronologically or by genre, which are more conventional in literary scholarship. This choice reflects our commitment to what Canadian interdisciplinary scholar Cecilia Chen calls "situated waters"—a framework that emphasises water's embeddedness in local ecologies, cultural practices, and material histories. "Waters take place", Chen writes, underscoring how "movements of water in our daily lives link us to places and to each other" (2013, 275). Instead of treating the sea as an "abstract and dislocated idea of resource" (2013, 276) or a universal symbol, we focus on its lived particularities. In developing the literary implications of this approach, we draw on Katie Ritson's book on the literary imagination of the North Sea lowlands (2019), which uses geography of the tidal basin as a starting point for her comparative study.

Each section of our book is thus anchored in a specific sea region and explores the literary responses these bodies of water have evoked across time and form. As Chen argues, the abstract idea of "the sea" often erases the very conditions through which water becomes meaningful. In response, this volume follows her call for a more sensuous, plural, and layered mapping of watery places that recognises multiplicity in meaning, temporality, and cultural resonance. Grouping our chapters around specific seas not only reflects the structure of the Nordic region as described earlier but also re-centres literary inquiry on oceanic spaces that have often remained peripheral. Such a shift challenges land-based hierarchies of knowledge and moves Nordic literature closer to oceanic imaginations, rather than merely orbiting the cultural capitals of the "World Republic of Letters" (Casanova 2007). Our contributors draw on various forms of knowledge—scientific, historical, literary, and embodied—to trace how Nordic seas have been inhabited, imagined, and contested. The result is "a deliberate plurality of maps", including

> (1) more than one understanding of water . . .; (2) an awareness of the same watery place or event over time . . .; (3) a selective layering of information to give depth to watery place . . .; (4) a deliberately synaesthetic use of many senses.
>
> (Chen 2013, 293)

The sections in our volume thus explore a wide array of Nordic oceanic zones, each marked by distinct ecological and cultural characteristics. In the Arctic Ocean, perennial ice cover, extreme seasonal shifts in light, and fragile polar ecosystems form the backdrop for intensifying climate change and geopolitical tension. Adjacent to it, the Greenlandic Sea is defined by cold currents, drifting icebergs, and nutrient-rich upwellings that sustain marine biodiversity and enduring Inuit traditions—now under mounting environmental pressure. The Icelandic Sea, situated at the confluence of warm and cold currents, is geologically active and biologically rich, but increasingly shaped by ocean warming, acidification, and resource extraction. The Norwegian Sea, deep and storm-swept, links Arctic and Atlantic waters and is central to migratory marine species, energy infrastructure, and the tensions between ecological conservation and industrial development. The shallower North Sea represents one of the most industrialised maritime zones in the world—criss-crossed by trawlers, oil rigs, wind farms, and shipping routes—yet recent decades have also seen renewed ecological protections and coastal restoration.

Further south and closer to land, the Wadden Sea stretches across tidal flats and barrier islands, offering a globally unique biosphere recognised by UNESCO for its role in bird migration and biodiversity, but increasingly threatened by rising sea levels. The Baltic Sea, a semi-enclosed and brackish body of water, is ecologically sensitive and historically overburdened by pollution, dead zones, and eutrophication; its limited water exchange makes it particularly vulnerable to anthropogenic impact. Moving outward, the Atlantic Ocean appears in this volume as a trans-oceanic expanse—both a historical arena of trade, migration, and colonial violence, and a modern site of ecological entanglement shaped by overfishing, warming currents, and vast logistical infrastructures. Conceptually, the volume also engages with the Global Sea—an abstract, planetary ocean that foregrounds the interdependency of Earth's water systems, challenging us to think hydrologically across scales and borders. Finally, there are also contributions that venture into the undersea—not a named sea but a vertical, immersive zone encompassing the benthic and bathypelagic depths, spaces often imagined as unknowable or sublime, yet increasingly mapped and mined in the service of techno-industrial desire.

Taken together, these maritime zones form a cold-water constellation distinctive to the Nordic world. Compared to other major oceanic regions, the Nordic seas are generally colder, less saline, and more seasonally variable. Ice, fjords, tidal flats, and brackish estuaries are common features. Their ecological productivity coexists with vulnerability: to warming currents, acidification, rising seas, and the industrial pressures of fishing, shipping, and resource extraction. In literature, these seas have long been imagined as both perilous and life-sustaining, spaces of separation and connection, and of migration and memory. Their particular materiality—dense, dark, often frozen or turbulent—conditions how stories are told, and which stories can be told, in Nordic sea literature.

Navigating blue humanities and critical ocean studies

Humanities scholarship that focuses on the ocean and other global waterways—often referred to as part of an oceanic turn—generally falls under the headings blue humanities or critical ocean studies. Of these, blue humanities has become the most widely used umbrella term, conceptually grounded in American literary scholar Steve Mentz's early formulation of "blue cultural studies" (2009a, 997) and American historian John R. Gillis's feature article "The Blue Humanities" (2013). However, this conceptual dominance has been questioned by scholars who advocate for the term critical ocean studies (DeLoughrey 2017), a label that emphasises the field's political, epistemological, and material stakes. In what follows, we briefly chart these two directions and some additional subfields, introduce key figures and approaches, and situate our volume within a wider scholarly conversation—bringing Nordic literary studies into dialogue with international oceanic thought.

Blue humanities is a wide-ranging, interdisciplinary field that places oceans, seas, rivers, and other bodies of water at the centre of human history, culture, and imagination. In Mentz's succinct formulation, it "names an off-shore trajectory that places cultural history in an oceanic rather than terrestrial context" (2018, 69). Key questions include the following: How does water shape narrative form and cultural meaning? What does it mean to think oceanically? Blue humanities emerged in close proximity to ecocriticism and the environmental humanities, drawing on literary studies, history, aesthetic theory, and cultural analysis. It typically emphasises the symbolic, affective, and aesthetic dimensions of water, engaging with concepts such as deep time, non-human agency, and planetary thinking.

By contrast, critical ocean studies bring an explicitly political, decolonial, and materialist lens to oceanic inquiry. As DeLoughrey puts it, this field "foreground[s] methodologies that examine the hydropower of militarism, empire, slavery, and extractivism to a greater extent" (2022, 149). It engages with questions such as: Who gets to know and navigate the ocean? How do power, empire, and knowledge intersect in oceanic space? Drawing from critical theory, postcolonial studies, science and technology studies, feminist theory, and Indigenous epistemologies, critical ocean studies centres issues of inequality, militarisation, maritime labour, and the ocean as a site of extraction, migration, and resistance.

While blue humanities and critical ocean studies often intersect and inform one another, each carries certain tendencies or blind spots. Early work in blue humanities, for instance, has at times been prone to aestheticising the ocean—approaching it as a space of metaphor, wonder, and poetic abstraction—while giving less attention to the material histories of extraction, militarisation, and structural violence that shape the world's oceans. The field is also characterised by a predominantly Eurocentric (Anglophone) scope and by insufficient engagement with Indigenous and decolonial epistemologies. In this regard, the risk lies in universalising the ocean without asking whose ocean is being imagined—and whose remains unacknowledged.

The political focus of critical ocean studies, in turn, is no doubt crucial, yet it also risks reducing the sea to a site of structural domination, occluding its affective, symbolic, and aesthetic dimensions. From the perspective of blue humanities, the ocean is not only a terrain of violence and circulation but also a space of imaginative excess—a fluid, unruly element that invites new ways of thinking about identity, time, and narrative.

While we have sought to distinguish blue humanities and critical ocean studies through conceptual contrasts, it is equally important to emphasise their points of convergence. Rather than being polemically opposed, the two fields exist in a state of productive tension—and often one finds elements of the former in the latter and vice versa. Together, they help build a pluralistic, ethically engaged, and aesthetically sensitive oceanic humanities.

Among the prominent figures in blue humanities are Steve Mentz, a specialist on early-modern literature and one of the founders of the field; Margaret Cohen, whose pioneering work on the novel and the sea has had a significant impact; the historian John R. Gillis, with his influential book on the human shore; political geographer Philip Steinberg, who has written on the social construction of the sea; and Serpil Oppermann, with her material ecocritical perspective on the sea. Key names associated with critical ocean studies include Elizabeth DeLoughrey, whose work fuses postcolonialism and environmental humanities; Melody Jue, who combines science studies and media theory; Christina Sharpe with her influential concept of "wake work"; Paul Gilroy, the early pioneer of Black Atlantic studies; and Stacy Alaimo, whose concepts of "trans-corporeality" and "exposure" have gained significant traction. In a Nordic context, literary scholars such as Christine Hamm, Jørgen Magnus Sejersted, Eirik Vassenden, Henning Howlid Wærp, Katie Ritson, Frits Andersen, Søren Frank, Magne Drangeid, and Malin Löf

Nyqvist have contributed significantly to introducing blue perspectives into Scandinavian literary scholarship and, in their publications in English, bringing Nordic insights into the wider blue ecocritical sphere.

The growing influence of ocean-centred inquiry is further evident in the emergence of related or hybrid subfields. Oceanic and archipelagic studies, closely tied to postcolonial and indigenous studies, focus on mobility, diaspora, and noncontinental modes of spatial thinking. Figures such as DeLoughrey and Tongan–Fijian writer and anthropologist Epeli Hau'ofa have been instrumental in articulating these perspectives. Hydrocriticism and hydropolitics analyse water as medium, infrastructure, and metaphor, often through the lenses of media theory and new materialism, with Canadian cultural theorist Astrida Neimanis being a key contributor to a specific hydrofeminist approach. Closely aligned with blue humanities and critical ocean studies, ecocriticism addresses also oceanic environmental crisis through literary and cultural forms, often from an environmental humanities perspective. Central figures in blue ecocriticism are Italian cultural and literary theorist Serenella Iovino, Turkish scholar of environmental humanities Serpil Oppermann, and American scholar Sidney Dobrin. Indigenous and decolonial oceanic thought, represented by scholars such as Hawaiian cultural theorist Karin Amimoto Ingersoll, emphasises relationality, lived knowledge, and decolonial critique grounded in indigenous epistemologies. Meanwhile, scholars like British–Iranian political sociologist Laleh Khalili (2020) have advanced the study of maritime labour and ocean capitalism, which investigates the political economy of global shipping, logistics, and seafaring through Marxist and critical logistics approaches.

Finally, we might also hint at terms such as "new thalassology" (Horden and Purcell 2006), which repositions the sea as central to historical analysis, especially in Mediterranean studies, highlighting connectivity and fluidity over fixed territorial boundaries; "terraqueous ecocriticism" (Brayton 2012), which emphasises the interdependence of land and sea in literary and ecological analysis, breaking down the binary between terrestrial and marine environments; "humanist oceanic studies" (Price 2017), which advocates for a renewed engagement with oceans in the humanities, attentive to both historical violence and the imaginative, affective potential of the sea; and "hydro-criticism" (Winkiel 2019), a term foregrounding water's fluid materiality in readings of literary form, politics, and embodiment, often linked to feminist, postcolonial, and queer theory.

There is another methodological axis worth mentioning, one between maritime history and oceanic history. While recent scholarship has emphasised a shift from the former to the latter, the chapters in our volume demonstrate their complementary relationship. Maritime history—with its focus on ships, trade, and coastal infrastructure—offers essential empirical insights into Nordic cultures' historical entanglement with the sea. Yet, as Mentz observes, it is a field "uneasily making space for a posthuman turn", as poets, literary critics, and environmental scholars seek to "place human histories in more-than-human contexts" (2018, 70). Blue humanities and critical ocean studies extend traditional concerns of maritime history by emphasising verticality, immersion, and ecological-elemental entanglement—often drawing inspiration from oceanography, marine biology, and other natural sciences to better understand the sea's dynamic materialities and multispecies networks. They invite us to consider not just the human activity *on* the ocean but also the material forces *within* it—currents, marine life, salinity, and sediment. As Sujit Sivasundaram, Alisan Bashford, and David Armitage put it, oceanic history turns the "longstanding historiography of humans, vessels, and exploration toward an analysis of complex relations between elements (winds, tides, currents), ocean life (mammals, fish, crustaceans, birds, plants), and human activity in and on the seas" (2018, 13). Rather than displacing maritime history, oceanic history deepens it and offers conceptual tools to read both the horizontal movements and the vertical entanglements of Nordic seascapes.

For all the possibilities offered by these theoretical insights, in our shared discussions during the Copenhagen workshop, we repeatedly found ourselves asking: Where is the sea, actually? For all the chapters' attention to maritime themes, we often found that the sea itself was elusive—polluted with metaphors, obstructed by metonymies. Despite the range of texts we had identified that were, we felt sure, deeply concerned with the sea, our discussions were often grounded on beaches; were caught up by boats or whales; descended into psychological, rather than watery, depths; or stayed stubbornly on the surface. While literature has long deployed the sea as a purely metaphorical space, a canvas for phantasmatic projections of all kinds—standing for the unconscious, the sublime, or the ungraspable—we were (also) interested in the sea as a real, physical body of water that had inspired literary texts over centuries: a saline expanse with its own histories, ecologies, and agencies. Yet this "real" sea remained difficult to access. Where was the real sea?

This difficulty, as DeLoughrey argues, is not accidental but structural. In *Allegories of the Anthropocene*, she suggests that the vastness, volatility, and scalar complexity of the ocean often resist direct representation, requiring literary allegory as a mode of "anchoring" meaning to materiality (2019, 135). Allegory, for DeLoughrey, is not escapist abstraction but a technique that allows readers to trace connections between representational figures (e.g. the boat, the shore, and the body) and planetary processes, especially in oceanic contexts where scale, movement, and opacity challenge referential clarity. In this sense, the metaphorical sea is not a distraction from the real sea—it is often the only way literature can access it.

Other scholars have similarly called for more robust theoretical approaches to the sea's material presence. Alaimo's "trans-corporeality" (2016) insists on the physical interchanges between bodies and environments, including toxic or oceanic flows. Steinberg and Peters's "wet ontology" (2015) and Peters and Steinberg's "more-than-wet ontology" (2019) reframe the sea as a dynamic medium of becoming, rather than a stable container. Sharpe's "wake work" (2016) forces attention to the sea as both a site of Black death and resistance, while DeLoughrey's notion of "interspecies worldings" (2019) highlights the ocean as a multispecies commons.

Such concepts were crucial as we turned to our own body of texts. While blue humanities and critical ocean studies offer divergent genealogies, their shared emphasis on theorising the sea as both material and symbolic proved vital. As Iovino and Oppermann write, the sea also requires "material ecocriticism" (2014)—a recognition that representation is never separate from substance. To grapple with this elusive, multivalent element, combining reflections on seas as ecosystems, habitats, sites of multispecies encounters, political violence, offshore industry and trade routes, literary sources and resources, imagined waterscapes, and problems of representation—to grapple with all this, our contributors draw on an array of methods, from blue ecocriticism and the poetic history of the oceans to postcolonial theory, the philosophy of the sea, archipelagic thinking, and environmental history.

Navigating colours

To call this field blue humanities or blue literary studies is not primarily an attempt to aestheticise the ocean or indulge in metaphor—though we insist that both gestures serve important critical purposes. Rather, the name registers a planetary fact: seen from space (on a cloudless day), Earth is overwhelmingly blue—its surface more than 70 per cent ocean. This global perspective is not meant to erase difference but to situate blue as a planetary colour, a visual shorthand for the interdependent watery systems that define and drive the Anthropocene. As Bengali–Indian historian Dipesh Chakrabarty reminds us, the humanities have historically emphasised "splitting"—a necessary resistance to universalist claims of power—but we are now also challenged to think in terms of planetary oneness: to hold together ecological-planetary commonality with historical-political difference (2021, 16–18). Blue, in this sense, is a chromatic metaphor for that dual task—it gathers without flattening. Interestingly, British literary historian Jeremy Davies defends the term Anthropocene along similar lines: not as a claim that humanity singularly dominates the planet, but as a stratigraphic marker of massive ecological change in which "human societies are only the most vigorous and distinctive among an irreducibly various array of altered forces" (2016, 73). "Anthropocene" is thus both a conceptual and geological condensation, just as "blue" is both a colour and a claim.

But blue also needs to be placed into a broader chromatic ecology. As American literary scholars and theorists Jeffrey J. Cohen (2013) and Margaret Cohen (2006) both argue, oceanic water is never just blue—it is brown, white, grey, and green. Blue is emphatic, but not exclusive. It connotes mobility, depth, and strangeness, but the sea's chromatic variability reflects sediment, storm, glacial melt, and pollution. Margaret Cohen's blue, brown, and white sea chronotopes remind us that these colours register ecological conditions (calm open water, river water, and stormy sea), not just aesthetic moods. This insistence on the material and ecological reality of colour is central to what American environmental humanist Heather I. Sullivan calls "the ecology of colors": "Colors are not just aesthetic or symbolic; they are material and ecological, entangled with chemicals, light, and perception" (2014, 81). In other words, blue is not merely a metaphor; it is a matter of entangled materialities, chemical interactions, and embodied perception; it even has its own colonial history (Ferwarda 2024). Recognising this entanglement helps us resist a purely symbolic reading of blue humanities and instead foregrounds the ocean as both sensuous surface and ecological process.

Moreover, blue stands in contrast to green—not as a rival, but as a necessary complement. Where green has long dominated environmental thought, tied to terrestrial metaphors of rootedness, harmony, and organic continuity, blue introduces a disorienting fluidity. It resists the fixity of green's pastoral visions, replacing equilibrium with turbulence and scale-defying depth. Green, as Jeffrey J. Cohen writes, "has become our synonym for sustainability", yet it often does so by evoking "a return, however belatedly, to the verdancy of an unspoiled world" combined with "a quiet faith in the totality envisioned by Deep Ecology" (2013, xx–xxi)—concepts that reproduce a nostalgic, terrestrial idealism rooted in the great bifurcation between nature and culture. Blue, too, has at times been romanticised—most notably in Romantic-era art and literature, where it signified longing, infinity, and spiritual depth, epitomised by Novalis's *blaue Blume* (blue flower), a symbol of unattainable desire and poetic transcendence. Yet the blue we invoke here is less about aesthetic yearning and more about ecological-geological unruliness. It resists resolution, clarity, and return. Blue, then, aligns more with "inhumanly vast or tiny scales of being and time", "the catastrophic, the disruptive", and "spaces inimical to homebuilding" (Cohen 2013, xxii, xxviii). And yet, as pigments remind us, green is born of blue and yellow—of sea and sun. This is not to initiate a competition between colours on our part; it is to emphasise a chromatic convergence. To invoke blue humanities is to embrace a prismatic planetary vision: one that balances abstraction and specificity, aesthetics and materiality, and unity and multiplicity.

Our contributors take blue as a starting point, but not as a constraint. They show that oceans and seas appear in many hues—literal and symbolic. Sometimes they are "blue" in Margaret Cohen's sense of the term:

> the realm of the open ocean containing immense and violent powers of weather, terrain (currents, tide, water depth), monstrous animals, aggressive warriors, as well as pirates and adventurers seeking gain in unpoliced zones beyond the control of sovereign and law.
>
> (2006, 650)

But we also read the ocean in its other colours. There is black water, stained by oil spills; there is the Black Atlantic, evoking histories of slavery and diaspora; there is white water, as when Amalie Skram describes the sea as "sun white" and "a shroud"; frozen water and grey slush in the Arctic; and in the Wadden Sea, brown waters churning with silt and history. Our blue humanities is attuned to this spectrum, resisting a too-rigid dichotomy between green and blue ecologies. The colour blue, like the sea itself, is both a surface and a depth, a unifying planetary image and a prism through which other ecologies—and other colours—are refracted.

The volume's structure allows the regional clusters to form a cross section of literary production that includes medieval, early modern, modern, and contemporary texts; prose, poetry, drama, and hybrid forms; as well as both canonical and marginal works. We believe this genre diversity and historical variety is crucial for revealing the *longue durée* of maritime imagination, while also showcasing how certain bodies of water have served as enduring sites of cultural meaning, poetic production, and political contestation. By placing well-known texts alongside lesser-studied or regionally specific works, we aim not only to expand blue humanities canon but also to suggest how marginalised texts may offer new insights into the entangled relations of land, sea, and literature.

Our contributors reflect a similarly productive mixture of disciplinary backgrounds and perspectives. Some are traditional Scandinavianists with expertise in particular authors or periods, but without prior engagement in blue humanities. We invited these scholars precisely because of their close familiarity with texts we believe deserve to be re-read through an oceanic lens. Others are researchers working from within blue humanities, bringing with them theoretical orientations from ecocriticism, environmental humanities, petroculture, and posthumanism. In several cases, these scholars have used this volume as an opportunity to explore new literary materials through the prism of the sea.

Thematically, *The Sea in Nordic Literature* is as varied as it is geographically, historically, and formally diverse. Several chapters engage directly with the Anthropocene and climate change, addressing the ecological upheavals that increasingly define the present and unsettle the future. Others explore pollution, environmental destruction, and the sea's role in extractive economies—whether through energy, industry, infrastructures, or the entangled histories of capitalism and resource exploitation. Colonial legacies and their ongoing reverberations are examined, especially with regard to Greenlandic and Caribbean contexts that complicate notions of Nordic unity. Questions of gender and embodiment emerge in texts where the sea becomes a site of both liberation and constraint. Other chapters focus on regionalism and place, anchoring their analyses in the distinctive geographies of islands, coasts, and archipelagos. Several chapters trace the shifting history of knowledge about the sea—through cartography, exploration, and scientific discourse—while others delve into the enduring force of religion and mythology in shaping the sea's cultural meaning.

Taken together, the chapters in this book do not chart a single narrative curve or impose a unifying theme, but a polyphonic map of Nordic literature's engagements with the sea. They enact a mode of *sympoiesis*—a collaborative "making-with" or even making-waves-with that reflects the ocean's own unruly plurality. Spanning a thousand years and a wide range of literary forms and concerns, the volume maps how Nordic literature has co-composed the sea as medium of exchange, material condition, metaphoric reservoir, and political space. By moving laterally across genres and periods, and by bringing together canonical and more marginal voices, the volume traces a maritime imagination that is at once continuous and contested—less a single current than a confluence of evolving tides and entangled waves.

Note

1 We didn't do all this on our own, of course. Thanks are due to Tanushree Ghosh and Alberte Ditte Steen Nemholt, doctoral student and student assistant, respectively, at the University of Copenhagen, who helped with the organisation of the workshop. We are furthermore deeply grateful to students Theresa Huber (in Munich) and Anastassia Zaltsberg (in Frankfurt), who were an invaluable support in copyediting and preparing the book manuscript for publication.

Bibliography

Alaimo, Stacy. 2016. *Exposed: Environmental Politics and Pleasures in Posthuman Times.* University of Minnesota Press.

Andersen, Frits. 2018. *Sydhavsøen: Nydelsens geografi.* Aarhus Universitetsforlag.

Andersen, Frits. 2024. *Wonders: A Literary History of the Deep Sea.* Aarhus Universitetsforlag.

Brayton, Dan. 2012. *Shakespeare's Ocean: An Ecocritical Exploration.* University of Virginia Press.

Casanova, Pascale. 2007. *The World Republic of Letters.* Translated by M. B. DeBevoise. Harvard University Press.

Chakrabarty, Dipesh. 2021. *The Climate of History in a Planetary Age.* University of Chicago Press.

Chen, Cecilia. 2013. "Mapping Waters: Thinking with Watery Places." In *Thinking with Water*, edited by Cecilia Chen, Janine McLeod, and Astrida Neimanis, 274–98. McGill-Queen's University Press.

Cohen, Jeffrey Jerome. 2013. "Introduction: Ecology's Rainbow." In *Prismatic Ecology: Ecotheory beyond Green*, edited by Jeffrey Jerome Cohen, xv–xxxv. University of Minnesota Press.

Cohen, Margaret. 2006. "The Chronotopes of the Sea." In *The Novel, Volume 2: Forms and Themes*, edited by Franco Moretti, 647–66. Princeton University Press.

Cohen, Margaret. 2010. *The Novel and the Sea.* Princeton University Press.

Davies, Jeremy. 2016. *The Birth of the Anthropocene.* University of California Press.

DeLoughrey, Elizabeth. 2007. *Routes and Roots: Navigating Caribbean and Pacific Island Literatures.* University of Hawai'i Press.

DeLoughrey, Elizabeth. 2017. "Submarine Futures of the Anthropocene." *Comparative Literature* 69 (1): 32–44.

DeLoughrey, Elizabeth. 2019. *Allegories of the Anthropocene.* Duke University Press.

DeLoughrey, Elizabeth. 2022. "Mining the Seas: Speculative Fictions and Futures." In *Laws of the Sea: Interdisciplinary Currents*, edited by Irus Braverman, 145–63. Routledge. https://dx.doi.org/10.4324/9781003205173-7.

Dobrin, Sidney I. 2021. *Blue Ecocriticism and the Oceanic Imperative.* Routledge.

Drangeid, Magne. 2022. *Jordomseilerne. Litterære reiser i naturkultur.* Scandinavian Academic Press.

Ferwarda, Susanne. 2024. "Blue Humanities and the Color of Colonialism." *Environmental Humanities* 16 (1): 1–18. https://doi.org/10.1215/22011919-10943081

Frank, Søren. 2022. *A Poetic History of the Oceans: Literature and Maritime Modernity*. Brill.

Gillis, John R. 2012. *The Human Shore: Seacoasts in History*. University of Chicago Press.

Gillis, John R. 2013. "The Blue Humanities." *Humanities* 34 (3). National Endowment for the Humanities. https://www.neh.gov/humanities/2013/mayjune/feature/the-blue-humanities.

Gilroy, Paul. 1993. *The Black Atlantic: Modernity and Double Consciousness*. Harvard University Press.

Hamelin, Louis-Edmond. 1968. *Nordicité canadienne*. Éditions Hurtubise HMH.

Hamm, Christine, Jørgen Magnus Sejersted, and Eirik Vassenden, eds. 2015. *Nordsjøen i norsk litteratur*. Fagbokforlaget.

Haraway, Donna J. 2016. *Staying with the Trouble: Making Kin in the Chthulucene*. Duke University Press.

Hau'ofa, Epeli. 1994. "Our Sea of Islands." In *A New Oceania: Rediscovering Our Sea of Islands*, edited by Eric Waddell, Vijay Naidu, and Epeli Hau'ofa, 2–16. University of the South Pacific.

Horden, Peregrine, and Nicholas Purcell. 2006. "The Mediterranean and 'the New Thalassology'." *American Historical Review* 111 (3): 733–36.

Ingersoll, Karin Amimoto. 2016. *Waves of Knowing: A Seascape Epistemology*. Duke University Press.

Iovino, Serenella, and Serpil Oppermann, eds. 2014. *Material Ecocriticism*. Indiana University Press.

Iovino, Serenella, and Serpil Oppermann, eds. 2017. *Environmental Humanities: Voices from the Anthropocene*. Rowman & Littlefield International.

Jue, Melody. 2020. *Wild Blue Media: Thinking Through Seawater*. Duke University Press.

Khalili, Laleh. 2020. *Sinews of War and Trade: Shipping and Capitalism in the Arabian Peninsula*. Verso.

Löf Nyqvist, Malin. 2023. *Hundraårsflöden: Översvämningsskildringar i svensk samtidsprosa*. Institutionen för kultur- og medievetenskaper. Umeå. https://urn.kb.se/resolve?urn=urn:nbn:se:umu:diva-213195.

Mentz, S. 2009a. "Toward a Blue Cultural Studies: The Sea, Maritime Culture, and Early Modern English Literature." *Literature Compass* 6 (5): 997–1013.

Mentz, S. 2009b. *At the Bottom of Shakespeare's Ocean*. Continuum.

Mentz, S. 2015. *Shipwreck Modernity: Ecologies of Globalization, 1550–1719*. University of Minnesota Press.

Mentz, S. 2018. "Blue Humanities." In *Posthuman Glossary*, edited by Rosi Braidotti and Maria Hlavajova, 69–72. Bloomsbury Academic.

Mentz, S. 2023. *An Introduction to the Blue Humanities*. Routledge.

Neimanis, Astrida. 2017. *Bodies of Water: Posthuman Feminist Phenomenology*. Bloomsbury Academic.

Oppermann, Serpil. 2023. *Blue Humanities: Storied Waterscapes in the Anthropocene*. Cambridge University Press.

Peters, Kimberley, and Philip E. Steinberg. 2019. "The Ocean in Excess: Towards a More-than-wet Ontology." *Dialogues in Human Geography* 9 (3): 293–307.

Price, Richard. 2017. "Afterword: The Last Universal Commons." *Comparative Literature* 69 (1): 45–53.

Ritson, Katie. 2019. *The Shifting Sands of the North Sea Lowlands: Literary and Historical Imaginaries*. Routledge.

Sharpe, Christina. 2016. *In the Wake: On Blackness and Being*. Duke University Press.

Sivasundaram, Sujit, Alison Bashford, and David Armitage. 2018. "Introduction: Writing World Oceanic Histories." In *Oceanic Histories*, edited by David Armitage, Alison Bashford, and Sujit Sivasundaram, 3–32. Cambridge University Press.
Södergran, Edith. 1992a. *Samlade skrifter. 1: Dikter och aforismer*. Edited by Holger Lillqvist. Svenska litteratursällskapet i Finland. https://litteraturbanken.se/f%C3%B6rfattare/S%C3%B6dergranE/titlar/SamladeSkrifter1/sida/42/faksimil.
Södergran, Edith. 1992b. *Complete Poems*. 2nd ed. Translated by David McDuff. Bloodaxe Books.
Steinberg, Philip E. 2001. *The Social Construction of the Ocean*. Cambridge University Press.
Steinberg, Philip E., and Kimberley Peters. 2015. "Wet Ontologies, Fluid Spaces: Giving Depth to Volume through Oceanic Thinking." *Environment and Planning D: Society and Space* 33 (2): 247–64. https://doi.org/10.1068/d14148p.
Sullivan, Heather I. 2014. "The Ecology of Colors: From Goethe to Ecocriticism." In *Material Ecocriticism*, edited by Serenella Iovino and Serpil Oppermann, 80–97. Indiana University Press.
Westerdahl, Christer. 1992. "The Maritime Cultural Landscape." *International Journal of Nautical Archaeology* 21 (1): 5–14.
Winkiel, Laura. 2019. "Introduction." *English Language Notes* 57 (1): 1–10.
Wærp, Henning Howlid. 2017. *Arktisk litteratur—fra Fridtjof Nansen til Anne B. Ragde*. Orkana Akademisk.

PART I

The Baltic Sea

1

THE SEA AS AN EMPTY SPACE IN EARLY MODERN LITERATURE

Johan Månsson's *Sea-book about Seafaring on the Baltic Sea*

Joachim Grage

In 1644, one of the most successful Swedish books of the seventeenth century was published in Stockholm. The *Siö-Book . . . om Siöfarten i Öster-Siön* (Sea-Book about Seafaring on the Baltic Sea)[1] came out in a total of twenty-two further editions until 1786, both in Swedish and in German, Russian, and Danish translations and adaptations (Meyer-Friese and Sauer 2020, 265–77)—making it not only a bestseller but also a veritable long-seller. The author, Johan Månsson, who was probably born in the early seventeenth century, had been chief helmsman ("ålder-styrman") in the Swedish navy from 1643 (Åberg 1973–75, 213). The places of publication and printing indicate where the book was in particularly wide circulation: Stockholm, Wismar, Lübeck, St. Petersburg, Åbo, and Copenhagen are all located on the Baltic Sea. It was here the book found its readers, who might be better described as users, as this is a text that could be called "functional literature" in modern terms.

The intended audience for this book is already named on the title page (see Figure 1.1): it is "for the service, instruction and correction of uninformed seafarers" (Månsson 1925–28, 3). In his dedication to high-ranking officers of the Swedish Navy, the author justifies "this little book on the correct description of the Baltic Sea" (Månsson 1925–28, 7) by stating that the books and nautical charts on the Baltic Sea available to date were too inaccurate and erroneous. He emphasises that his book is based on empirical experience, because as a helmsman he had "from the very beginning devoted my utmost diligence to observing, recording and noting all courses, coastlines, landmarks, entrances, harbours, dangerous shoals, cliffs, sandbanks and reefs" (Månsson 1925–28, 7). In 1645, a year after the publication of his sea book, Månsson also published a nautical chart of the Baltic Sea covering the same geographical area described in the book.[2]

DOI: 10.4324/9781003647638-3

Een Siö-Book /

Som innehåller

Om Siöfarten i Öster-Siön /

Jämwål om Koosar / Landkännningar / Streckningar / Inlopen / Banckar och Grunden.

The Enfaldigha Siöfarende til tiänst / Underwijßning och Rättelse /

Kortelighen sammansatt

Aff

Johan Månßon / Styrman /

medh sin eghen bekostnadt / och fins hoos honom til kiöps.

Tryckt i Stockholm aff Ignatio Meurer / Åhr 1644.

FIGURE 1.1 Johan Månsson, *Siö-Book . . . om Siöfarten i Öster-Siön* (1644), title page.

Permissions: Both images scanned from the 1925–1928 facsimile edition of the book Månsson, Johan: *Een Siö-Book / Som innehåller Om Siöfarten i Öster-Siön*. Ignatio Meurer, originally published 1644.

Månsson's description of the Baltic Sea for sailors is a text that can be made fruitful for the field of blue humanities from both a history of knowledge and a literary perspective—even though it is in the strict sense neither a literary text nor clearly a scholarly text. If the author had had an academic audience in mind, he would have written in Latin, the common language for scholarly texts in seventeenth-century Europe. We know as little about Månsson's educational background as we do about his origins. It is doubtful that he attended a Latin school, let alone studied at a university. Nevertheless, his book entered scholarly discourse. The famous Swedish polymath Olof Rudbeck the Elder quotes Månsson's sea book in the first volume of his monumental speculative-historiographical work *Atland eller Manheim* (1677) in connection with his remarks on the depth of the Baltic Sea (Rudbeck 1937, 104).

In terms of the history of knowledge, Månsson's sea book coincides with an "empirical turn" in scientific engagement with the sea (cf. Deacon 1971, 69–92), which was stimulated by physical experiences of the sea: "Much early modern ocean science took shape from the deck of a ship, or in the disorienting moments when human bodies immersed themselves in salt water" (Mentz 2019, 433). Stacy Alaimo emphasises the necessity of a historical approach to knowledge about the sea within the blue humanities:

> While science studies has been vital for the environmental (green) humanities, it is even more essential for the development of the blue humanities, since most aquatic zones, species, and topics exist beyond human domains, requiring the mediation of science and technology.
>
> (2019, 429)

However, science studies are important not only in terms of the technology used to explore the oceans but also in terms of knowledge about the sea, which varies from culture to culture and changes over time—and which always determines the perception of maritime space. Since the field of blue humanities itself claims to be scientific, a look at historical texts concerned with the maritime world also serves scholarly self-reflection and highlights the historical contingency of one's own approach.

A literary perspective on Månsson's sea book allows for a more precise analysis of how the sea is represented, because in the book, the maritime space appears as text. The present chapter will examine how Månsson organises the knowledge about the Baltic Sea that he compiles in his book and transfers it into the text, and which traditions he draws on in doing so. Of particular interest here is the perspective on the sea that manifests itself in the text and the framing of this knowledge through paratexts. This is because the information about the Baltic Sea in the book is framed by reflections on the technology and ethics of seafaring and by thoughts that place seafaring in a religious perspective. Finally, my aim is to show the extent to which the sea book itself is part of the maritime world it describes.

The Baltic Sea from the perspective of a helmsman

When Månsson calls his sea book a "description" in the dedication (see p. 19), he is using a term that has a long rhetorical tradition in Latin as *descriptio*. Cicero says the description should give a "clear explanation and almost visual presentation of events as if they are actually taking place" (De or. III, 53, 202; cf. Cicero 1982, 161). Description thus transforms information into text with the aim of giving the listener or reader an accurate impression of the object. What is not currently visible should be reconstructed in the text as if it was present. Numerous sub-types of *descriptio* can be distinguished, depending on the object being described. Månsson's description of the Baltic Sea would be classified as topography; that is, the description of a geographical space. Topographical texts were popular in Scandinavia in the early modern period: In Norway, Peder Claussøn Friis (1545–1614) had worked on a description of the country, which was published posthumously in 1632 as *Norriges . . . sandfærdige Bescriffuelse* (Norway's accurate description), while in Denmark, Peder Hansen Resen (1625–1688) collected information for the large-scale topographical encyclopaedia project *Atlas Danicus,* which was never realised.[3] Topographies such as these served to collect knowledge about geographical features, natural resources, and meteorological conditions, and make it available to an interested audience.

In terms of genre history, Månsson's book also follows in the tradition of Dutch and Low German sea books, which had been in existence since the fifteenth and sixteenth centuries and were used as navigational aids (cf. Richter 1925–28; Meyer-Friese and Sauer 2020, 11–23). The unique selling point of Månsson's book was the subject of his description: until then, there had simply been no sea book about the Baltic Sea. Sailors had to rely on inaccurate maps and the knowledge of local pilots if they wanted to navigate in unfamiliar areas of the Baltic Sea. Given the importance of the Swedish navy and merchant shipping on the Baltic Sea, the need was great. In the seventeenth century, Sweden was a major European power that possessed Finland and numerous territories in what are now the Baltic states, Poland, and Northern Germany. The *Mare Balticum* was almost a Swedish inland sea.

Månsson's sea book is a "literary representation of [a] specific bod[y] of water", which is the subject of this volume (see the Introduction to this volume, p. 2), but the "body of water" itself is only indirectly described. Anyone expecting a vivid *descriptio* of the Baltic Sea that brings the maritime space to life for readers, as Cicero demands, will be disappointed. "The essential part of Johan Månsson's sea book is a description of the coasts of the Baltic Sea as they appear to the sailor approaching from the water" (Meyer-Friese and Sauer 2020, 25). In doing this, Månsson anticipates, almost five hundred years before the establishment of blue humanities, its central epistemological shift in perspective, which Steve Mentz describes as an "offshore trajectory": "[It] means that you arrive in this placeless place as soon as you step off land. To offshore oneself—to move from land to sea—defines this practice in terms of location, method, and direction" (2024, 26). On the one hand, Månsson consistently takes an offshore point of view, describing the Baltic Sea from the fluid perspective of a ship. On the other hand, however, he does not focus his gaze on the sea, but on its boundaries, on the land that surrounds it.

Månsson's method of describing the Baltic Sea based on its coastline has a contemporary parallel in one of the most complex works of Scandinavian literature in the seventeenth century, the epic poem *Hexaemeron* by Anders Arrebo (1587–1637), a literary description of God's work of creation, which he was working on about ten years before Månsson published his sea book. In his description of God's works on the third day of creation, Arrebo talks about the oceans. In his description, he stages a fictional voyage around the world, has his narrator board a ship that sails along the coasts of the known world, and primarily lists geographical names of islands, ports, and countries (Arrebo 1965, 124–29; cf. Grage 2000, 122–29). Månsson could not have known Arrebo's epic, as it was not published until 1661, posthumously. The essential difference between Arrebo's description of the world's oceans and Månsson's description of the Baltic Sea lies in the fact that Arrebo's work is an artfully crafted didactic poem, a verse epic written in rhyming hexameters and alexandrines, while Månsson's book is essentially a prose text. Arrebo's great poem is a hymn of praise to God's creation, which demands a sublime form and pulls out all the rhetorical stops that are available. Månsson, on the other hand, is concerned with the very concrete question of how humans can make use of creation through technology and the art of navigation; that is, how they can sail the Baltic Sea without suffering shipwreck. He is focused on providing information rather than on artistic presentation.

Nevertheless, Månsson's work is not formless. The form of the description is the list—the ordered sequence of geographical information that is not structured hierarchically but in which all aspects claim equal value. The list is an essential basic form of premodern texts (cf. Contzen 2017; Contzen and Simpson 2022), not only in natural history and texts for practical use but also within literary, especially narrative texts, here mostly in the form of the so-called catalogues, enumerative descriptions of certain animal or plant species, and of geographical and spatial conditions. Månsson compiles the geographical information in a prose text that comprises one hundred printed pages in its Swedish first edition. The text is structured by a total of thirty-eight subheadings, each naming a harbour, an island, or a section of coastline, the surroundings of which are described in the respective section. These thirty-eight sections also contain thirty-two tabular lists that show the routes from the location described to other locations, indicating the direction to be taken and the distance in miles.

The first section on the port of Sandhamn, located on an island in the eastern part of the Stockholm archipelago, gives an impression of this combination of list-like descriptions of a port and its surroundings and a list of courses to other locations[4]:

About Sandahampn

If you want to sail from Sandahampn in Sweden, the course is south-southeast, then you should have two parts of water on the port side towards Grönskier and the third part on the starboard side towards Swartebodhan. This is because there is a cliff called Södherkubben off Grönskier in the south-southwest, and southwest of the cliff, about half a quarter mile away, there is a shoal at 5 fathoms. Outside Sandahampn, one and a half miles into the fairway, there is a sandbank at 12, 13 or 14 fathoms and coarse sand, where you can run aground in offshore winds if necessary.

From Sandahampn to Uthön, you leave Nyskieran on your port side, which you should watch out for in dark weather, as it lies 12 miles off Sandahampn to the northeast-east.

Courses from Sandahampn

	Uthön, east-northeast and northeast to east	24	
	Hangö Udd, east-northeast fairly eastern	30	
	Odensholmen, east, then north	30	
	Simpersnäs, east to south	24	
	Dagerort, east-southeast, then east	22	
To	Loserort, southeast to south	34	Miles
	Window, southeast to south, then south	35	
	Lijbo, south-southeast, then south	44	
	Memell, south to east	5	
	Gottsche Sandö, south, then east	12	
	Karssöyiar, southwest to south	30	
	Norre Udden, southwest	3	

(Månsson 1925–28, 11)

Already in this first sentence of the description, we see that the perception embodied in the text is technologically determined. For Månsson, the sea is only of interest as a space for sailing ships and thus for "the world's most complex technological machine during the four centuries spanning from the mid-fifteenth century to the mid-nineteenth century" (Frank 2022, 206). Typical for Månsson's description is its syntactic simplicity and stylistic austerity, which stand in sharp contrast to the dedication to the officers in the service of the Queen, in which Månsson utilises every stylistic tool at his disposal in this genre. The paratactic sequence of topographical information, especially the large number of place names, dispenses with any variation in vocabulary: apart from verbs denoting movement with the ship ("löpa" and "setta Koos"), the localising verb "ligga" (to lie) and the simple "wara" (to be), which states existence, dominate.

In terms of content, the description is essentially based on three points. Firstly, cliffs and rocks in the sea that must be navigated around are mentioned, but at the same time, they serve as recognisable and identifiable points on the water surface. Secondly, water depths, measured in fathoms, which are important for navigation insofar as the ship is in danger of running aground if the depth is too shallow. And thirdly, in some places, prominent landmarks on land, for example, mountains or large buildings, which are useful for orientation at sea. The knowledge manifested here is that of the pilots who are familiar with the respective coastal section and who assist foreign sailors in navigating close to the coast. Art and maritime historian Meyer-Friese suspects that Månsson obtained much of his information from other pilots (2020, 198). The sea book is thus a compendium of knowledge that was originally distributed among several experts, who passed it on orally, and which has now been collected in written form and made publicly available.

The thirty-eight sections in Månsson's sea book are arranged in geographical order: beginning with Sandhamn and the Stockholm archipelago, the book first covers the coastline of the Gulf of Finland, from its western entrance to the easternmost point at Nyenskans, the former Swedish fortress in what is now St. Petersburg (which was not founded until 1703). This is followed by important coastal sections of present-day Estonia from the north-east to the south-west, to the Gulf of Riga and further south to the coasts of present-day Latvia and Lithuania. Next, Månsson describes the coastlines around the island of Gotland before turning his attention to the southern archipelago of Stockholm. The description then follows the Swedish south-east coast southwards to the end of the Kalmar Sound. The last large area is the southern Baltic Sea, first the Danish coast, then the Mecklenburg and Pomeranian coasts, the latter receiving relatively little attention. If the sections are understood as places on a travel route, the journey could be described as a grand Baltic Sea cruise. It is interesting to note that the northern part of the Baltic Sea—the Gulf of Bothnia and the Åland Islands—are omitted entirely. This part of the Baltic Sea is also not shown on Månsson's nautical chart from 1645. The reason for this is probably that the Gulf of Bothnia was less heavily trafficked in the seventeenth century than the southern and eastern parts of the Baltic Sea, where the important trade routes were located.

Even though the arrangement of the individual sections follows a fictional itinerary, the text is anything but a travelogue or guidebook, as it contains no narrative passages or historical digressions about individual places, as was common in seventeenth-century topographical literature. Reading the description from beginning to end would therefore be rather tedious. Instead, the intention is that readers will selectively read those sections (or parts thereof) that are of interest to them because they are currently at the location in question or want to find out about the route there (or away from there).

The description of the Baltic Sea is linear in its focus on the coasts and in the sequence of the individual sections. Here, the focus within the text is primarily on the coast. However, the inserted tables showing the routes from a port of departure to various destinations shift the perspective to the sea or, more precisely, beyond the sea to the space behind the horizon, to those coastal locations that are not visible from the point of departure due to their distance. The routes stretch from one place to another like the threads of a spider's web. Although they are also arranged in sequence in the table, they can be read almost simultaneously due to their condensation. By crossing and intertwining with routes from other starting points, they form a network that spans the Baltic Sea. The network of recorded sea routes turns the sea into a space of transit and a connecting element between the coastal towns. Although the marine space is bounded by a detailed description of the coastline, and although a network of routes spanning the Baltic Sea is stretched between the individual ports, the sea itself remains an empty space of description.

Sidney Dobrin has pointed out that "a sense of place becomes problematic in marine environments" because "[a]ny 'sense of place' can only be established in fleeting glimpses pinned to specific historical moments and specific habitations" (2021, 58). While seventeenth-century topographical literature characterises places on land in terms of their human settlement and history, the sea is uninhabited and is usually perceived as devoid of history, as an eternal element. A "sense of place" can therefore only develop on ships, where sailors live together in a defined social situation. The sea itself, however, is a "placeless place" (Mentz 2024, 26).

The fact that the coast plays such an important role in Månsson's description of the Baltic Sea is also due to the fact that sailors in the seventeenth century were unable to determine the exact location of their ship if they did not have a coastline in sight that they could identify (Meyer-Friese and Sauer 2020, 197). After a long voyage on the high seas, it was only possible to determine the ship's exact location when land was sighted. Until landfall, sailors could not know exactly where they were at sea, neither with a nautical chart and a compass nor by astronomical observation—this only became possible with the invention of the sextant, almost one hundred years after the publication of Månsson's sea book. If "place" is understood as a specific location that can be differentiated from its neighbouring spaces, then only the sea near the coast is a place that can be defined as such. Only near the coast can the sea be described topographically, and even then only in relation to the specific conditions of the coast.

The technology and ethics of navigation

As was common in many printed books of the seventeenth century, Månsson's sea book contains a number of paratexts, some of which provide an important framework for describing and understanding the Baltic Sea. In addition to the title page and frontispiece, the royal privilege granting Månsson exclusive rights to his book for twelve years, the dedication to high-ranking officers of the Swedish navy, and a topical poem to critics ("Försmädarens Rijm"), it is above all the "Förmaning til Styremän" (Admonition to Helmsmen) and a spiritual song that deserve special attention.

The "Admonition to Helmsmen" comes immediately before the description of the individual coastal sections. It consists of a short didactic poem, followed by a paragraph in prose on the importance and use of the compass in navigation. The didactic poem comprises twenty-two verses in rhyming couplets, each with three stresses, but otherwise quite irregular in structure: their length varies between eight and fourteen syllables. Formally, it is a rather unsophisticated poem, but it summarises the virtues of a good sailor in concise form in its first verses:

> Those who wish to be good helmsmen
> must not skimp on the plumb line,
> but should also be sober and alert.
> Diligence and attentiveness are the best adornments.
> Pay attention to your compass, then make your calculations,
> then, with God's help, everything will go well.
> I truly advise you to worship God diligently,
> then you will have good fortune on water and sand.
> If you do this and heed my teaching,
> then God will grant you happiness and blessings. (Månsson 1925–28, 9)

The poem mentions two nautical instruments that are also depicted on the frontispiece of the book (see Figure 1.2). This shows a helmsman standing on the shore, against the backdrop of a handful of sailing ships and a rocky coastline that is quite untypical for the Baltic Sea. In his hand, he holds an hourglass (for measuring time) and the plumb line mentioned in the poem, which he can use to measure the depth of the water, and at his feet lies a compass, which he uses to check the ship's course.

The good helmsman is therefore first and foremost a technician, who has to make calculations based on his measurements. Only those who know how to use the instruments correctly are able to draw the right conclusions from the resulting geographical and hydrographical information. How to use the compass is described in more detail in the prose paragraph that follows the poem: It must be checked regularly and aligned with the North Star or the rising and setting sun. This is apparently based on the experience of the compass needle being deflected and influenced by the ship's magnetism (Meyer-Friese and Sauer 2020, 188–89).

In addition to mastering nautical techniques, the poem emphasises the helmsman's attentiveness as another important virtue in the verses quoted earlier and throughout the rest of the poem. In order to remain "alert" and "attentive", he must remain "sober"—in other words, he must not get drunk and allow alcohol to cloud his senses. He must also pay attention to safety and be aware of his responsibility

FIGURE 1.2 Johan Månsson, *Siö-Book... om Siöfarten i Öster-Siön* (1644), frontispiece.

for the ship, its cargo, and his own life as well as that of the rest of the crew ("Remember what is sailing with you"). In addition, the necessary diligence is repeatedly mentioned; that is, a high level of commitment to the task entrusted to the helmsman. The dangers posed by the sea, on the other hand, are only indirectly addressed by stating that many a sailor comes to grief. This, in turn, can be understood both literally and metaphorically: sailors can also stray from their course in a moral sense.

Finally, the helmsman must not lose his religious compass. Without God's help and protection, the poem suggests, all art and virtue are in vain. Therefore, the helmsman should pray to God, for He decides the happy outcome of the sea voyage. At the same time, it is clearly stated that God's blessing depends on the observance of the other virtues.

The "Admonition to Helmsmen" places humans in an ethical relationship with the sea. It prescribes rules of conduct that must be observed in order to make the sea usable as a temporary habitat. These rules are the basis for the successful use of the information provided in the description. Anyone who is unable to use their instruments, sharpen their senses, and entrust themselves to God is putting themselves in danger and has no business being at sea.

Spiritual framework: Christian seafaring and seafaring life

The fact that Månsson's extremely factual, sober, and dry description of the Baltic Sea is firmly anchored in a religious worldview is evident not only in the "Admonition to Helmsmen" but also in the paratext at the end of the book, which closes the religious framework opened up by the didactic poem. This is the hymn "IJ HErrans Nampn nu reese wij" (Now we travel in the name of the Lord), which comprises nine stanzas of six verses each and is to be sung to the melody of the better-known song "Kommer hijt til mifh sägher Gudz Son, &c." (Come to me, says God's Son, etc.) (Månsson 1925–28, 71–72). The name of the author is not mentioned. In 1695, half a century after Månsson's sea book, the song was included in *Den svenska Psalmboken* as number 335, with reference to the same melody as in Månsson. According to Lars Högmarck's *Psalmopoeographia* (1736), the song was written by the translator and lexicographer Ericus Schroderus (1609–1639) (Högmarck 1736, 99, 203–4; cf. Odelman 2000–2). It was therefore written relatively shortly before the publication of Månsson's book. It is not known where Månsson knew the song from.

The song is narrated in the first-person plural. The speakers ask God for guidance on a journey they are about to embark on, just as he once helped the people of Israel when they left Egypt. There is no explicit mention of a sea voyage, but there is a reference to particular dangers on the journey. The prayer for the journey is extended to include a general request for protection and blessing on the journey of life, of which the journey now being undertaken is only a part. Just as they hope for a safe arrival at their destination, they also ask for a safe arrival at the destination of their life's journey. At the end, the text returns to the specific departure and asks for protection and mercy for those who are left behind.

The function of this song at the end of a navigation manual is not limited to referring back to the introductory didactic poem, which presents God as the one who blesses seafaring. Nor is it to be understood as a mere reverence for God as the highest authority in human life. The inclusion of this song probably also has a pragmatic purpose. The intended performance of hymns is singing—and in this case explicitly: the singing of a congregation turning to God as a group. The situation described in the song corresponds to that of the crew or passengers of a ship embarking on a sea voyage. Similarly, the intended location of the book, that is, the place where it is used, is the ship on which the helmsman sails the Baltic Sea and where it serves as his *vademecum*. By printing this song at the end of his navigation guide, Månsson provides his readers with a text that they can sing together as a ship's community when they set sail or when they are travelling.

This song is not found in the translations of Månsson's sea book, but other spiritual texts are, clearly showing that Månsson's book was not intended solely as an aid to navigation on the Baltic Sea, but could also be used as a songbook, devotional book, or prayer book. The Danish translation of 1735 offers a whole selection of "sighs, prayers, and songs" to be used at sea (Monsen 1735, 81), whether as morning or evening prayers, as prayers at the start of a voyage, or as pleas for help in a storm. In the eighteenth century, songbooks and devotional books written specifically for use at sea were also published in Scandinavia. Their songs and spiritual texts were based on the everyday life of sailors, for example, the volume *Sebulon eller Søe-Havn* (Sebulon or Sea Harbour, 1720) by Jens Colding, who was a ship's chaplain in the Danish navy at the time, or the book *Besværede Søe-Mænds Søde Siæle-Roe eller Aandelige Lyst-Hauge* (Sweet peace of mind for troubled sailors or Spiritual pleasure garden, 1730) by the Norwegian cartographer and merchant captain Johan Heitman, which he claimed to have written during long sea voyages and which was intended to serve as pious entertainment for sailors (Heitman 1730, 17–18).

During the seventeenth century, maritime metaphors and allegories also increasingly found their way into Scandinavian spiritual poetry. In these, the sea voyage usually symbolises the course of human life, which in turn—as in the song printed by Månsson—is itself conceived as a journey. The various stages of a sea voyage—setting out on the high seas, the dangers of storms and running aground on rocks, and arrival in a safe harbour—offered the opportunity to illustrate religious content (cf. Grage 2000, 23–69). The sea itself was a symbol of the world in which people live in constant uncertainty. The rich maritime imagery in hymns and other spiritual poetry of the seventeenth and eighteenth centuries is an indication of the high status of the sea and seafaring in the Scandinavian cultures and of how the sea shaped the lives of the people. At the same time, it shows how deeply rooted the Christian view of humanity and the world was in everyday culture.

On the other hand, God is mentioned exclusively in the paratexts, not in the description itself. This can, of course, be explained by the list-like structure described earlier and the focus on geographical and hydrographical conditions. However, a look at a contemporary description of a sea voyage that follows a different literary pattern confirms this finding. In the logbook published by the Danish sea captain Jens Munk about his failed expedition in search of the North-West Passage in 1619–20 (Munk 1624), God also hardly appears, while He (as well as the king) is very present in the paratexts, as Søren Frank has shown (Frank 2022, 78–89). The fact that the logbook as a literary form "dictate[s] a particular empirical style and anthropocentric world picture" (Frank 2022, 87) can also be applied to the sea book and its mode of description. In both cases, however, it is also evident that—in the wake of the "empirical turn" of the early modern period—the sea is primarily described as a space for human action. There was no doubt that God was the creator of the sea, but this fact recedes into the background when it comes to the specific actions by which humans take possession of this space.

Månsson's sea book was practical literature for sailors. It was part of maritime life, read by sailors on ships as they navigated the Baltic Sea. The many editions of the book over a period of 150 years show its widespread distribution. It probably sailed across the Baltic Sea on many ships and saved them from disaster. The bookending of the navigational information by spiritual paratexts indicates that the respective sea voyage was understood as part of a larger, metaphorical journey, as a section of the journey of life, the goal of which is eternal life in heavenly bliss, in contrast to the tribulations of earthly life.

Notes

1 Unless otherwise stated, all translations are by the author.
2 A reprint of the map from the collection of the Royal Library in Stockholm is available digitally: Johan Månsson, Passcort öfüer Östersiöön (Stockholm: A.L. Normans boktr., 1725), http://urn.kb.se/resolve?urn=urn:nbn:se:kb:dig-10924949.
3 Only Resen's description of the island of Samsø was published during his lifetime and gives an impression of the concept behind the entire work, Resen 1675. Cf. Grage 2000, 136–40.
4 In the translations from the original Swedish text, the historical place names have been retained in their original spelling and have not been adapted to today's common names.

Works Cited

Åberg, Alf. 1973–75. "Johan Månsson." In *Svenskt Biografiskt Lexikon*, Vol. 20. https://sok.riksarkivet.se/sbl/Presentation.aspx?id=12108

Alaimo, Stacy. 2019. "Introduction: Science Studies and the Blue Humanities." *Configurations* 27 (4): 429–32.

Arrebo, Anders. 1965. "Hexaemeron Rhythmico-Danicum: Det er: Verdens Første Vges Sex Dages prægtige og mæctige Gierninger." In *Samlede Skrifter*, edited by Vagn Lundgaard Simonsen, Vol. 1, 1–264. Munksgaard.

Cicero. 1982. *De Oratore: De Fato. Paradoxa Stoicorum. De Partitione Oratoria.* Translated by H. Rackham. Cicero in Twenty-Eight Volumes 4. Harvard University Press.

Claussøn Friis, Peder. 1632. *Norriges oc omliggende Øers sandfærdige Bescriffuelse.* Marzan.

Colding, Jens. 1720. *Sebulon eller Søe-Havn for de aandelige Søe-Farende*. Kongl. Maj. privil. Bogtrykkerie.

Contzen, Eva von. 2017. "Grenzfälle des Erzählens: Die Liste als einfache Form." In *Komplexität und Einfachheit. DFG-Symposion 2015*, edited by Albrecht Koschorke, 221–39. Metzler.

Contzen, Eva von, and James Simpson. 2022. *Enlistment: Lists in Medieval and Early Modern Literature*. The Ohio State University Press.

Deacon, Margaret. 1971. *Scientists and the Sea 1650–1900: A Study of Marine Science.* Academic Press.

Dobrin, Sidney I. 2021. "Unearthing Ecocriticism." In *Blue Ecocriticism and the Oceanic Imperative*, 1–87. Routledge.

Frank, Søren. 2022. *A Poetic History of the Oceans: Literature and Maritime Modernity.* Brill.

Grage, Joachim. *Chaotischer Abgrund und erhabene Weite: Das Meer in der skandinavischen Dichtung des 17. und 18. Jahrhunderts*. Vandenhoeck & Ruprecht.

Heitman, Johan. 1730. *Besværede Søe-Mænds søde Siæle-Roe eller Aandelige Lyst-Hauge.* Kiøbenhavn.

Högmarck, Lars. 1736. *Psalmopoeographia: En utförlig Beskrifning öfwer them, som hafwa 1. sammansatt, 2. öfwersatt och 3. förbettrat Psalmarna*. Joh. Laur. Horrn.

Månsson, Johan. 1925–28. *Een Siö-Book, som innehåller om Siöfarten i Östersiön: Med efterskrift, ortnamsregister, ordlista och sjökort.* Edited by Herman Richter. Namn och bygd 13–16, bilaga B. Blom.

Mentz, Steve. 2019. "Experience Is Better Than Knowledge: Premodern Ocean Science and the Blue Humanities." *Configurations* 27 (4): 433–42.

Mentz, Steve. 2024. *An Introduction to the Blue Humanities*. Routledge.

Meyer-Friese, Boye, and Albrecht Sauer. 2020. *Johan Månssons Seebuch der Ostsee von 1644: Eine – fast vergessene – Erfolgsgeschichte*. Oceanum Verlag.

Monsen [= Månsson], Johan. 1735. *En nyttig Søe-Bog, som klarlig udviser den rette Søefart i Øster-Søen*. Joh. J. Hopffner.

Munk, Jens. 1624. *Navigatio Septentrionalis. Det er: Relation eller Bescriffuelse, om Seiglads oc Reyse, paa denne Nordvestiske Passagie, som nu kaldis Nova Dania, igiennem Fretum Christian at opsøge*. Heinrich Waldkirch.

Odelman, Eva. 2000–2. "Ericus J Schroderus." In *Svenskt Biografiskt Lexikon*, Vol. 31. https://sok.riksarkivet.se/sbl/Presentation.aspx?id=6398.

Resen, Peder Hansen. 1675. *Descriptio et illustratio Samsoæ insulæ Maris Balthici*. Gödian.

Richter, Herman. 1925–28. "Efterskrift." In Johan Månsson: *Een Siö-Book, som innehåller om Siöfarten i Östersiön*. Med efterskrift, ortnamsregister, ordlista och sjökort, edited by Hermann Richter, 73–98. Namn och bygd 13–16, bilaga B. Blom.

Rudbeck, Olof. 1937. *Atland eller Manheim [. . .]*. Edited by Axel Nelson. Vol. 1. Almqvist & Wiksell.

2

GULF OF FINLAND TEMPORALITIES

Petroculture, geopolitics, and Tove Jansson's resonant island

Pia Ahlbäck and Mia Österlund

During a nearly thirty-year period extending from 1965 to 1993, the world-famous Finland-Swedish writer Tove Jansson (1914–2001)[1] and her partner Tuulikki Pietilä (1917–2009) spent holidays on the tiny island of Klovharun off the Finnish south coast, close to the main waterway of the Gulf of Finland, not far from the territorial border.[2] Those are the vicinities we will visit in this chapter, Klovharun and adjacent islands in the Gulf being the setting for Tove Jansson's novel *The Summer Book* (1974; *Sommarboken* 1972) and short story "The Island" (2019; "Ön" 1961). *The Summer Book* is one of Tove Jansson's most well-known and cherished works along with her popular Moomin series (1945–1970), which has been translated into some forty languages and developed into TV series and films. Today, the Moomins and their world are part and parcel of children's popular culture in large parts of the world, and *The Summer Book* is not far behind. In 2025, *The Summer Book* was made into an international film with the same title.[3] Along with these works, Jansson also published novels, satirical sketches, essays, and polemical writings; in addition, she was also an accomplished painter.

"Visiting", in the case of *The Summer Book* and "The Island", means tracing the material environments undergirding these works, opening up a number of temporalities within which they are situated, and reading them counter-intuitively and against the popular traditions of moominological fun and idyllicism.[4] In other words, the task is to read these particular works by Tove Jansson as literature-*cum*-life and *landscape* in the temporal contexts of the material hazards of their expanding petrochemical and geopolitical environments.[5] Highlighting this context in Jansson's work enables it to be read anew in our historical moment of climate emergency and recurring Russian totalitarianism and European warfare.

DOI: 10.4324/9781003647638-4

Initially, we pay attention to the temporal reclusion of the island itself, making use of Hartmut Rosa's concept of "resonance" to understand the interactions between the island and the human visitors (Rosa 2017, 39–55).

Island resonance

Rosa defines the concept of resonance—as opposed to acceleration—as being about human affection, transformation, and uncontrollability in the interactive relationship between humans and the physical world. More importantly, he also acknowledges physical entities as being resonant and agential (Rosa 2023, 1–11). We understand the material island of Klovharun as being inwardly "resonant": it has a temporality of its own defined by natural forces, life forms, and cycles, of which the humans become part when they arrive in early spring and depart in autumn. The humans are thus immersed in the circular temporality of the island's various natural forms of life and must both adapt to and interact with them. Once they have set foot on it, the humans are formed by the active agency of the island. Island agency consists of natural forces, wildlife, and geographical forms—waves, winds, and animals—including the island's very barrenness. Similar phenomena are pictured in *The Summer Book*, which highlights the natural forces and events typical of the outer archipelago and the many ways in which they shape the humans. During the course of their yearly summer holiday on the island, Grandmother, father, and the girl Sophia repeatedly adapt to these agencies.

In Jansson's sensuous portrayal of autonomous island dwelling in her short story "The Island", the narrator gives expression both to a sense of unity with the island and an act of communication with it. The following two quotes depict an immersion into the material world:

> The mutating sea, the beach that rises and sinks and changes shape, everything that grows and dies and grows again in a new and surprising place, the way in which trees and shrubs withstand the storm, decay takes its natural course on everything one has built, and the pleasure of recognition and repetition.
>
> (Jansson 2019, n.p.)

> After being alone for a long time, one starts to listen differently, to perceive the organic and the unexpected all around, to brush against all the incomprehensible beauty of the material.
>
> (Jansson 2019, n.p.)

The material world, however, also correlates in an unexpected way with a twang of threat in *The Summer Book*, each of the chapters in the novel providing the tingling insight that there is no obvious way out of the dangers met, that things really can go wrong. Despite this, the characters are *in resonance*: exactly because the natural threats are tangible, they are possible to understand and thus not quite so frighteningly overwhelming. Grandmother, an island-whisperer of sorts, is more than well-versed in the natural environment of the outer archipelago, and Sophia, being a daring storm-lover, confides in her. In other words, in *The Summer Book*, there is something different from the natural dangers of the island that is *really* threatening, something one cannot "resonate" with.

Independent of genre, Tove Jansson is a master of the unspoken. *The Summer Book* and "The Island", too, show a remarkable high presence of voids, thus creating a creeping feeling of hazard, maybe even of fear. This situates Jansson in the modern mimetic tradition described by Erich Auerbach in his famous work *Mimesis: The Representation of Reality in Western Literature* (1946). Here, he uncovers the two major literary traditions that have shaped Western literature since antiquity—the Homerian and the Biblical, Old Testament styles, respectively (Auerbach 2013, 3–23). Auerbach shows that the latter, by means of its very exposition of temporality, is more modern than Homer's. In the Old Testament, background, depth, and void are introduced into the narration, and so is suspense. This happens in contradistinction to the slow and all too apparent narration of *The Odyssey*, where the spelling out of every detail and event means that the reader can never be taken by surprise. In the biblical horror story of Abraham's—happily uncompleted—sacrifice of his son Isaac, threat and suspense are everything: the outcome is increasingly clear to the reader until there is a release of tension almost as frightening as the initial divine demand on Abraham's loyalty.

Voids, consequently, are not empty, but pregnant with the unexpected, yet-to-appear. Surprisingly perhaps, a great deal of Tove Jansson's work draws on this very tradition and *The Summer Book* is typical in that respect. Instead of burning bushes, we get boats appearing in the horizon, creating the odd impression that a vessel turning up at sea is something quite unexpected, a bit like a black swan. There is, consequently, nothing quite as sudden as the suddenness of Tove Jansson's archipelagic stories, their spaces filled with phenomena of potential disaster. By means of the absence turned into presence, exemplified by the ship appearing out of the seeming nowhere of the vast sea, a poignant sense of insecurity is communicated, the absences themselves returning to the island those very things, phenomena, and people that have been shut out. Or rather, all that which Grandmother has done her utmost to exclude.

This undefined "something" is never explicitly spoken in the works we study, but it has left a trace in them of oil spill, plastic, and, as said, suddenly appearing ships. A very different presence than that of the natural dangers on the island thus appears: this is the, despite everything, powerful proximity of the modern world and its many hazards connected with the particular far-reaching temporality of invasive objects, substances, and infrastructure.

Whereas the island resonates inwardly, it becomes a temporal parenthesis in the wider sixties and seventies contexts of the Gulf. The pragmatic middle generation of the novel represented by the unnamed father of the girl Sophia—son of Grandmother—has one foot in the Finnish mainland temporality of oil and strongly emerging plastic commodities, the other in the temporality of the island, which is defined by natural cycles.[6] In 1972, when the novel was first published, both plastic discourse and the fear of an energy crisis were at their height in the Western world. With the toxicity, invasiveness, and persistence of the miraculous material of plastic not yet having been fully realised, the conviction was that plastic was a blessing for all time. In hindsight, the irony of this belief is striking. What is interesting in the context of the novel, moreover, is that there are many instances where Jansson treats plastic ironically; in fact, plastic itself becomes a major irony in *The Summer Book*. We return to this subject further on, but for now, starting in the present, we will focus on the petrochemical and geopolitical history of the Gulf. That is the concrete physical and geographical context in which we wish to read Tove Jansson's work.

The chemical corridor of the Gulf

The Gulf of Finland is a chemical corridor of a kind, both shallow and narrow, a Baltic back alley, which in its eastern direction has always ended in a particular kind of totalitarian nowhere. This is the no-go area represented first by the Communist dictatorship of the Soviet Union and, as of 2022, of warmongering Russia. In a European geopolitical perspective, the Gulf ends in the St Petersburg bay, where the river Neva has its outflow, bringing with it the sewage water of millions of people. In addition, there are large petrochemical plants situated in Primorsk and Ust-Luga to the north-west and south-west of the city, respectively, oil and gas being Russia's major export goods (Rodin 2009). As a consequence, the Gulf was highly polluted for a long time, although clear improvements were noticed during the last decade due to the early twenty-first-century investments in three sewage water treatment plants in the St Petersburg area (John Nurminen Foundation 2024). However, the most recent reports are alarming, showing huge environmental problems both in the Gulf and in the Baltic main basin, where large bottom areas have died (Piepponen et al. 2024, 9–11).

In the sixties and seventies, the Gulf was already filled with oil tankers, transport vessels, warships, and submarines (*Helsingin Sanomat* 2004). Today, traffic is even more intense due, among other things, to the increased freedom of mobility achieved in the northeastern Baltic Sea as of 1991. The problem of traffic noise has increased correspondingly (Piepponen et al. 2024, 9–11). The dense traffic in the Gulf after 2022, however, is partly due to the heavily increasing military presence in the Gulf. A particular danger is posed by the oil tankers and the Russian so-called shadow fleet: oil tankers in very poor condition, moving in the Gulf without following proper regulations (MarineFinland.fi, n.d.). Whether they have other purposes than just the transportation of oil is not entirely clear, but there is a strongly suspected connection between these tankers and the increasing number of accidents in which underwater communication cables between Finland, Estonia,

Sweden, and Germany have been cut. In addition, despite the Western commercial blockade of Russia, Russian oil traffic is still growing (Russia Fossil Tracker 2025).

Up until 1991—two years before Jansson and Pietilä left Klovharun for good—with the fall of the Soviet Union and simultaneous sovereignty of the Baltic states of Estonia, Latvia, and Lithuania, the southern shores of the Gulf were also a forbidden territory. What free movement there was in the Gulf was restricted to the international waterways and Finnish territorial waters. Regarding the impact of the oil industry on the Gulf, however, it was not merely of Russian origin. Not that far from Klovharun, to the north-west, lies the petrochemical plant of Finnish Neste,[7] allegedly the largest oil refinery in the Nordic countries. Tove Jansson and Tuulikki Pietilä built their cottage on Klovharun in 1965; it was completed in the same year. From their tiny island, really not much more than a skerry, it must have been impossible to *not* notice the intense traffic further out at sea. In the same year as they initiated their cottage project, the Neste petrochemical plant was also beginning to be constructed in the small village of Sköldvik outside the historical town of Borgå (Fi. Porvoo). To avoid public and political protest, the refinery had secretly been prospected in a woodland area of the Sköldvik village already in 1962 (Vennonen 2006, 12; Katavisto 1990, 94–95), the advantages of the village being a natural, deep harbour and a way out into the Gulf that was smooth enough (Katavisto 1990, 94). By means of Neste's arrival, however, the Finland-Swedish village Sköldvik was practically deleted: exactly one house would remain of the original village after the area had been exploited by the petrochemical company. The rest, including the eighteenth-century Sköldvik manor, was destroyed, and the entire area thus exploited for petrochemical purposes (Katavisto 1990, 100). The Neste refinery was opened in 1966, one year after the completion of Jansson's and Pietilä's cottage on Klovharun, and further expanded throughout the sixties and seventies. In other words, the construction and successive expansion of the largest oil refinery in the Nordic countries took place in the very same region where the "paradise island" of one of the world's most famous writers of children's stories was situated. Klovharun—one of the utmost islands of the Pellinge archipelago outside Borgå—and Sköldvik both belong to the Borgå municipality, Sköldvik lying on the shore somewhat west of the town of Borgå and north-west to Klovharun. Strikingly, almost like an ambiguous lighthouse itself, the eternal petrochemical flame from the Neste plant can be witnessed also from Klovharun (Brunow 2010).

Petrified waters, plastic irony

In Jansson's short story "The Island", the inhabitant walks around the island again and again, satisfied, letting her sight go looking for what is increasingly farther away, ultimately towards the open horizon. She is convinced that her eyes will meet only open space, owning it, liberty and safe seclusion at the same time. Then, suddenly, there is the burning bush, the boat out of nowhere, and she comments on her feeling of being exposed: "It's a rickety structure, one pays for it with fear of darkness and sudden panic—a stirring in the dark, a boat on the horizon" (Jansson 2019, n.p.). The short story was published four years before the construction of the cottage on Klovharun had begun, but Tove Jansson had visited Klovharun many times before, from at least as early as 1958. Even before that, she and her family had spent many summers on the nearby island of Bredskär, eventually leaving it because of the rising number of visitors (Westin 2007, 370).

Neither Grandmother in *The Summer Book* nor the narrator of "The Island", it is apparent, are keen on visitors. It is a well-known fact in Jansson's research that she was remarkably private and protective of her space (Westin 2007, 373, 377, 397). For the purposes of this chapter, we read the quote earlier as an emblem of intrusion, ascribing relevance to it also in the future context of the Gulf. We suggest that there is another epistemological level to that "stirring in the dark, a boat on the horizon" than merely potential visitors. What arrival is it that the narrator fears? Not only intruders but perhaps also *that which intrudes* makes itself felt. This is the idea that the short story provides: the very materiality of intrusion. It therefore becomes meaningful to situate *The Summer Book* and "The Island" within a materialist epistemology.

In her book *Exposed: Environmental Politics and Pleasures in Posthuman Times* (2016), Stacy Alaimo highlights the concept of "trans-corporeality" (2016, 112), understanding it as something that "traces material interchanges across human bodies, animal bodies and the wider material world" (2016, 112). The following passage from the short story, again contributing to the understanding of the later novel, underscores the seashore as being in a state of material flux: "The beach—the narrow border between land and sea, fickle and treacherous, shaped with heedless violence and strewn with peculiar objects that the sea has smoothed, giving them their softness and strength" (Jansson 2019, n.p.). The materialist perspective thus implies the arrival of the more-than-human in Jansson's novel, be it by means of things or phenomena: the darkness of oil; flotsam washing up on an island; and the vessel as a synecdoche of the oil tanker. Tove Jansson's nearly thirty-year stay on the minute island of Klovharun coincided with the high tide of oil and its refinement into plastic and other petrochemical products. During this period of what we suggest is a particular kind of *petrification* of the Gulf, it would be continually coated in a fluttering dark cloak of oil spills and orange spots of plastic in accordance with the discourse of the age. We claim that it is no coincidence that the swiftly growing oil infrastructure of the Gulf and its various and frequent side effects also communicate with the occasionally petrified human emotions in "The Island" and *The Summer Book.*

In the chapter of the novel ominously titled "Dead Calm" (Jansson 1974, 41–48), the family is on their way to a cairn even further out at sea. The sea is described as being so still that it is "as smooth as oil" (1974, 42), whereas Grandmother is sitting under a violet parasol in the boat—violet being a colour she hates. Oil is black, but when floating on the sea, it takes on various shades of violet. Consequently, in this instance, Jansson offers a refined environmental criticism, connecting the death of the calm in the title with the oil on the water. Describing the smoothness of water in terms of oil is also rather far-fetched. A word for smoothness closer to hand would be silk. The choice of "oil", consequently, works to defamiliarise the story of the chapter, reinforcing its critical perspective.

Previously, in the chapter "The Scolder",[8] the debris of the beach is perceived through the eyes of the girl Sophia one early morning in May:

> No one had had time to gather driftwood yet, and the whole shore was a tangle, a broad bulging mass of tumbled planks and seaweed and reeds along with posts and fragile wooden boxes that had turned inside out around their steel frames, and on top of everything lay an enormous heavy log that was black with oil-spill. (1974, 17)

Sophia continues to observe her immediate environment: "Out on a flat rock in the water lay a scolder. It was wet and dead and looked like a wrung-out plastic bag" (1974, 18). Oil spill and plastic bags, indeed. In the seventies, the naive optimism of eternal progress, which had been so characteristic of the fifties and sixties, would remain remarkably strong. Tove Jansson's construction of the similarity between the scolder and the plastic bag is striking and ironic, considering our knowledge today of the dangerous dispersion of plastic in all life forms: fish, seals, birds, and humans. The marine biologist Karen Scales notes that a "[p]lastic fog in the ocean is thickening and now comprises hundreds of trillions of particles" (2024, xii), whereas Alaimo points out that plastic bags look like prey, that is food, to many sea animals (2016, 137).

An ingenious instance of Jansson's ironical use of plastic in *The Summer Book* can be found in the chapter "The Neighbour" (1974, 76–87), where Grandmother and Sophia make a secret and unauthorised visit to a luxurious, modern summer house on a nearby—relatively speaking—island. The chapter highlights a lack of authenticity, the most remarkable of all being the presence of plastic—a phenomenon described by Serenella Iovino as "the other side of our presence in the world, our absence made visible" (2009, 6). Having landed without permission, inspecting the place critically, and finding obvious fault with it, the artifice of the whole place is ridiculed by Grandmother and Sophia. Plastic takes centre stage: the plastic is what they explicitly laugh at, and through the ironic repetition, the artifice of the material is emphasised. At the same time, the iteration recasts plastic in language as a generic product, a kind of mimicry.

> "What we are now doing", Grandmother said, "is a demonstration. We are showing our disapproval. Do you understand?"
> "A demonstration", her grandchild repeated, adding, loyally, "This will never make a good harbour."
> "No", Grandmother agreed. "And they have the door on the wrong side of the house. They'll never get it open in a south-wester."

> "And look at their water barrels. Ha-ha. Plastic, of course."
> "Ha-ha", Sophia said. "Plastic, of course." (Jansson 1974, 77–78)

The entire chapter titled "The Enormous Plastic Sausage" (1974, 95–106), on the other hand, is a hilarious celebration of the power of the natural forces by means of the exposition of a pathetic plastic commodity: a tall and thick orange hose. Having been transported by Father from the mainland with much trouble and finally installed for its purpose—to channel water from the sea up to the flower beds—there is a thunderstorm and the island is drenched. As Grandmother does not have much respect for fragile modern infrastructure but counts on the forces of the island to care for its own, the whole project is associated with her son, Sophia's father. The male middle generation, consequently, is conflated with the artefacts of petrocultural society when trying to cultivate the island by planting flowers that do not belong there, and making it dependent on plastic in order to make it thrive artificially.

The sound of diesel engines

The overwhelming risk of oil disasters was and is today increasingly present in this part of the Baltic. Accidents of different degrees were indeed plenty in the Gulf throughout the seventies, eighties, and nineties, even if few of them were really catastrophic. What needs to be acknowledged, however, were threats in the form of the geopolitical power moves of the Soviet Union as they once again need acknowledging in the case of Russia today. Oil, its infrastructure, and the warships were all part of the same pattern. Tove Jansson's "stirring in the dark" could have been—can be—the shadow of a submarine or an elite soldier arriving from underneath the water; the arrival of the enemy.

In 1979, seven years after the original edition of *The Summer Book* was published, the infamous Soviet oil tanker Antonio Gramsci caused a major oil disaster in the Baltic Sea with effects for the Gulf and as far west as the Åland Islands and Sweden (Sinivaara 2020, 17). There would be another accident caused by the same tanker in 1987, this time in the very port of Sköldvik (Sinivaara 2020, 17). In addition, there were, among other things, the problems of Neste's early "supertankers" Tiiskeri and Enskeri, the latter of which tried to dump its cargo of arsenic in the Atlantic in 1975. After international protests, the cargo had to be reloaded onto a different tanker, which brought it back to Finland. At the time these events unfolded, Tove Jansson and Tuulikki Pietilä had many summers left on Klovharun. What threats there were, many of them definitely came from "the outside".

In the final chapter of *The Summer Book*, "August" (1974, 146–52), summer turns into autumn, and the family is preparing to return to the city for winter. The arrival of autumn makes the boats seem more intrusive: "There was an unbroken chain of navigational lights and beacons along the horizon, and sometimes big boats thumped by in the channel" (1974, 148). Equally clear is that, just as the family had entered the temporality of the island when they arrived in spring, they now have to extract themselves from it. Removing the debris from the island and restoring it to its primaeval condition is a particularly important project for Grandmother: "She picked up everything that had to do with human beings. She gathered nails and bits of paper and cloth and plastic, pieces of lumber covered with oil-spill, and an occasional bottle top" (1974, 147). Only the cottage will remain. In their last night on the island, Grandmother wakes up and walks out to pee. She is feeling her way carefully, watching and listening into the night, paying attention to a big ship passing. It is a cargo ship on its way east, to Kotka.[9] She recognises the different kinds of engines and varieties of fuel: a diesel engine here, a petrol engine there. Finally, she hears the sound of a small boat approaching and passing closely by the island only to head straight out to sea, the sound of the engine never stopping. Suddenly, she realises this is no boat but her heart beating, and she decides to remain sitting on the chopping stump for a while.

The ending of the last chapter is open and can accordingly be interpreted in many ways. In the recent film *The Summer Book* (2025), there is a clear emphasis on death, which becomes obvious as Grandmother's heart stops beating while the camera keeps following the imagined boat further and further out at sea until it disappears. We, on the contrary, suggest that the boat is exactly that mirage which Grandmother tells the reader it is. She hears her heart, not a boat, and decides to remain sitting for a while before returning to the house. Nevertheless, as we indicated previously, if things were that straightforward, this would not be Tove Jansson writing. In such case, the suspense and absences would be obliterated. There is something more to this scene after all.

It is night, autumn, there is darkness. People are leaving. These tropes are traditional literary devices to indicate death. The boats, too, can be interpreted in this metaphorical way. But from the material perspective of petrocultural modernity, the boats, as we have seen previously, are intrusive: unwelcome containers of a different, aggressively accelerating, toxic temporality, pressing upon the island and Grandmother. Let us finally return to Rosa's idea of resonance and Alaimo's of the trans-corporeal. We claim that Grandmother, by remaining sitting on the chopping stump, persists in her resonance with the island, insisting on being a material part of it, melting into it in the darkness of night. Not returning to the house means refusing to eventually re-enter the temporality of the over-intense modern world with its commodities of oil and plastic. Just before, Grandmother has asked herself: "How can I ever leave this room?" (1974, 150), indicating the guest room in the cottage where she sleeps, so full of material fragments of pre-plastic history. In our reading, the room transgresses its walls and becomes a metonymy for the space of the island itself. When Grandmother gives expression to her inability to leave the room, it is, finally, the island she is referring to. This is the space she has become one with over a long life.

Conclusion

In this chapter, we have reinterpreted Tove Jansson's novel *The Summer Book* and short story "The Island" by relating them to the petrocultural and geopolitical material environments of the Gulf of Finland between 1962 and 1993, with outlooks to some of the present-day conditions of the Gulf. We have paid attention to the simultaneous temporalities of Jansson's presence on the geographical island of Klovharun and the surrounding, swiftly developing petrochemical infrastructure around the Gulf. The oil refinery of Sköldvik together with the tanker traffic on the Gulf, the Cold War presence of warships, and the Russian oil industry outside St Petersburg all made their mark on and in the waters of the Gulf. Reading the novel and short story from a temporal perspective related to a material one using Rosa's idea of resonance and Alaimo's of trans-corporeality, we have shown how oil and its refinement into plastic permeate these texts from the sixties and seventies. Nevertheless, these modern and destructively polluting, permanent materials did not just permeate but were also critically negotiated in Tove Jansson's work through her artistic genius.

Notes

1 "Finland-Swedish" denotes a Finnish citizen with Swedish as their native language. The Swedish-speakers in Finland date back to the thirteenth century. Swedish is acknowledged as an official language in Finland, and the native Swedish speakers amount to 550,000 people, which is about 5 per cent of the entire population.
2 Tove Jansson and Tuulikki Pietilä have described their comings and goings and the process of building the cottage in the memoir *Notes from an Island* (2021).
3 The American actress Glenn Close acted the role of Grandmother.
4 By idyllicism in this case, we primarily indicate the widespread *public* conception that much of her work is idyllic. In Jansson's literary research, the concept of idyllicism is strong, but there is a scholarly consensus that the idyll in her work is threatened (Laakso 2017; Westin 2007; Rehal-Johansson 2006, 63; Holländer 1983, 57–59).
5 The cultural historians Otto Latva and Maarit Leskelä-Kärki have written about Tove Jansson and her relation to the sea in what they characterise as an "environmental biography", *Meri ja Tove* (2022; The Sea and Tove). Focusing on the particular surroundings of the Pellinge archipelago (to which Klovharun belongs) and the Janssons' relation to this, they do not, however, pay further attention to the wider petrochemical and geopolitical implications of the Gulf. Nevertheless, in the section "The Pollution and Protection of Marine Nature" (Latva and Leskelä-Kärki 2022, 151–156), they take note of the problem of pollution, including an oil-spill. They interestingly highlight one of four pictures with environmental topics from the seventies, all of them signed by both Lars and Tove Jansson. The picture shows the Moomin Troll trying to clean the sea from oil, while a ship resembling a tanker is seen further out (2022, 154). Latva and Leskelä-Kärki also point out a letter from Jansson to Pietilä, in which she, in 1962, critically comments on a previous oil-spill from the Swedish cargo ship M/S Coolaroo, which sunk outside Helsinki in 1961, affecting the Pellinge group of islands one year later.
6 Compare, for instance, Margaret Cohen's discussion of the island as a chronotope in her article "Chronotopes of the Sea" (2006).
7 The Finnish word "neste" means liquid in English.

8 The Swedish original gives "strandskata" (oystercatcher) for scolder.
9 The port of Kotka ("HaminaKotka") is the major port in Finland today, lying even further east on the Gulf, close to the Russian land border.

Works cited

Alaimo, Stacy. 2016. *Exposed. Environmental Politics and Pleasures in Posthuman Times.* University of Minnesota Press.

Auerbach, Erich. 2013. (1946). *Mimesis. The Representation of Reality in Western Literature.* Translated by Leonard R. Trask. Princeton University Press.

Brunow, Stefan. 2010. "Tärnornas krig på Tove Janssons ö." *The Finnish Broadcasting Company YLE,* August 17. Accessed June 27, 2025. https://yle.fi/a/7-515799.

Cohen, Margaret. 2006. "The Chronotopes of the Sea." In *The Novel, Volume 2: Forms and Themes,* edited by Franco Moretti, 647–66. Princeton University Press. https://doi.org/10.1515/9780691243740-039.

Helsingin Sanomat. July 22, 2004.

Holländer, Tove. 1983. *Från idyll till avidyll: Tove Janssons illustrationer till muminböckerna.* Finnish Children's Literature Institute.

Iovino, Serenella. 2009. "Naples 2008, or the Waste Land: Trash, Citizenship, and an Ethic of Narration." *Neohelicon* (36): 335–46.

Jansson, Tove. (1972) 1974. *The Summer Book.* Translated by Thomas Teal. Hutchinson.

Jansson, Tove. (1961) 2019. "The Island." Translated by Hernan Diaz. *The Paris Review.* Accessed June 25, 2025. https://www.theparisreview.org/blog/2019/02/13/tove-janssons-the-island/.

Jansson, Tove, and Tuulikki Pietilä. (1996) 2021. *Notes from an Island.* Translated by Thomas Teal. Sort of Books.

John Nurminen Foundation. 2024. Accessed June 28, 2025. https://johnnurmisensaatio.fi/en/our-work/projects/wastewater-treatment-plant-of-st-petersburg/.

Katavisto, Anja. 1990. *Vanha Sköldvik.* Neste.

Laakso, Maria. 2017. "Idylleistä ironiaan: Saarten monet merkitykset Tove Janssonin tuotannossa." In *Lintukodon rannoilta: saarikertomukset suomalaisessa kirjallisuudessa,* edited by Maria Laakso, Toni Lahtinen and Merja Sagulin, 147–66. Suomalaisen Kirjallisuuden Seura.

Latva, Otto, and Maarit Leskelä-Kärki. 2022. *Meri ja Tove.* John Nurminen Säätiö.

MarineFinland.fi. n.d. "A Major Ship Oil-spill Is the Most Significant Environmental Threat in the Baltic Sea." Accessed June 25, 2025. https://itameri.fi/en/state-of-the-baltic-sea/harmful-substances/oil-spills/.

Piepponen, Hanna, Leena Laamanen-Nicolas, Korpinen Samuli, Meri Back, Jan Ekebom, Janne Suomela, Titta Lahtinen, Pekka Paavilainen, and Henna Rinne. 2024. *The State of Finland's Marine Environment.* Reports from the Finnish Environment Institute: (35).

Rehal-Johansson, Agneta. 2006. *Den lömska barnboksförfattaren: Tove Jansson och muminverkets metamorfoser.* Makadam.

Rodin, Johnny. 2009. "Finska viken—Rysslands val." *Nordisk Östforum* (23): 3, 217–46.

Rosa, Hartmut. 2017. "Available, Accessible, Attainable: The Mindset of Growth and the Resonance Conception of the Good Life." In *The Good Life Beyond Growth: New Perspectives,* edited by Hartmut Rosa and Christoph Henning. Routledge.

Rosa, Hartmut. 2023. "Resonance as a Medio-passive, Emancipatory and Transformative Power: A Reply to My Critics." *The Journal of Chinese Sociology*: 1–11. https://doi.org/10.1186/s40711-023-00195-4.

Russia Fossil Tracker 2025. Centre for Research on Energy and Clean Air, CREA. Accessed June 28, 2025. https://www.russiafossiltracker.com/.

Scales, Karen. 2024. *What the Wild Sea Can Be: The Future of the World's Ocean*. Grove Press.
Sinivaara, Erkki. 2020. *Ympäristövahinkouutinen Antonio Gramsci: Neuvostotankkerin öljypäästöt ja torjuntatoimet suomalaisessa mediassa 1979 ja 1987*. Turku University Press.
The Summer Book. 2025. Directed by Charlie McDowell. SF Studios and Music Box Films.
Vennonen, Pekka. 2006. *Viikinkisatamasta öljysampoon: 40 vuotta öljynjalostusta Porvoossa*. Neste.
Westin, Boel. 2007. *Tove Jansson. Ord, bild, liv*. Albert Bonniers förlag.

3

OBSTRUCTED PASSAGES

Archipelagic thinking and planetary reach in Tomas Tranströmer's *Östersjöar*

Dan Ringgaard

Atlantic prologue

This chapter attempts an archipelagic and planetary reading of Swedish poet Tomas Tranströmer's 1974 long poem *Östersjöar* (*Baltics*). But—water being the great connective—it begins somewhere else on the blue planet. Here is Emily Dickinson's poem #1302:

> I think that the Root of the Wind is Water –
> It would not sound so deep
> Were it a Firmamental Product –
> Airs no Oceans keep –
> Mediterranean intonations –
> To a Current's Ear –
> There is a maritime conviction
> In the Atmosphere—(2016, 567)

The first verse articulates a notion or perhaps even a conviction. The rest of the poem is symmetrical: The three verses in the middle are isolated from each other by dashes and enclosed by coupled verses, the first of which is an argument, the last a conclusion that confirms the initial verse. Meteorologically speaking, the poem makes sense. Anyone living in a temperate coastal climate, as Dickinson did, can confirm that the weather most often comes with the winds from the sea, in her case the Atlantic Ocean. So, the water is present in the winds. It is "more-than-wet" (Peters and Steinberg 2019). But it's not just about meteorology, is it? It is about the "deep sound" of the wind, about the "Airs", a noun that means both winds and songs, about "intonations" and about "a Current's Ear"—all musical or sonic terms

DOI: 10.4324/9781003647638-5

pointing to the poem itself. But it is obviously not just about the poem either. The musical metaphors make the sound of the wind and the water come alive. The poem is about the power, depth, and omnipresence of "Water", the "Oceans", about the "Mediterranean" sea and all things "maritime". The "Current's Ear" rhymes with "Atmosphere", a word for a mood that is in the air, but also the sphere in which the winds circle the Earth in currents making the figure of the current one of wind as well as water supporting the initial insight. It is a planetary poem, insisting on the immensity of a planet beyond reach and its exchanges with and influence on what is going on right here, on the invisible wind that touches our bodies and reaches our ears. There is a sense of wonder to it all.

But then the word "Root" stands out. As a vehicle of the ground and of immobility, it is a disturbing presence within the free-floating exchange of wind and water. The syntactical fragment "Airs no Oceans keep –" can be interpreted in context as the departure of the wind from the water, but read in isolation, as the dash suggests, it contradicts the initial verse. Is there a hint of metareflective irony in "It would not sound so deep" directed towards the bold opening of the poem? And how about the collision of the poetic vocabulary of wind and water with the abstract "Firmamental Product"? The meaning of "conviction" is also unruly. The connotation to verdict may associate to something doomed lurking in the atmosphere, a haunting since the wind speaks of something that is not there. And what exactly is a "maritime conviction"? There is, as often in Dickinson, a teasing tone to it all that makes you wonder about the sense of wonder.

The concepts

What in the poem appeared to be a seamless planetary exchange is interrupted by ambiguity. And that, of course, is what the dashes are about. They too interrupt. They diverge from the seamless syntax, the regular flow of words. They make you wonder. The singularities of metaphor, syntactic fragment, irony, mixed vocabularies, and connotation together with each particular dash obstruct easy meaning. The poem tells the truth, but tells it slant. Instead of the image of a coastline where the wind sweeps effortlessly in from the sea, or of a planet with constant meteorological circulation between sea and air and land, the poem is rather like an archipelago: a sea of interruptions, detours, redirections, and stops—currents rather than convictions. Which makes it truly planetary, since planetarity according to Gayatri Spivak is letting the particularities of the lived world together with the incomprehensible vastness, the sublimity of the planet, disrupt the global order of thought (2003, 70–102). This is why the planetary must be grasped by shifting and twisted scales. The incompatibility of the planetary means that the scales cannot be subsumed into one another. They can be neither systemic (global) nor synthetic (holistic). The shifting and twisted scales call for the same kind of thick description that Zach Horton recommends for the Anthropocene (2019, 6), as does David Ferrier under the heading of "thick time" (2019). Understanding planetarity is all about the reflective construction of assemblages.

The focus on the particular and the critique of the global also lie at the heart of the archipelago according to Edouard Glissant:

> *Archipelagic thought*, the thought of the attempt, of the intuitive temptation, which could be apposed to continental thoughts, which would first be system thoughts. With continental thought the mind sprints with audacity, but the fact makes us think that we see the world as a bloc, taken wholesale, all-at-once, as a sort of imposing synthesis, just as we can see, through the window of an airplane, the configurations of landscapes or mountainous surfaces. With the archipelagic thought, we know the rivers' rocks, without a doubt even the smallest ones.
>
> (1997, 113)

Archipelagic thought favours the particular and the unforeseeable, and it is a geographical thinking. The geography of the archipelago is one of neither oceans nor isolated islands but of groups of islands in the sea, of neither boundlessness nor borders but of connections and trajectories, and flows and interruptions. Archipelagic thinking, Michael Wiedorn writes commenting on Glissant, is about "radical interconnection within difference", and about being at once "rooted and adrift" (2017, xvi–xvii).

Returning to Dickinson's dashes and her inclination to scribble her brief and seemingly loose and fragmented poems on whatever paper that was within her reach, it is tempting to think of them in the image of the archipelago: each word, each line as a world of its own in the currents of sound and meaning that connects them within the poem, and each poem scribbled down on unremarkable surfaces floating around in the poet's environment, "rooted and adrift". Even as her poems are arranged on the page according to the conventions of lyric poetry, with a shift in optics, they remain archipelagos: instead of a dense, self-reliant island floating on the blank sea of the page, what we see is a sea that floats in currents, in and out between the lines and the words of the poem. It is exactly the same shift in optics that can make the continents on a map of the world look like one big archipelago. A truly planetary analogy where the only difference seems to be one of scale. But of course, poems and places are not the same. They can, however, as we shall see, exchange properties as they cross.

The sea

Now, let's return to Tomas Tranströmer. Tranströmer published thirteen books of poetry between 1954 and 2004, and he is translated into more than sixty languages. In 2011, he received the Nobel Prize in literature, because—the committee writes—"through his condensed, translucent images, he gives us fresh access to reality" (Svenska Akademien 2011). In his Award ceremony speech, Kjell Espmark speaks of how the metaphors, so characteristic of the poet, lend a "sensuous precision" to the "huge breathing" in Tranströmer's poems (Espmark 2011). If there is one particular place where this "huge breathing" can be sensed, it is in his long poem *Baltics*.

Tranströmer begins the second movement out of six in *Baltics* with an observation similar to Emily Dickinson's: "The Wind is in the pine forest. Sighing heavily and lightly. / The Baltic is sighing in the middle of the island also, far within the forest you are out on the open sea" (1997, 107).[1] The sea is present on the island not only on a representational level but also on the level of sound. The see-saw motion of opposites in "heavily and lightly" echoes a number of similar phrasings in the poem such as "understanding and misunderstanding" and "great doors opening and great doors closing" (1997, 107, 108). The latter closes the second movement and connects the doors to the wind and the sea, reusing the verb "sighing" (in Swedish *susar*: blows or whistles). The phrasings mime the rocking motion of the waves that is present already in the title. The primary meaning of *sjö* is lake, but the word can also mean "sea" and "large wave". The uncommon plural form suggests that there are many versions of the Baltic Sea and underscores the image of large waves, since these are commonly referred to in the plural as *sjöar* (Ringgren 1997, 12–13). The waves make themselves heard in the two trochaics of the title, emphasised by assonance: *<u>Öst</u>e<u>r</u>s<u>jö</u>ar*. In a similar fashion, the opening of the second movement, quoted earlier, rhymes "ön" (the island) with "sjön" (the sea). The poem ends with yet another see-saw (or should I say sea-saw?) motion uniting the dead with the waves and the wind: "The hovel is lit up / with all those who were driven *by a certain wave, by a certain wind* / right out here to their fates" (1997, 114, my italics). Death prevails as the wind and the waters efface all things human.[2]

In *Baltics*, Tranströmer writes about life around the Baltic Sea from the point of view of Runmarö, a tiny island in the Stockholm archipelago. It is a poem about memory and place. The place in question is, as the plural form of the title suggests, overlapping versions of the Baltic waterscape as it is experienced by the poet through time and made known to him by way of various kinds of mediations: almanacs, anecdotes, photographs, diaries, family history, and place-specific objects. The Baltic waterscape connects life with little regard to national borders and turns it into a region, and it does so from the local and apparently marginal point of the coast-near archipelago (the site-specific Swedish term is *skärgård*). The poem speaks from the archipelago, and it thinks with the archipelago. I will end the chapter by suggesting that the archipelagic thinking, together with the planetary reach of the poem, can serve as a way of doing comparative regional literary studies in particular and perhaps literary studies in general. I will get there by looking at the aquatic elements of the poem, starting, as I have done, with the omnipresence of the sea and continuing with the planet, the aquarium, and the submarine before returning to the archipelago.[3]

The planet

> Here is the opening of the poem:
> It was before the time of the radio masts.
> Grandfather was a new-made pilot. In the almanac, he wrote down the vessels he piloted –
> Names, destinations, draught.
> Examples from 1884:
> SS Tiger Capt. Rowan 16 ft. Hull Gefle Furusund
> Brigg Ocean Capt. Andersen 8 ft. Sandöfjord Hernösand Furusund
> SS St. Petersburg Capt. Libenberg 11 ft. Stettin Libau Sandhamn (1997, 106)

The notes of the grandfather offer a set of trajectories within the Baltics and beyond, connecting the archipelago to global trade flows. They create a web and make up a map of movements between dots. This changes as the poem follows the pilot into "the marvellous labyrinth of islands and waters" (1997, 106). We are no longer above, instead "we're-just-here" (1997, 106). The immediacy of being in-the-middle-of has replaced the outside-of-time panoptics of the map. But then it never was a map, it was a mapping mediated by words written in a tactile historic object: the handed-down notebook. Furthermore, the map is internalised and a matter of experience: "(Had he the labyrinth in his head?)" (1997, 106). The grandfather is guiding ships through the labyrinth and the "thick fog" towards the open sea (1997, 106). If it all seemed accessible to the distant gaze, the maze and the haze moves the poem further into the invisible: "At one single stride the cape came out of the invisible" (1997, 106). The grandfather's "eyes read straight into the invisible" (1997, 106). The global flows and systems are substituted by the particulars of time and place, the map by the labyrinth, and positive knowledge by riddle and wonder as the human world is absorbed by the environment. Globalisation becomes planetary.

The opening line of the poem, "It was before the age of the radio masts" (an echo perhaps of the opening of Herman Melville's *Billy Budd*: "In the time before steamships"), contains a similar tension. The radio masts, this potent symbol of globalisation, electrifying and connecting the human world, are set aside as the poem moves back in time to a world untouched by distanced and speeded-up communication. By the end of the first movement, the abstract connectivity of the radio masts is substituted by the substantial beauty of the boat's "compound machine" (1997, 106). Its stamina is compared to the human heart, its movements to acrobats, and it smells of kitchen. Throughout the poem, technologies are necessary media for connecting to memory and place.

The effacement of things human by the planet is evident in a passage from the third movement where the speaker of the poem looks at an old photograph. The people in the photo are repeatedly "in the process of being rubbed out", and their boat seems foreign to the landscape (1997, 109). What is left are the "ripples of the water", the "sighing" of the wind in the spruces and the tactile "rough rock-faces" (1997, 109). In conclusion: "The waves are topical" (1997, 109). What is present is what was always present. Deep time outlives human-scaled time. After this shift in optics between figure and ground, the scene changes and the poet walks the beach at night and notices a planetarium and above it the night sky "full of ciphers" (1997, 110).[4] He compares the power of the optical instruments to "a locust swarm that denudes the acreage of Somaliland in half an hour" (1997, 110). This leap of thought from technosphere to biosphere and to a distant part of the planet by way of the planetarium and the night sky leads to a planetary conclusion: "The summing-up can't be done, the summing-up is impossible" (1997, 110). The planetary is incompatible.

The aquarium

Shortly after the "The great current" of sea and wind of the second movement, these enigmatic lines occur:

> You go on, listening, and then reach a point where the frontiers open
> or rather
> where everything becomes a frontier. An open place sunk in darkness.
>   The people stream out from the faintly lit buildings round about.
>   Murmuring.
>
> A new breath of wind and the place lies desolate and silent again (1997, 107).

The paradoxical first sentence encourages several conjectures regarding the sea and borders: the border between the sea and the island that is overrun by the wind, the coast-near archipelago as a border between the mainland and the sea, the borderless expanse of the sea, the sea as a place between borders, the idea of the border as a meeting place, the sea as the border of the land, the borderline of the horizon, and, not least, at the time of the poem's publication, the hard border between Eastern and Western Europe epitomised in one of the most frequently quoted lines of the poem: "But it's a long way to Liepaja" (in Latvia) (1997, 111). This points to an archipelagic mode of thinking. The Baltics are a place neither of definite borders like the island nor of open space like the sea, instead they are presented through the dynamic image of the pilot navigating a labyrinth of pathways and obstructions.

The archipelagic space of border/not border is directly followed by another kind of space: the "open place".[5] The place is delimited as it is "sunk" as if in water, the people move like a "stream", and the passage ends with a third reference to the sea: "Murmuring", in Swedish: "Det sorlar" (it murmurs)—the "it" referring to the sea. "Det sorlar" echoes "det suser" ("it sighs"), repeated four times in the previous lines, also referring to the sea. The "it" of "Det susar" and "Det sorlar" gives agency to a nature that is more enduring than things human. The dimly lit square (there are buildings around it) is sunk in water metaphors, confirming the omnipresence of the sea. The enigmatic force of the sunken square comes from its lack of connection with what comes before and after in the poem, making it a singular poetic image; it comes from the very long verse that spans two lines and prolongs the image; but it comes first of all from a sense of looking into an aquarium. An aquarium is a deep-sea simulation in a magically diminished scale. It is the limitless within the limited, a visual machinery of wonder. It pulls the beyond of the planet within reach and makes it transparent (Andersen 2023, 62–74, 135–60). The wonder in this case, though, is uncanny. There are people, not fish, in the aquarium. They spill out on the empty, shadowy square. Do they react to some kind of emergency? Do they seem lost?

A key quality in Tranströmer's poetry is its condensed imagery. The sunken square is one such example of a *Lichtung*. It is characterised by a sensibility to place—in this case, the exposed towns and rocky islands of the archipelago at nightfall or sunk in winter darkness—and its metaphorical extension into what can hardly be sensed and into wonder. What makes *Baltics* stand out from this Tranströmer commonplace is its proximity to prose, the unusually long verses that border on and sometimes spill over and into prose. In *Baltics*, prose is that which brings the immense and contingent realities of the lived world and of history into intelligible form. Poetry is the condensation and incantation of the world, often in the form of a fragmentary image, such as the sunken square. It seems that the "huge breathing" (to quote Espmark) of the sea thrives not only on the incantation of poetry but also on the duration of the long poem and the world of prose.

The submarine

Baltics is a montage poem in the tradition of T. S. Eliot and Ezra Pound and at the same time a symphonic poem in six movements. As a two-sided structure of *opsis* (montage) and *melos* (symphony), *Baltics* is both a fragmented space and a flow of time—or in keeping with the image of the archipelago: a structure of flows through fragments. The fourth movement changes the pace of the poem, introducing four short pieces of prose, rather like a scherzo that interrupts the major motifs of a symphony. (Music is also thematised in the poem, especially in part three, and Tranströmer has on several occasions used musical forms in his poetry.) It also changes the scale of the poem as it moves to close-ups of the small things living in or by the sea. There are the "weed forests" of bladderwrack "that holds itself up with air-bladders" (1997, 110). Someone might wish to "sink to a certain depth" through his mirror image and down into that submarine forest (1997, 110). There is the bullhead seen through water and disentangled from a net, leaving nothing but gleaming slime on one's hands. There are the insects that "scurry" on the "sun-warmed lichens" and the grass on "the lee-side" growing with "a faint drumming from underneath" (1997, 110).

This change of scale from the human world of history and the natural world of wind and sea in movements two and three slows down the pace of the poem. After leaping through history in the third movement (I shall return to this), the reader is now asked to closely watch the life under or just above the tiny waters. The deceleration also changes the quality of time. The insects are like "second-hands" on a watch, the pine shadows on the lichens are like "an hour-hand—inside me time stands still, an infinity of time" (1997, 110). The slow time becomes a *durée*, and the *durée* becomes the antipode to the historical time set by the opening line of the poem. It is also a different kind of time than the deep time of the sea and the wind. It is the time of human consciousness. Incompatible types of time coexist. Likewise, space comes in different scales, underlining the incompatibility of the planetary. The maritime close-ups are replaced by "the width of water, without doors, the open frontier . . . the Baltic is a calm endless roof" (1997, 111). In a similar manner, the sea that the pilot steers towards in the first movement is counter-scaled by a much smaller sea as the difficulties of navigating the archipelago are likened with "carrying a brimful pail without spilling a drop" (1997, 106). After the omnipresence of the sea has been established in the second movement, the third movement opens with another tiny sea, this time within a Middle Age baptismal font in a church on Gotland (a large island on the east coast of Sweden). Soon after, a hospital drip that drips "drop by drop" adds to the assemblage of watery scales (1997, 109). The aquarium of the second movement is of course another example of the change in water scales.

The maritime close-ups of the fourth movement spill over into the fifth that begins with a diary entrance arranged like the prose pieces that proceed it. Here, the jellyfish "pump themselves forward" and

> drift like flowers after a sea burial, if you take them out of the water all their form vanishes, as when an indescribable truth is lifted out of silence and formulated to an inert mass, but they are untranslatable, they must stay in their own element.
>
> (1997, 111)

The jellyfish or *Aurelia* are followed by aphasia, dreams, and music, all versions of the untranslatable, of "Something which can't be said", of "fragments of the high nocturnal style that drew past?" (1997, 111). Something—a truth, a wonder—is out of reach of language but perhaps within grasp of the poetic image and of the sound and the rhythm of the poem.

The paradox of out-of-reach-but-within-grasp is present by the end of the third movement. I quoted the conclusive line: "The summing-up can't be done, the summing-up is impossible." It is followed by: "The summing-up is the mandrake –", and by a parenthesis that informs the reader that this "miracle-working plant" screams so appallingly when torn from the ground that "a man would drop dead. A dog had to do it" (1997, 110) (the analogy to the jellyfish and the bullhead taken out of water is clear). The Swedish poem uses "ALRUNA" for mandrake, a word that literally means the magic letter or the cipher that evokes the whole (Tranströmer 1974, 22). This explains why the phrase "The summing-up is a mandrake –" comes the second time around without its contradiction: "the summing-up is impossible." Furthermore, "ALRUNA" in capitals at the end of the third movement mirrors the beginning of the movement where the name of the mason that made the aforementioned font, "HEGWALDR", is also written in capitals. His name on the font is "like a row of teeth on a mass grave", and it is followed by the most chaotic and destructive passages of the poem, ending with the uncanny rotation of the font "like a slow rumbling merry-go-round" (Tranströmer 1997, 108). The deadly mandrake confirms the presence of death that lies heavily on the third movement especially, but the cipher of wholeness, "ALRUNA", defies the disintegrating power of history.

The archipelago

The tension in *Baltics* between the excessive and devastating immanence of history as prose and the transgressive experiences captured in poetic images, and in the represented as well as the actual rhythm of the poem, is analogous to the tension in the concept of planetarity between a surplus of close to life particularities and the sublime beyond of the inconceivable planet. *Baltics* ends with the image of a dilapidated fishing hut. As often in Tranströmer, the rowdy particulars of place slowly evolve into a figure:

> So much wood crouching. On the roof, ancient tiles which have slipped downways
>   and crossways over each other
> (the original pattern deranged over the years by the rotation of the earth)
> it reminds me of something . . . I was there . . . wait: it's the old Jewish cemetery in
>   Prague
> where the dead live more packed than they were in life, the stones packed packed.
> So much love encircled! The tiles with their lichen-script in an unknown tongue
> are the stones in the ghetto cemetery of the archipelago folk, the stones raised and tumbled.—
> The hovel is lit up
> with all those who were driven by a certain wave, by a certain wind
> right out here to their fates (1997, 114).

The hut is marked by deep history ("an unknown tongue") and planetary involvement ("earth's rotation") as it connects to the human history of the Jewish cemetery in Prague. The awry gravestones of the Jewish community visually as well as semantically echo the "row of teeth in a mass grave" at the beginning of the third movement. Time and place implode in the gradual discovery of a resemblance between the distorted tiles and the tumbled gravestones, and holocaust and death are given the penultimate word of the poem—the last word going to the wind and the waves. The blending of the tiles and the gravestones, of surfaces overlapping, criss-crossing, sliding, and tossing in an ultra-slow pattering, ultimately gives way to the image of an archipelago. The dilapidated roof of the fishing hut becomes a metaphor for the poem, itself archipelagic in its montage form, challenging the idea of a symphonic composition progressing smoothly in time. As a poem of place and memory, *Baltics* takes the spatial form of the archipelago, of flows obstructed by fragments.

The image that the poem offers of the Baltic region is one of a vast archive of particulars spread out in time and place and held together by the sea and the trajectories that it makes possible. The poet navigates the labyrinth connecting isolated incidents, or in the words of the poem: "People, beasts, ornaments. / There is no landscape. Ornaments" (1997, 109). Ornaments being the dynamic architectonic equivalent of trajectories, whereas landscape may be seen as the totalising and immobilising gaze that the poem disavows. As always, Tranströmer is place-specific. The coast-near archipelago does not qualify as a landscape. It is too spread out, too difficult to survey from the viewpoint of the pilot in his boat, too monotone, too remote from any traditional art historical appreciation of landscape. And, literally, it's not a landscape. It's a waterscape. The water is what binds the region together, it is omnipresent, as deep time it effaces human history, it negates terrestrial distinctions like the one between border/not border, it comes in all scales from the tiny waters of everyday life via human-scale aquariums to planetary immensity, it comes with the wind, it comes as fog, and as the all-encompassing life rhythm of the waves it's surface can be "a calm endless roof", and its depths a place of wonder.

The obstructions

Baltics speaks from the archipelago, it takes the shape of the archipelago, and it thinks with the archipelago. What can literary scholarship learn from that, when it comes to matters such as place, region, and sea? I suggest that we regard the archipelagic thinking of the poem as obstructions in the sense of Lars von Trier in his and Jørgen Leth's movie *The Five Obstructions* (2003). The purpose of an obstruction is to derail habits, habits such as organising texts in national or supernational containers, homogenising texts within the containers, and obliterating texts that don't fit the prevailing narratives. And also habits such as thinking the planet from the continent, more or less ignoring the most planetary element of the Earth and of our bodies: water. And even more generally: habits such as systemic and anthropocentric, in two words: *global thinking*. Here are eight obstructions. They do not offer an alternative system. They do not oppose. In Glissant's terminology, they appose. Their purpose is to deflect and redirect:

1. *Reverse figure and ground.* Put place and all that constitutes it (geology, climate, fauna and flora, architecture, everyday objects, etc.) centre stage and let human actions and characters join the place instead of using it as a backdrop.
2. *Write spatial history*, not historical narrative. History is central to *Baltics* but not as chronology and linear causality. Instead, history makes itself visible in each place as remnants of lives lived and as a vertical axis crossing the horizontal axis of geography.
3. *Make it personal. Baltics* is situated knowledge. Like the sea, the speaker of the poem has a discrete but absolute presence. He tells of a place that is important to him. The choice and distribution of subjects and the voice of the poet are what hold the poem together.
4. *Do decentred mapping.* Circumscribe a space, don't try to be comprehensive, and avoid thinking from the centre, instead think from places such as Runmarö. There is no centre in an archipelago, everywhere is centre. This is a way of doing power geography.
5. *Compare, don't contain.* Compare incidents of many kinds (texts, objects, images, events, and ideas) and scales, as long as they are approximately within the circumscribed space and make up meaningful worlds. It's a kind of thick description. Let's call it Comparative Regional Literary History.
6. *Think with the planet.* Exploit the tension between the infinite particulars and the excesses of the sublime. Make the singularities of the sensible world and of texts supplement meaning and postpone truths. That's how literature works. Although not always with the planet as horizon.
7. *Write in water*, not in stone. And not only *of* water and stone. See where it takes you. Archipelagic thinking is unpredictable. Let the writing be "rooted and adrift". Redirect attention. Make connections sometimes by way of analogy or association. Be alert. Navigate the archipelago.
8. *Write prose.* Prose is the discourse that strains to contain and make sense of the excess of the phenomenal world, and in doing so, it not only reaches beyond literary texts, it also provokes its counter-image: poetry as the vehicle of the equally untranslatable subjective truth.

Notes

1 I will be using Robin Fulton's translation. Although at some points perhaps slightly archaic, Fulton catches the rhythm and tone of Tranströmer well compared to the more modern choices made by Patty Crane (Tranströmer 2023).

2 For the workings of the waves and the wind, see also Nielsen 2002, 63–65.

3 In a previous essay (Ringgaard 2021), I read Tranströmer's poem within the theoretical framework of world literature, including the concepts of the archipelago and planetarity. There are some overlaps between the two essays, but here I will be reading the poem through another lens, that of the blue humanities.

4 I have exchanged "numbers" in Robin Fulton's translation with "ciphers".
5 See Svenska Akademiens Ordbok, https://www.saob.se/artikel/?unik=P_1053-0096.fa03&pz=3.

Works cited

Andersen, Frits. 2023. *Underværker: Fortællinger om havets vidundere*. Aarhus University Press.

Dickinson, Emily. 2016. *The Complete Poems*. Edited by Thomas H. Johnson. Faber and Faber.

Espmark, Kjell. 2011. "Academy Award Speech." *NobelPrize.org*. Accessed June 27, 2025. https://www.nobelprize.org/prizes/literature/2011/ceremony-speech/.

Ferrier, David. 2019. *Anthropocene Poetics: Deep Time, Sacrifice Zones and Extinction*. Minnesota University Press.

Glissant, Éduard. 1997. *Poetics of Relations*. University of Michigan Press.

Horton, Zach. 2019. "The Trans-Scalar Challenge of Ecology." *ISLE: Interdisciplinary Studies in Literature and Environment* 26 (1): 5–26.

Leth, Jørgen, and von Trier, Lars. 2003. *The Five Obstructions*. Zentropa.

Nielsen, Birgitte Steffen. 2002. *Den grå stemme: Stemmen i Tomas Tranströmers poesi*. Arena.

Peters, Kimberly, and Philip Steinberg. 2019. "The Ocean in Excess: Towards a More-than-wet Ontology." *Dialogues in Human Geography* 9 (3): 293–307.

Ringgaard, Dan. 2021. "The Region as an In-between Space: Tomas Tranströmer's Baltics and the Making of an Archipelagic Nordic Literature." In *The Cambridge History of World Literature*, edited by Debjani Ganguly, 495–510. Cambridge University Press.

Ringgren, Magnus. 1997. *Det är inte som det var att gå längs stranden: En guide til Tomas Tranströmers Baltics*. Bokbandet.

Spivak, Gayatri Chakravorty. 2003. *The Death of a Discipline*. Columbia University Press.

Svenska Akademien. 2011. "Press Release." Accessed June 27, 2025. https://www.nobelprize.org/uploads/2018/06/press-16.pdf.

Tranströmer, Tomas. 1974. *Baltics: en dikt*. Albert Bonniers Förlag.

Tranströmer, Tomas. 1997. *New Collected Poems*. Translated by Robin Fulton. Bloodaxe Books.

Tranströmer, Tomas. 2023. "Östersjöar/Baltics." In *The Blue House: Collected Works of Tomas Tranströmer*, translated by Patty Crane. Copper Canyon Press.

Wiedorn, Michael. 2017. *Think Like an Archipelago: Paradox in the Work of Éduard Glissant*. State University of New York Press.

4

WRITING THE DELUGE

Drachmann and the 1872 Baltic Sea flood

Søren Frank

"A strange landscape!"—with this brief, unelaborated exclamation, Holger Drachmann registers his bewildered encounter with the aftermath of one of the most destructive storm surges in Danish history. The sentence is striking less for what it conveys than for what it withholds. Coupling the nondescript adjective "strange" (*besynderligt*) with the emphatic finality of an exclamation mark, the utterance encapsulates the challenge that structures this chapter: how can literature represent the disorienting aftermath of oceanic catastrophe, a reality that defies symbolic containment and overwhelms narrative form?

The Baltic Sea flood of November 1872 claimed about 300 human lives. It caused massive material losses, drowned villages, and led to long-term engineering responses, including dike construction and land reclamation. Literary responses also emerged soon after: poems by Hans Christian Andersen and Christian Richardt; prose reportage by Drachmann; and Friedrich Spielhagen's German novel *Sturmflut* (1877). Later reflections include Sophus Claussen's memoir "Fabius' Stamtavle" (1927) and recent novels such as Sven Halse's *Og stormfloden kom* (2022) and Peder Frederik Jensen's *Rans vilje* (2023), which revisit the flood in light of contemporary climate anxieties.

Drachmann, a 26-year-old painter and literary debutant who would later become a prominent artist of the Modern Breakthrough, was dispatched by the newspaper *Dagbladet* to report on the devastation. His three dispatches, "Efter Orkanen" (1872; After the hurricane), stand as both vivid literary records of disaster and poignant documents of aesthetic struggle. They are rhetorically sophisticated and sharply observant, yet Drachmann was navigating multiple mimetic complexities inherent in his subject matter and confronting substantial rhetorical challenges—challenges that are not incidental but constitutive. The result aligns with what Steve

DOI: 10.4324/9781003647638-6

Mentz has labelled "Salt Aesthetics": an artistic practice attuned to the "disorienting pressure of the inhuman sea" (2018, 72). Rather than seeking narrative mastery or formal containment, Drachmann's prose attempts to stay with the turbulence, rendering not just the material consequences of the storm but the instability it imposes on perception, form, and voice.

This chapter considers how literary responses to historical disasters like the 1872 flood continue to resonate in the present. Such texts help sustain the collective memory of storm surges—even if the 1872 event itself has faded from public view and must be periodically reactivated, as in Jensen's *Rans vilje*. More broadly, storm surges tend to be forgotten from one generation to the next, making the literary archive an essential resource for cultivating both an emergency alertness and a cultural preparedness, particularly in light of the Anthropocene's escalating risks (cf. Ritson 2019, 27–28). The Baltic Sea flood prefigures the climate-related calamities of the twenty-first century, exacerbated by global warming and compounded by persistent patterns of residential development along vulnerable coastlines. Drachmann's reports, moreover, embody deeply embedded assumptions about nature, agency, and human–environment relations—assumptions that are central to our efforts to understand and inhabit an increasingly unruly planet.

Writing after

If we consider the 1872 Baltic Sea flood as a "black hole"—a traumatic event that challenges artistic representation to its very limits and threatens to absorb or annihilate any attempt at narrative—then one mitigating factor in Drachmann's case is his temporal distance from the event. He arrived belatedly, more than a week after the flood. His reports are thus marked by what Dominick LaCapra has termed "belated effects": "Trauma indicates a shattering break or cesura in experience which has belated effects. Writing trauma would be one of those telling aftereffects" (2014, 186). Drachmann's task is not to document the catastrophe as it unfolds, but to write trauma retroactively. His reliance on secondary witnesses further reflects the complex temporality of trauma, which, as Cathy Caruth argues, is not simply the experience of a violent event but rather the way in which that event was not fully grasped in the moment and thus returns, belatedly, to haunt the survivor (1996, 4).

This double belatedness—the temporal delay between catastrophe and arrival, and the deferred temporality inherent in trauma itself—poses a dual difficulty. On the one hand, trauma, as LaCapra notes, "is a disruptive experience that disarticulates the self and creates holes in existence; it has belated effects that are controlled only with difficulty and perhaps never fully mastered" (2014, 41). On the other hand, literature that seeks to represent trauma must contend with what Caruth calls "the story of a wound that cries out, that addresses us in the attempt to tell us of a reality or truth that is not otherwise available" (1996, 4). Caruth insists that this truth "cannot be linked only to what is known, but also to what remains unknown in our very actions and our language" (1996, 4). Writing trauma thus entails a paradox: it attempts to bear witness to something that resists comprehension and representation.

On a personal, existential level, trauma continues to destabilise the individual who experienced it, and he or she may never achieve a state of control or resolution. But, as LaCapra makes clear, writing trauma may itself constitute an after-effect of the traumatic event. If that is true, then writing trauma faces the same limits as the traumatised self. The literary form and language used to represent and process trauma are also vulnerable to distortion, fragmentation, and opacity—that is, difficult to control or master for the writer. Caruth notes that trauma involves "a crisis of truth" (1996, 6), one that must be explored "not in a straightforward way", but in a language "that defies, even as it claims, our understanding" (1996, 5). This is precisely what characterises "Efter Orkanen": a form of aesthetic vulnerability that becomes a productive strength. Literature may help render the "holes in existence" caused by natural disasters visible, understandable, and partly processable through narrative. Yet in doing so, it does not simply repair or restore but reflects critically on the ways trauma is (or fails to be) integrated into individual and collective identity and memory.

Mediated voices

A defining feature of Drachmann's reportage is its reliance on mediated testimony. He cannot document the catastrophe as it unfolds, but must instead reconstruct it through conversations, overheard accounts, written traces, and retrospective narratives. The text is thus shaped by a complex relay of voices, each marked by temporal distance and subjective filtering. This polyphonic texture is not merely a documentary necessity; it is a rhetorical and epistemological strategy that underlines trauma's belated effects—what Drachmann renders as the difficulty "to tell again second hand" (23 November).[1]

Drachmann is a collector of fragmented stories. In Roskilde, he listens in on "a group of sailors" (21 November), whose speech reveals the narrative logic of catastrophe as dispersed and accumulative, privileging the trivial over the traumatic and thereby mirroring the representational challenge of disaster itself: "Characteristically for their way of communicating, it was almost always the relatively insignificant events that were at the forefront of the longer development of danger, loss, exhaustion, and finally catastrophe" (21 November). In the evening, at Hotel Prinsen, a

> cloud of tobacco smoke hung over the whole scene, and through this smoke I heard stories of shipwrecks from fore and aft, from starboard and port, until I myself felt like a stranded ship that had to dispose of its cargo.
>
> (21 November)

The drifting smoke evokes the opacity and friction of post-disaster testimony, as both survivors and narrator struggle to articulate the inarticulable; submerged in this fog of voices, the narrator internalises the shipwreck, becoming himself a foundering vessel burdened with narrative cargo. These mediated voices introduce an indirect, deferred transmission of an experience not fully assimilated when it occurred.

The use of secondary narrators also dramatises the instability of narrative authority. In one instance, Drachmann reports, “my coachman told me”, and “the coachman now described” (24 November), placing narrative agency in the hands of a local coachman. Onboard a shipwreck, he discovers “the draft of a letter” (24 November), which he quotes at length, treating epistolary remnants as documentary fragments. These insertions displace the author’s own voice and signal a distributed authorship of disaster. This narrative dispersal reflects the ontological destabilisation in Drachmann’s prose, which we will examine more closely in the following section. The catastrophe unmoors not only the physical world but also the structures of representation: experience becomes fragmentary, voice circulates around the black hole of disaster without clear origin, and truth resides in the sediment of multiple, contingent tellings. People bear witness, not to a coherent event, but to the frayed edges of memory and perception. In this way, “Efter Orkanen” renders the catastrophe not as a single story but as an archipelago of testimonies floating in the wake of deluge.

Fluid boundaries and watery forms

The storm surge confronts Drachmann not only with how to know and narrate catastrophe but with what kind of world emerges in its wake. So far, I have approached the event through an epistemological lens, probing belatedness and mediated testimony. Yet as the landscape is reshaped by water—blurred, widened, and liquefied—epistemology proves insufficient. What is at stake is not just the delayed grasp of a past event but a hydro-infused destabilisation of ontological categories: land and sea, solid and liquid, and subject and object begin to dissolve. Drachmann’s turn to fluid boundaries marks a shift from knowing disaster to inhabiting a radically altered world.

The storm surge is rendered as an ontological breach. Saltwater spills beyond its habitual boundaries, flowing inland, mingling with mud and ruin, and with freshwater rain. This opens a space for rethinking the coast not as a liminal line between two stable entities—land and sea—but as what Philip Steinberg and Kimberley Peters call “a dynamic environment of flows and continual recomposition where, because there is no static background, ‘place’ can be understood only in the context of mobility” (2015, 258). To help conceptualise such transformations, they advance the notion of “wet ontology”. Against terrestrial ontologies that privilege spatial fixity, they advocate for an attentiveness to the volumetric and fluid dynamics of the ocean. The sea, they argue, does not merely occupy space but incessantly reshapes it, making space itself an unstable category.

At the heart of Steinberg and Peters' intervention is their critique of static geographies. In 2019, they extended "wet ontology" by arguing that the ocean exceeds wetness, demanding a "more-than-wet ontology" attuned to fluidity's relational, atmospheric, and sensory entanglements. "The ocean", they argue, "is not simply liquid; it is not simply wet. It is solid (ice) and air (mist); it generates winds, which transport smells and tastes that permeate senses and imaginations, emoting the 'marine' and the 'maritime' miles inland" (2019, 294). Drachmann's flooded coast is exemplary in this regard, his narrative being attuned to an ontological excess where the sea widens the coast, reclaims land, penetrates homes, and together with constant rain and enveloping fog liquefies the conceptual binaries that structure terrestrial life. The reports stage an encounter with a world in which boundaries do not sit still. The storm is not a discrete event. Its effects linger in bodies and landscapes, blurring distinctions between weather and writing. The sensory imprint of the storm—its smells, moisture, animal and human debris—is not contained by time or topography.

This ontological disturbance is captured in a telling passage:

> the landfall with the stationary mills, with the factory chimneys that did not smoke, with the town that looked like a shapeless clump of fog around a church tower, fell exactly into the tone of the leaden, stagnant water surface.
>
> (21 November)

The mechanisms of industrial society have stalled, suggesting an order of malfunction and purposelessness. But more radically, the town dissolves into the fog and becomes indistinguishable from the sea. Similar images of spatial dislocation abound. Firstly, as a reciprocity between land and sea: "Masts and ropes mingled in a companionable manner among the beech trees of the forest. The trunks jutted out over the deck, and the masts nodded in between the trees" (25 November). Secondly, as a world turned on its head: "monstrous tree roots twisted in fantastic twists and turns high up in the air" (24 November). Each scene captures not just spatial dislocation but categorical confusion through inversions. These images are not just surreal juxtapositions but representational strategies for articulating a world undone.

The culmination of liquefaction appears in a passage that strongly resonates with a more-than-wet ontology:

> All around was a layer of sand, mixed with gravel, stones, and seaweed. It had penetrated the houses; it stretched over fields, meadows, and gardens; it was no longer the southern tip of the fertile island on which one walked; it was the bottom of the ocean over which one drove, the bottom of the ocean, covered with everything that belongs to the ocean: wreck, destruction.
>
> (25 November)

What was once solid and domestic is now marine and wrecked. The ocean is no longer distant or delimited. It is here, and it is everywhere. This ontological dissolution of categories prefigures Peters and Steinberg's argument that oceanic thinking invites us to consider how "the ocean exists far inshore, above ground, underground, in our senses, and as part of fantasy" (2019, 295). Drachmann's reportage performs precisely such an inland oceanic infiltration.

The experience of elemental saturation is not only thematic but formal. Drachmann's narrative reflects a shift in medium under the pressure of environmental immersion:

> Out on the stone pier, . . . we sat down to draw. The water seeped up from below and the rain sprinkled down from above. The leaves of the sketchbooks were soon curled and wet. We grabbed the paint boxes. They soon became full of rain, and we had to give up this attempt to capture what we saw before our eyes. But what we saw remained in our memory, and here it is.
>
> (21 November)

Drachmann repeatedly insists on wetness: foggy, watery, humid, misty, and thickening, saturated with sand and silt. These conditions thwart his attempt to respond visually to the ruined coastal landscape: the visual medium dissolves; drawing becomes prose; and sight becomes memory. This too is a fluidisation: of genre, of medium, of perceptual modality. The scene exemplifies yet another delay—between the perception of catastrophe's after-effects and artistic composition. But even the verbal descriptions remain haunted by the impossibility of full capture.

In sum, Drachmann's dispatches can be read as a literary exercise in thinking *with* the sea specifically, *with* water generally. His text responds to a world rendered unstable not merely by destruction, but by ontological overflow. It is a world in which land becomes oceanised, where roots float and vessels ground, and where language must shift from depiction to attunement. If, as Steinberg and Peters argue, a wet ontology challenges us to think *from* the sea and water's fluidity rather than merely *about* it (2015, 261), then Drachmann's narrative offers a remarkable nineteenth-century attempt to do just that. It registers not only the force of water but its capacity to undo and recompose the world—a capacity we are only beginning to fathom in the Anthropocene.

From contact zone to force field

Mary Louise Pratt's original concept of the contact zone described "social spaces where cultures meet, clash, and grapple with each other" (1991, 34) under conditions of asymmetrical power, particularly in colonial contexts. The coast is emblematic of such encounters. In *Planetary Longings* (2022), Pratt reorients her concept for the environmental humanities, warning against domesticating it into a metaphor for solutions to situations of heterogeneity and conflict, because this would deny the concept its critical force and jump "over the necessary step of thinking through the chaotic, uncontrollable energies that are in play" (2022, 130). Instead, the contact zone "is a zone of hazard and unpredictability" (2022, 133)—a space where destructive forces and material entanglements unfold in turbulent relation.

In this expanded environmental framework, the coast remains central but now hosts a broader range of actors: not just imperial powers and colonised subjects, but human and more-than-human entities, including the non-living. In what may be her most striking theoretical shift, she argues that even the multispecies frame of the contact zone may be inadequate for capturing the agency of elemental, non-sentient forces—fire, wind, darkness, and, crucially, water. These are not life forms but powerful actors that "trigger chains, combinations, and arrays of events" (2022, 135). For such forces, Pratt proposes the concept of the force field: a zone shaped by the raw capacity to act. "How can lakes, rivers, and seas that rise, fall, flow, sustain, and destroy be imagined as inert?", she asks (2022, 135). These geoforces do not desire life or death; "they do not face extinction" (2022, 136). And yet they define the conditions of survival, belonging, and meaning. Just as the original contact zone displaced Eurocentrism by refocusing analysis on colonial peripheries, the environmental contact zone displaces anthropocentrism by asking us to "redefine the contours of our sphere of attention, observation, engagement, and care" (2022, 131). Here, the work of curiosity is reengineered; attention must be trained on non-human agencies and altered ecologies.

Drachmann's storm-ravaged landscapes vividly evoke such a site of planetary entanglement, documenting a world where traditional coordinates—cultural, spatial, and ontological—begin to dissolve. Consider his depiction of the inundated coast: "I looked around. Shattered dinghies, flooded gardens, wreckage, and an unbearable muddle" (23 November). This is a coastal contact zone not only in a literal sense but in Pratt's extended meaning: a site where history and geology, and human structures and elemental forces collide. It is not simply a backdrop for disaster but an active agent of material encounter and ontological confusion. Drachmann often describes the flood not as an isolated calamity but as a distributed field of forces—shaping not only geography but sensory experience, social rhythm, and the limits of narration. One report mentions "the muddy and almost impassable roads", continuing to stress that "it was swampy and morass-like here", finally pointing to an "insane, furious, corruptible mixture of water, sand, clay" (24 November). What Pratt names a force field—an assemblage of death-dealing and life-defining powers—finds literary form here in the thick descriptions of marine encroachment and disorientation. If the contact zone emphasises encounter, the force field emphasises atmospheric immersion and submission to non-human temporality.

Figuring disaster

In his 1956 essay "Two Aspects of Language and Two Types of Aphasic Disturbances", Roman Jakobson proposes a foundational distinction between metaphor and metonymy as the two basic poles of language, cognition, and artistic production. Metaphor operates through substitution, founded on similarity-in-difference, and organises meaning along the vertical axis of selection. Metonymy, by contrast, is based on contiguity—closeness in time, space, or causality—and works through contextual association along the horizontal axis of combination. "The development of a discourse may take place along two different semantic lines", Jakobson explains: "one topic may lead to another either through their similarity or through their contiguity. The metaphoric way would be the most appropriate term for the first case and the metonymic way for the second" (1971, 254).

These two tropes are not merely rhetorical devices but reflect fundamentally different modes of thought. Metaphor is paradigmatic, centripetal, and inclined towards symbolic condensation, coherence, and analogy. It underpins poetic and idealist worldviews that seek resonance and unity; its generic form is poetry. Metonymy, on the other hand, is syntagmatic, centrifugal, and generates meaning through displacement, fragmentation, and adjacency. Jakobson associates metonymy with realism, narrative prose, and impressionistic styles in which cause and effect are obscured, and details in the form of sensory overload accumulate without resolving into a whole. Ultimately, metaphor implies a coherent universe of symbolic correspondence, a world interpretable through resemblance, while metonymy suggests a reality fractured by contingency and resistant to synthesis, a world marked by opacity and disconnection.

Drachmann's reports oscillate between these two poles. Metaphoric strategies help him make sense of the chaos and epistemological confusion by reorganising the disfigured landscape into recognisable figures. When two ruined houses resemble "a fat man crouching" and "a thin man humbly bowing to the fat one" (25 November), he reorders disfigured material into anthropomorphic forms. Metaphor here is a cognitive anchoring—a way of seeing the wreckage not just as broken, but as legible through figuration. Such metaphors do not merely aestheticise catastrophe; they perform the labour of comprehension by transforming perceptual rupture into recognisable shapes.

It is in this context that Drachmann introduces his most explicit statement of literary intention, invoking the metaphor of a previous Danish military campaign, to frame the catastrophe:

> Forgive me for seeing, in the present destruction, . . . the outlines of a sad campaign in our history . . . . But I intend to convey a mood, not to list one by one all the losses that the last catastrophe brought; I intend to reproduce in brief, sketchy strokes the landscape where the accident happened, and I must have a mood to start from that is known to all and which, with all its eeriness, can form the background for my picture.
>
> (25 November)

This declaration is not only a reflection on rhetorical method but a defence of metaphor as a mode of historical and affective representation. By evoking "a mood" rather than enumerating empirical details, Drachmann claims aesthetic license to shape the disaster through analogy—linking the present scene to the military defeat at Ditmarsken in 1500. The deluge becomes decipherable through historical metaphor; the flooded fields are like battlefields, the retreating Danish army like a displaced rural population. Yet his insistence on a "background mood" also signals an awareness of metaphor's limitations: it provides form, but not facts. His use of metaphor thus alternates between assertive framing and epistemic humility. Drachmann's "mood" implies a poetics of catastrophe—one that relies on analogy to process trauma, while conceding that any representation must begin from a shared (in this case: national) mood, not total knowledge.

Drachmann's most ambitious attempt to render the catastrophe's aftermath intelligible occurs through a syntactically and structurally complex passage in which linguistic form mirrors material fracture, and the scene of ruin is elevated to the status of witness:

> But just as it is not the storm and the ship tossed about between the waves that speak loudest, but rather the desolate coast the day after, with the dull, lapping water and the broken, abandoned wreck, so too did everything I now saw speak: these undermined buildings, the rooves lying on the ground, the beach sand across the fields, the willow trees with seaweed and hay, the dead animals and the distraught people, louder than on the day of the disaster itself, and the desolate, dull calm that rested over the landscape hid a silent, heart-rending lament that gathered within itself all the individual scenes from that stormy night and day—and this common mute language of desolate, mistreated nature I shall not hastily forget.
>
> (25 November)

The delay between "speak" and "louder" acts as a syntactic deferral mirroring the struggle to comprehend catastrophe. By inserting a dense catalogue between verb and intensifier, Drachmann postpones closure, just as disaster defers comprehension. The sentence accumulates wreckage before releasing its claim: that ruins speak louder than the storm. Oxymoronic phrases—"mute language" and "silent, heart-rending lament"—dramatise the paradox of witnessing the unspeakable. Language here stretches to its representational limit. Crucially, the passage also vindicates belatedness. Absent during the storm, Drachmann frames the aftermath as more eloquent: a delayed chorus that speaks with greater authority than immediate chaos. The silent landscape becomes a medium of testimony, and the narrator, through retrospective attentiveness, asserts interpretive privilege by shifting epistemic weight from action to aftermath, from noise to residue.

The descriptive catalogue that separates "speak" from "louder"—"these undermined buildings, the roofs lying on the ground, the beach sand across the fields, the willow trees with seaweed and hay, the dead animals and the distraught people"—is metonymically structured. Its paratactic syntax establishes contiguity, but without causal sequence or visual hierarchy; the reader is offered no connective tissue in the sense of narrative or explanatory logic. Buildings, roofs, sand, trees, animals, and people are placed side by side, yet their relation remains unresolved. This rhetorical disjunction enacts disaster's aftermath, where normal functions are suspended, coherence collapses, and the components seem ontologically untethered. Buildings and bodies, sand and sorrow, and seaweed and roof beams are arrayed without narrative scaffolding. Jakobson's notion of metonymy as a structure of adjacency is vividly enacted here: the storm is not represented by a symbol but indexed by its scattered traces—debris, residue, and silence. The brokenness of the physical world mirrors the disruption of narrative order itself. Here, metonymy becomes more than a figure; it structurally echoes catastrophe itself.

Climate change, residential patterns, and philosophies of nature

Climate change and contemporary settlement patterns have placed coastlines at the forefront of environmental vulnerability. As rising sea levels and intensified storm activity reshape the edges of continents, the implications of human habitation along these thresholds become increasingly urgent. John Gillis describes the past decades as marked by "an unprecedented surge to the sea", noting that societies across the globe "have been turned inside out. . . . We are all now creatures of the edge" (2012, 1). Recent data support this claim: nearly a billion people now live within 10 kilometres of a coastline (Reimann, Vafeidis, and Honsel 2023, 3).

This growing coastal presence coincides with escalating environmental risk. The IPCC's 2022 report indicates that floods increased fourfold in the tropics and 2.5-fold in northern mid-latitudes between 1985 and 2015, with a 20–24 per cent increase in global flood exposure from 2000 to 2018 (2022, 573). Their 2019 special report projects sea-level rise of 43–84 centimetres by 2100, possibly up to 2 meters, with ascents continuing for centuries (2019, 324). Shorelines, once symbols of opportunity, leisure, and conquest, have become the frontlines of the Anthropocene. The clash between permanent residence and planetary instability forces us to reconsider how and where we build.

It is worth noting that if anthropogenic climate change has already triggered irreversible processes, there is also an understanding of the entire Earth System being caught in an irreversible transition between the Holocene and the Anthropocene (Bonneuil and Fressoz 2017). The current epoch of increasingly freakish weather events is not going to go away soon, meaning that settlement pattern represents our most urgent arena for transformative action. We should reconsider where and how we build our homes.

While the causes of climate change have been well theorised, less attention has been paid to the historical and cultural drivers of settlement along risky coastlines. Amitav Ghosh attributes the global gravitation towards water-adjacent residences to a colonial-modern vision of mastery, in which proximity to the sea signified power and commerce. Enlightenment attitudes towards nature promoted an exploitative view of the earth's resources, leading people to disregard risks posed by environmental forces. A critical cause identified by Ghosh is a cognitive tendency to think in discontinuities, ignoring wider environmental contexts: "it is a perspective that renders the interconnectedness of Gaia unthinkable" (2016, 56). Over time, and in line with the Enlightenment's conviction of gradualism as geological paradigm, humans developed a misplaced belief in environmental stability that reflected "the uniformitarian expectations that are rooted in the 'regularity of bourgeois life'" (2016, 35). Despite historical warnings, societies repeatedly ignored ancestral advice, choosing short-term convenience and monetary priorities over long-term safety. Together, these causes illustrate a historical, cultural, and cognitive shift towards the irresponsible settlement of exposed, vulnerable coastal regions.

In light of this history, the Baltic Sea flood, though not directly attributable to anthropogenic climate change, resonates as a foreshadowing event. Literary responses to the disaster serve as test cases for how earlier societies imagined, resisted, or succumbed to environmental volatility. As flood memories, they offer insight into a (near) future of wetter weather and increasingly violent storm surges. If we temporarily set aside debates over the origins of the Anthropocene and instead attend to its effects, the 1872 flood becomes a training ground for emergency preparedness: a site from which to learn how to live, narrate, and imagine amid the rising tides of a bluer world.

Through a range of historical and literary sources, the 1872 deluge elucidates the inherent constraints in humanity's capacity to control the unpredictable dynamics of nature, particularly those associated with saltwater. Literary representations serve as valuable resources in addressing pivotal contemporary questions: Which strategies might we employ in adapting to our bluer future? Should we persist in our combative stance of living *against* the sea, seeking to exclude seawater at any cost? Or should we instead adopt an accommodating approach, a living *with* the sea, and permit greater interaction between land and sea by inviting saltwater into our terrestrial territories? Alternatively, are we compelled to accept that the sea will inevitably prevail, necessitating strategic retreat from flood-prone regions?

In his third report, Drachmann turns to the large-scale infrastructural response to the catastrophe, describing the damming efforts undertaken by local inhabitants and the Danish state:

> [T]he great wave had receded again, like an army withdrawing from a ravaged country. But it had left a crew behind, and it would take years of persistent

> fighting, of tenacious, late-strength energy to drive the enemy out completely—if he could be driven out at all. And it was with a doubt repeated from all sides of this last that I once more set out and worked my way northward across the border of prosperity and impoverishment, security and uncertainty.
>
> (25 November)

The martial metaphor—nature as a retreating army—reflects a worldview in which the elements are cast as adversaries, and the appropriate human response is one of conquest and endurance. At the same time, the rhetoric evokes a broader national imaginary, shaped by military defeat and territorial loss. In 1864, Denmark suffered a traumatic blow with the loss of one-third of its territory to Bismarck's Prussian forces in the Second Schleswig War. The flood, in this light, becomes not only a natural but a national crisis, echoing anxieties about the erosion of borders—both geographic and symbolic.

In the decades following 1864, Denmark entered a period of introspective nation-building, shifting its identity inward through agricultural modernisation and landscape reclamation. Drachmann's account implicitly participates in this post-defeat logic: the enemy is no longer Prussia but the sea; victory lies not in territorial expansion but in infrastructural endurance. Yet his tone remains sceptical rather than triumphalist. The "enemy" may never be vanquished, and the narrator's return journey northwards reimposes a spatial and symbolic geography of control. What was ontologically destabilised by the flood is reinscribed as a national project of resilience and repair. The border holds.

Earlier in the same report, Drachmann offers a more intimate parable of nature's reversal of human ambition:

> behind the sunken house stretched a clear lake the size of Lake Peblinge. It was Noret, which some years ago had been reclaimed from the Baltic Sea with Grønsund as an intermediary. Now both the former owner and his commission agent had broken their contract without prior notice. Noret was Noret again. The pastures and wheat fields formed the bottom of the lake.
>
> (25 November)

Here, the contract metaphor—an anthropocentric fiction of mutual agreement and control—collapses in the face of water's indifferent agency. The return of Noret is not just spatial but temporal: a suspension of historical progress and a return of geological cadences. Human claims prove revocable at the whim of the elements. While Drachmann rarely theorises nature explicitly, his metaphoric and narrative choices repeatedly cast doubt on the viability of large-scale interventions aimed at disciplining the sea.

These passages offer a latent philosophy of nature that cuts against the grain of nineteenth-century confidence in technological mastery and territorial consolidation. The young Drachmann, later to become an important figure in the Modern

Breakthrough launched by Georg Brandes the previous year, is not anti-modern, but he is wary of its hubris. The flood is not merely a disaster to be managed; it is a challenge to the ontological structures that underwrite modern life—ownership, linear time, and infrastructural control. In this sense, his reports suggest an alternative ethics: one of humility, responsiveness, and ecological attunement. The sea, for Drachmann, is not a passive object to be subdued but a dynamic force that unravels human certainties. To think nature here is to recognise the limits of mastery and to imagine forms of dwelling attuned to volatility rather than permanence.

Conclusion

In the wake of the 1872 Baltic Sea flood, Drachmann's "Efter Orkanen" grapples not only with the limits of witnessing but with the destabilisation of form, space, and voice. His writing, shaped by the disaster's temporal delay, epistemological gaps, and the "disorienting pressure of the inhuman environment of the sea", becomes a site where narrative breaks down and begins again—fluid, porous, and provisional. By mobilising trauma theory, wet ontology, and a rethinking of the contact zone as elemental force field, this chapter has shown how Drachmann's dispatches do more than document—they enact the very disorientation they seek to name through a salt aesthetics. If the flood marks a historical catastrophe, it also offers a proto-Anthropocene scene in which landscape, agency, and narration become unstable. Drachmann's literary struggle thus becomes ours: to narrate from within a world where distinctions collapse, where the sea breaches every boundary, and where literature is asked not to resolve but to inhabit the turbulence of planetary change.

Note

1 The reports were published on 25 November, 28 November, and 2 December, but the dates referred to here follow Drachmann's own internal dating of his observations.

Works cited

Bonneuil, Christophe, and Jean-Baptiste Fressoz. 2017. *The Shock of the Anthropocene: The Earth, History and Us*. Verso.

Caretta, M.A., A. Mukherji, M. Arfanuzzaman, R.A. Betts, A. Gelfan, Y. Hirabayashi, T.K. Lissner et al. 2022. "Water". In *Climate Change 2022: Impacts, Adaptation and Vulnerability*. Contribution of Working Group II to the Sixth Assessment Report of the Intergovernmental Panel on Climate Change, edited by H.-O. Pörtner, D.C. Roberts, M. Tignor, E.S. Poloczanska, K. Mintenbeck, A. Alegría, M. Craig, S. Langsdorf, S. Löschke, V. Möller, A. Okem, and B. Rama, 551–712. Cambridge University Press. doi:10.1017/9781009325844.006.

Caruth, Cathy. 1996. *Unclaimed Experience: Trauma, Narrative, and History*. Johns Hopkins University Press.

Drachmann, Holger. 1872. "Efter Orkanen: Pennetegninger fra Kysten." *Dagbladet*, 25 and 28 November and 2 December.

Ghosh, Amitav. 2016. *The Great Derangement: Climate Change and the Unthinkable*. University of Chicago Press.
Gillis, John R. 2012. *The Human Shore: Seacoasts in History*. University of Chicago Press.
IPCC. 2022. "Water." In *Climate Change 2022: Impacts, Adaptation and Vulnerability (Contribution of Working Group II to the Sixth Assessment Report of the Intergovernmental Panel on Climate Change)*, edited by H.-O. Pörtner, M.A. Caretta, A. Mukherji, M. Arfanuzzaman, R.A. Betts, A. Gelfan, Y. Hirabayashi et al., 551–712. Cambridge University Press.
Jakobson, Roman. 1971. "Two Aspects of Language and Two Types of Aphasic Disturbances." In *Selected Writings II: Word and Language*, 239–59. Mouton.
LaCapra, Dominick. (2001) 2014. *Writing History, Writing Trauma*. Johns Hopkins University Press.
Mentz, Steve. 2018. "Blue Humanities." In *The Posthuman Glossary*, edited by Rosi Braidotti and Maria Hlavajova, 69–72. Bloomsbury Academic.
Oppenheimer, M., B.C. Glavovic, J. Hinkel, R. van de Wal, A.K. Magnan, A. Abd-Elgawad, R. Cai et al. 2019. "Sea Level Rise and Implications for Low-Lying Islands, Coasts and Communities." In *IPCC Special Report on the Ocean and Cryosphere in a Changing Climate*, edited by H.-O. Pörtner, D.C. Roberts, V. Masson-Delmotte, P. Zhai, M. Tignor, E. Poloczanska, K. Mintenbeck, A. Alegría, M. Nicolai, A. Okem, J. Petzold, B. Rama, and N.M. Weyer, 321–445. Cambridge University Press. https://doi.org/10.1017/9781009157964.006.
Peters, Kimberley, and Philip Steinberg. 2019. "The Ocean in Excess: Towards a More-than-Wet Ontology." *Dialogues in Human Geography* 9 (3): 293–307. https://doi.org/10.1177/2043820619872886.
Pratt, Mary Louise. 1991. "Arts of the Contact Zone." *Profession* (91): 33–40.
Pratt, Mary Louise. 2022. *Planetary Longings*. Duke University Press.
Reimann, Lena, Athanasios T. Vafeidis, and Lars E. Honsel. 2023. "Population Development as a Driver of Coastal Risk: Current Trends and Future Pathways." *Cambridge Prisms: Coastal Futures* 1 (e14): 1–12. https://doi.org/10.1017/cft.2023.3.
Ritson, Katie. 2019. *The Shifting Sands of the North Sea Lowlands: Literary and Historical Imaginaries*. Routledge. https://doi.org/10.4324/9780429490644.
Steinberg, Philip, and Kimberley Peters. 2015. "Wet Ontologies, Fluid Spaces: Giving Depth to Volume through Oceanic Thinking." *Environment and Planning D: Society and Space* 33 (2): 247–64. https://doi.org/10.1068/d14148p.

PART II

The Wadden Sea, the North Sea, and the Norwegian Sea

5

COASTAL WRITING

The Wadden Sea in the works of Steen Steensen Blicher, Peter Seeberg, and Dorthe Nors

Anders Ehlers Dam

The Wadden Sea is one of Europe's most distinctive and dynamic coastal environments. Formed in an ongoing process that started approximately twelve thousand years ago, it stretches in an unbroken littoral band from Den Helder in the Netherlands, across the Frisian and North German coasts, to the south-western edge of Denmark near Esbjerg and the desolate Skallingen peninsula. It is the largest tidal flat system in the world, and a UNESCO World Heritage Site. Twice daily the tide withdraws, exposing expanses of sand and mudflats; twice daily it returns, submerging them once more. Marshland—situated slightly higher—is only flooded in storms or unusually high tides and is protected by dikes, sluices, canals, and pumping stations. The coast is thus a zone of constant transition, in which the boundary between land and sea is never fixed but rhythmically renegotiated.

This littoral region embodies both a sharp tension and entanglement between nature and culture. While the tidal movements and shifting sediments reveal the elemental forces of the sea, the landscape is also profoundly shaped by human engineering: dikes built through centuries of labour, agricultural land reclaimed from the water, harbours, and systems of drainage. In a blue humanities perspective, the Wadden Sea region is interesting as both a geophysical and cultural territory: a site where deep geological time intersects with human history, and where natural and cultural processes are inseparably intertwined.

Despite its vastness and environmental significance, the Wadden Sea remains relatively marginal within literary history. It lacks the dramatic mountain ranges traditionally associated with the Romantic sublime or the idyllic groves of pastoral tradition. Instead, its flatness, apparent monotony, and rhythmic dissolving of boundaries render it a challenging landscape for conventional literary representation. With its constantly shifting edge—neither fully land nor fully sea—it may be described as a kind of non-place, as a liminal zone, and a "no-man's-land" between

DOI: 10.4324/9781003647638-8

elements, difficult to situate within traditional landscape aesthetics. This does not mean, however, that no one has engaged with the Wadden Sea in literature. A number of interesting texts from the Netherlands, Germany, and Denmark confront this dynamic landscape, and the result is that the Wadden Sea is attributed diverse meanings by different authors. Katie Ritson has made several groundbreaking contributions to the study of the connection between the Wadden Sea and literature, for example, the foundational reference work *The Shifting Sands of the North Sea Lowlands* (2019), which analyses works across national borders, including the most famous Wadden Sea text, Theodor Storm's *Der Schimmelreiter* (1888). A recent comprehensive introduction to the Wadden Sea as both a natural and cultural region can be found in John Frederiksen's *Vadehavet* (2022), which also includes a chapter on the Wadden Sea in literature by Johs. Nørregaard Frandsen.[1]

Wolfgang Behschnitt has argued that the nineteenth-century Danish west coast in its full extent from the south to the northern tip of the country functioned as a "liminal landscape": once regarded as "a wild, uncivilised, adventurous, and rather exotic space, a desert-like region with wide sand dunes, roaring storms, shipwrecks, and utterly poor inhabitants", yet increasingly romanticised by writers such as St. St. Blicher, Hans Christian Andersen, Meïr Aron Goldschmidt, Holger Drachmann, and Henrik Pontoppidan. This region, he suggests, became a projection surface for fantasies, anxieties, and desires—an "other" space that promised literary riches despite its material austerity (Behschnitt 2017, 80–82; specifically on Blicher: Behschnitt 2006, 341–89).

By turning attention to the literary representation of a specific landscape and seascape and examining the concrete connection between the Wadden Sea coast and writing, this chapter situates itself within the broader context of the spatial turn and literary geography (see, for example, Collot 2014). I consider the Wadden Sea as a site where space, nature, and culture are mutually constitutive. The landscape's dynamism—its tides, erosion, sedimentation, storms, and anthropogenic interventions—disrupts linear temporality, juxtaposing human time with deep time, as Katie Ritson and Eveline de Smalen have argued (2021). What emerges is an amphibious cultural landscape in which the human and the non-human cannot be neatly separated. The Wadden Sea landscape is thus a perfect example of what Ursula Kluwick and Virginia Richter call a "littoral space":

> the zone *between* the sea and the land, the beginning and end of which are indeed hard to define and which is constantly reconfigured in a material as well as a symbolic sense, depending on whether it is approached from the sea or the land.
>
> (2015, 4)

Against this background, I examine three works of Danish coastal writing that traverse the western shoreline from the Netherlands or Northern Germany to northern Jutland: St. St. Blicher's *Vestlig Profil af den Cimbriske Halvø fra Hamborg til Skagen* (1838/1839; Western Profile of the Cimbrian Peninsula from Hamburg to Skagen), Peter Seeberg's *Vestlig profil af den jyske halvø* (2000; Western Profile of the Jutland Peninsula), and Dorthe Nors's *En linje i verden* (2021; A line in the world, published in English translation, first under the title *North Sea* 2021 with Henrik Saxgren's photographs, like in the Danish original version, and again the following year as *A Line in the World: A Year on the North Sea Coast* 2022). These texts constitute a case of intertextuality—or more precisely hypertextuality in Gérard Genette's (1997) sense—where later works write "on top of" earlier ones, reconfiguring their forms and imaginaries. In generic terms, the three works are a kind of non-fictional prose that can be defined as coastal writing. In essayistic form, the coastline is examined through the author's eyes. It is thus a form of travel writing in which a route along the coast is followed, and the Wadden Sea constitutes part of that route. The texts all attend to the sea and the coastal landscape as a border zone between land and water, solid and fluid, and culture and nature. The coast dictates the itinerary, and specific places along the way provide inspiration.

I examine the type of writing engendered and inspired by the coastal landscape, as well as the intertextual connections between these texts, beginning with Blicher's work, which serves as a topographical hypotext. How are the Wadden Sea and its coastline written and rewritten across the examined works? What role does the Wadden Sea play in understanding the fluid, ever-changing border between sea and land? In all three, the Wadden Sea plays a central role, though the authors' engagements differ markedly: from Blicher's Romantic fascination with geological force and cultural alterity, through Seeberg's phenomenological attention to the unspectacular and atmospheric, to Nors' contemporary, introspective, and psychologically inflected coastal writing. By analysing what these authors write about, how they do so, and why, I seek to illuminate the different literary functions of the Wadden Sea as a shifting borderland and a generative imaginative space, as well as the intertextual relation between the texts. As we shall see, coastal literature is not merely writing *about* the coast; it is a mode of writing in which the underlying imagination is inseparable from the coast—in this case, specifically, the Wadden Sea.

Romantic imaginaries of the Wadden Sea: Steen Steensen Blicher

Steen Steensen Blicher's writings belong to what the historian Steen Bo Frandsen has termed the nineteenth-century "exploration of Jutland": A cultural moment in which Jutland appeared geographically and culturally distant from the national centre in Copenhagen. The author who set out into this perceived wilderness—this "desert" of the Danish kingdom—performed the role of explorer, encountering unfamiliar landscapes, unfamiliar dialects, and unfamiliar customs (Frandsen 1996, 109–17).[2] *Vestlig Profil af den Cimbriske Halvøe* is a hybrid work, combining historical reflection, natural description, anecdotal reportage, political commentary, and quasi-ethnographic portraits of local people. Although the text is filtered entirely through Blicher's perspective, it is not a personal or confessional account; he does not foreground his inner life, but appears instead as an informed and curious observer.

Blicher writes simultaneously from the inside and the outside. He was himself from Jutland and lived there as a parish priest, yet he wrote for an audience belonging to the educated elite in Copenhagen. He did not address the rural communities he encountered on his travels, but instead mediated their world to metropolitan readers. The same holds for his more famous short stories set on the Jutland heath. As the first Danish author to systematically represent Jutland's landscape, culture, and dialect, including the West Coast, Blicher played a formative role in expanding the literary map of Denmark. For readers in the capital, his depictions held a distinct exotic appeal.

Vestlig Profil af den Cimbriske Halvøe fra Hamborg til Skagen tagen paa Reiser af S.S. Blicher was published in 1839, describing a journey of roughly six hundred kilometres undertaken over six weeks in late summer 1838. A few years earlier, Blicher had produced two short travel books about Sweden; his west coast journey was supported financially both by liberal circles in Copenhagen and by the king, attesting to a metropolitan appetite for accounts of the "exotic" western periphery. As Johs. Nørregaard Frandsen notes in his study of the Danish province, with his work on the coastline, Blicher created "an original piece of ethnology and a topographical work that circles in the character of exotic 'otherness' of the province", rendering the region "interesting and dramatic" and, crucially, "literary" (Frandsen 2023, 97). Blicher's method combines realistic interest in how people survive in the marginal environments *through which he travels* with a Romantic fascination for the sublime and unusual landscapes. Whereas Jonas Ullerup Dejgaard (2026) focuses on the realism of *Vestlig Profil*, I emphasise the text's Romantic features.

Throughout *Vestlig Profil*, Blicher appears as a fascinated observer of the coastal landscape. North of the Wadden Sea, at Blåbjerg, he admires the horizon's vast reach. What particularly draws his attention is the transitional zone between land and sea. In "Tepperne", a lush marshland, he describes the area "between the yellow barren desert and the bitter-salty water" (Blicher 2013, 30). In "Klitboerne" (The dune-dwellers), this no-man's-land becomes embodied in the West Jutland fishermen: "They are a strange people, these amphibians of the west coast, who live in the sand and off the water" (Blicher 2013, 34). The Jutlandic west coast is thus portrayed as an amphibious zone—neither fully land nor fully sea—and its inhabitants as liminal figures suspended between the two elements in what Michael N. Pearson calls a "Littoral Society" (2006).[3]

Several passages reveal Blicher's Romantic fascination with strangeness and alterity. In Chapter 11, describing the remote town of Agger, he imagines a metropolitan visitor transported there by modern means of travel:

> If a Copenhagener who had never been beyond Gefion's Isle were suddenly, by airship or steamship, transported to Agger; saw the inhabitants' strange clothing

> and strange faces, and heard their tongue, entirely foreign to him—one might perhaps persuade him that he had arrived in a completely foreign country.
>
> (Blicher 2013, 44)

Here, the coastal inhabitants are exoticised: to the imagined Copenhagener, they appear as foreign as Africans or Asians might to a European traveller of the era. Yet Blicher also ironicises the metropolitan perspective. If Agger seems strange, it is partly because the imagined visitor has never left his familiar cultural orbit, in a kind of inverted provincialism. Blicher claims a third position: neither local nor provincial urbanite, he observes from an intermediate standpoint as a literary traveller.

In the chapters concerning the Wadden Sea itself, Blicher's interest shifts to geological time and the forces that have shaped the coast. In "Marsken" (The Marsh), he describes the formation of the marshlands through storm surges and tidal incursions:

> At the strait that separates Britain from Europe these additions from the sea begin—whether we choose to regard them as gifts, or as spoils won back. . . . The peculiar bed of the sea is sand and stone; the clayey ground points to gained possessions—conquests wrested from the land. . . . The mighty North Sea rolls its salt-laden heavy waves, piled up and driven forward by the storms from the west, with a force against which human steadfast effort can scarcely set limits. Compared with this wild sea, the Baltic is but a pond.
>
> (Blicher 2013, 14)

The Baltic ("a mere pond") is contrasted with the violent power of the North Sea, and by extension, the cultivated calm associated with Copenhagen and eastern Denmark versus the western part of the country. Blicher continues by celebrating the resilience of coastal communities, whose dikes he compares to the world's great monuments:

> When this reconquest began, we do not know . . . . [T]hese giant works, which remind us of the Chinese barrier against the sand sea . . . Belgians, Batavians, Frisians, Ditmarsians—all have defended their fertile marshes with the strength of mountain dwellers. In their utmost distress, they have even called upon the old enemy, *the sea*, for help against the new enemies from *the land*.
>
> (2013, 14)

For Blicher, the origins of both the landscape and the dikes recede into obscurity, inviting speculation across deep time. The text reflects the intellectual climate of the period, in which geological discoveries—by figures such as the Scottish geologist James Hutton—had expanded conceptions of the earth's age far beyond biblical chronology. Until the mid-eighteenth century, the prevailing assumption was that the world was around 6,000 years old, as calculated from the Old Testament (Heringman 2023). The dikes are, of course, man-made, yet their origins also dissolve into Blicher's sense of what the French historian Fernand Braudel (1980) would later define as a *longue durée*, an extended, slow process that transcends recorded history.

Blicher's evocation of immense natural forces and ancient engineering also aligns with Romantic aesthetics of the sublime. He likens the marsh soil to Egyptian fertility—"this Egyptian richness" (Blicher 2013, 15) —and compares the exposed tidal flats at low tide to "ice fields": "from the dikes one sees the flats at ebb tide, white and gleaming like ice fields" (Blicher 2013, 15). In Blicher's imagery, the sublime and overwhelming blend with the foreign and exotic in his description of the coastal landscape.

In *The Lure of the Sea* (1994), Alain Corbin describes how the coast, once perceived primarily as a place of utility or a frightening and dangerous place, from the mid-eighteenth century and into the nineteenth was transformed into a destination people sought out in order to encounter both nature in its magnitude and otherness and authentic cultures distinct from those of the civilised metropolis. The very presence of slow geological forces and deep time contributed to this encounter with aspects that transcended the human world:

> People came there to wonder about the earth's past and the origins of life. There, more than anywhere else, it was possible to discover the multitude of temporal rhythms, to sense the duration of geological time, and to observe the indecisiveness of biological borderlines, the vagueness of different kingdoms, and the surprising transitions that linked them to one another.
>
> (Corbin 1994, 97)

Blicher is among the first authors in Nordic literature to describe exactly this type of perception of the coast and the sea.

In sum, Blicher depicts the Wadden Sea as a frontier shaped by a titanic struggle between land and sea, framed within deep geological time and populated by hardy, amphibious inhabitants. His figurative analogies exoticise the region while conferring a certain dignity upon it. His representation is quintessentially Romantic: a geography of sublime forces and cultural otherness.[4]

Similar exoticisations of the Wadden Sea landscape can be found in the works of Blicher's poet-colleague Hans Christian Andersen, in the novel *The Two Baronesses* (1848) and in his diaries. In Andersen's novel, which partly takes place on the small Frisian island, or *hallig*, of Oland, Katie Ritson identifies parallels between "the shifting fortunes of Andersen's protagonists, the unstable landscape, and the nation [Denmark] struggling with its recent history and sense of itself on the European stage" (2019, 29). Andersen's text displays a pronounced interest in the dynamics of the landscape, and Ritson encapsulates the "defining features" of the Wadden Sea as "its constant reshaping, both by slow processes (erosion, siltation) and sudden events, such as sandstorms, the appearance of floating islands, and storm surges" (2019, 34). In his reading of Andersen's representation of the Wadden Sea, Henrich Detering extends this interest in the coast's dynamically hybrid liminal landscape, situated between land and sea, incorporating diary passages and interpreting it as a "queer landscape" (Detering 2026). Andersen's texts, like Blicher's, similarly attest to a sustained fascination with the Wadden Sea and its literary possibilities (cf. also Klaus Müller-Wille's chapter on Andersen in this volume).

Phenomenologies of the Wadden Sea: Peter Seeberg

While Blicher focused on deep time sublimity and the exotic aspects of the landscape and the locals in the middle of the nineteenth century, Peter Seeberg (1925–1999) takes a markedly different approach more than 150 years later in *Vestlig profil af den jyske halvø* (2000). Seeberg's book explicitly positions itself as a modern counterpart to Blicher's work, as stated on the cover. Known primarily for his modernist prose and existential concerns, Seeberg had already depicted the Wadden Sea in the novel *Ved havet* (1978), offering a detailed portrayal of a summer's day on Rømø. In the posthumously published *Vestlig profil*, he mixes genres—memoir, anecdote, topography, and observation—tracing the West Coast from the Danish–German border at Siltofte to Skagen. The book is accompanied by photos by the photographer Katrine Ussing.

The coastal areas, and especially the Wadden Sea, have a whole other function in Seeberg's text than in Blicher's. In sharp contrast to Blicher, Seeberg does not seek out historical sites or pursue exotic otherness. Instead, he turns towards the seemingly insignificant: fragments of memory, minor places, and objects without obvious symbolic weight. In one extended recollection from Hjerpsted, he dwells on ordinary houses, a cauliflower seller, a wedding, a deserted white coastal house, and a plastic boat:

> In Hjerpsted there are a couple of unusual Jugendstil villas in brick, in Hjerpsted one summer there was a man who sold cauliflower from a rubber cart, I bought two, in Hjerpsted, one day as I drove past, there was a wedding at the church, many cars and many people, many children and pretty girls. There were extraordinarily many cars, I was astonished. In such a small town as Hjerpsted. But I especially remember that summer with the cauliflowers, and the little man who sold them, who looked as if he had just gotten out of bed. Yet even more clearly I remember the small white house that lies right down by the Wadden Sea. You don't get the impression that anyone lives there. It lies a little south of a very, very modest parking place where a boat is sometimes pulled ashore. Of plastic. The little boat is of plastic. From the road to the parking place there is

> an overgrown path leading to the white-washed house. No matter how close you come to this house, it looks just as deserted and empty, yet it is well maintained. That house fascinates me tremendously, more than anything else in Hjerpsted. I don't know anyone in Hjerpsted. But once there was a man who was a baker in Hjerpsted, there isn't one anymore. The baker had a secondary profession. He was a woodcarver, he carved bridal chests, for which he became famous and enjoyed a good profit. He was truly famous.
>
> (Seeberg and Ussing 2000, 13)

Nothing here is ostensibly important. The scene is composed of banal objects and fleeting impressions and memories—yet Seeberg treats them with reverence. The plastic boat, the abandoned house by the Wadden Sea, and the forgotten baker: these fragments constitute a quiet phenomenology of place. Where Blicher sought the sublime and the foreign, Seeberg privileges what remains in memory despite its ordinariness. The fascination lies not in grandeur or history but in the persistence of minor details and the mere appearance of things.

In another passage, Seeberg articulates how the Wadden Sea landscape affects perception—offering a kind of littoral phenomenology:

> The fundamental aesthetic structure of the area is one of distance, remoteness, and it takes on the character of everything that withdraws. Above all, it withdraws because it stretches, like the dike, into the infinite, or repeats itself monotonously like the barbed-wire fences, whose monumental dominance makes them a sight in themselves. But even more powerful is the atmospheric element: the greyness that optically removes Rømø from the mainland in degrees of visibility that can be torn away within minutes, whereupon the vault of sky over the island spreads out and bursts into light, visibility becomes sharpness and nearness—especially in the early and late hours of the day.
>
> (2000, 22)

This is a neutral, almost meditative description of visibility and distance. The Wadden landscape withdraws and reappears; Rømø vanishes in grey haze only to return in sudden light. The focus is not on dramatic nature or local life but on perception itself. The "aesthetic structure" is one of distance, monotony, and atmospheres—an experience of subtle dissolution and sudden return. In his work on the colour grey, Peter Sloterdijk quotes the painter Paul Cézanne's statement: "You're not a painter, if you haven't painted gray." Sloterdijk takes up this challenge within philosophy, developing a thinking of greyness from the premise that "anyone who proposes to take seriously the weather reports of the soul as an ongoing language game, or to consider it as its own genre of bulletin, can hardly avoid making the gray explicit" (Sloterdijk 2025, 1–2). Seeberg's coastal book, especially the passages on the Wadden Sea, is, as with much of his other work, an attempt to capture this atmosphere of greyness and everydayness.

Seeberg's coastal writing thus foregrounds attentiveness, surface, and phenomenological presence. Compared to his Romantic predecessor, he is unconcerned with exoticism, dramatic forces, and political and historical questions. Instead, he attends to the everyday and the anonymous, to non-places and fleeting impressions. His Wadden Sea is quiet, unspectacular, and yet profoundly affecting: a landscape of neutrality where meaning emerges not from symbolic grandeur but from the delicate interplay of perception, memory, and atmosphere.

Psychological figurations of the Wadden Sea: Dorthe Nors

Moving from Blicher and Seeberg into contemporary literature, we encounter in Dorthe Nors's *A Line in the World*, another essayistic work structured around the West Coast. Like her two predecessors, Nors is primarily known as a writer of fiction, yet here she, too, travels along the shoreline, tracing the coast in the opposite direction: from Skagen southwards all the way towards the Netherlands. Her book participates in the same coast-following literary gesture inaugurated by Blicher, but its scope is explicitly transnational, extending beyond Danish borders into German and Dutch coastal territories.

Although all three works may be called essayistic, Nors's writing is the most overtly personal. She describes different places and people who live along the West Coast with great sensitivity, but her book is shaped by introspection and symbolic thinking; the coast becomes a space of self-encounter, and the Wadden Sea a site of psychological and existential reflection. The shoreline functions as a metaphorical threshold, where questions of mortality, disappearance, and belonging are negotiated. While the book contains vivid chapters about individual places, its true subject is the narrator's identity and her reckoning with loss, memory, and familial ties. The borderland between land and sea is also a border between interior and exterior worlds, between not-knowing and self-understanding. This is also true of the parts of the book that deal with the Wadden Sea region.

In the chapter "Wadden Sea Suite", Nors recounts living on the island of Fanø for a year. Standing on a dike, she experiences a profound silence—yet this emptiness is revealed as internal rather than external. The landscape acts as a mirror of the psyche:

> The starry sky formed a dome above the island, vast, curved and infinite. As I stood there, I sensed the depth of this curve. Up and down became relative terms, and when I shut my eyes, I could hear the silence of the Wadden Sea, like some kind of resonance. . . . [T]he Wadden Sea became a huge basin beneath the vaulted stars, where everything fertile grew, including death. Perhaps the cello's sound was coming from the deeps, from the tideways. Perhaps I'd brought it out there myself. Perhaps it couldn't be any other way.
>
> (Nors 2022, 138–39)

Where Seeberg phenomenologically analyses how the outer world appears to the perceiving subject and how the landscape finds its place in memory, Nors suffuses the outer world with inner meaning. The Wadden Sea becomes a site where inner and outer experience fold together, the landscape revealing the movements of the mind.

In "Magnets", she puts this directly: "The Wadden Sea is, in many ways, a piece of psychology. The year I lived in the middle of it, I was fascinated by how it affected me and the people I met in the area" (Nors 2022, 190). A similar sentiment appears in Nors's short story "The Wadden Sea" from *Karate Chop* (2008), where she calls the Wadden Sea "an image in the mind's eye" (2008, 88).

Entering the tidal landscape draws forth memories and vulnerabilities:

> I lost my layers of protection. I felt grief for the losses that have been and the losses to come. . . . But we can't stop time. We can't escape the force that draws us through the world. We have magnets in our bodies.
>
> (Nors 2022, 197)

Here, the Wadden Sea becomes an emblem of the existential condition: A place where layers of self are shed and where one confronts grief and temporality. The border between Denmark and Germany acquires symbolic weight, representing a threshold between life and death; the physical journey thus mirrors a psychological one.

Nors at times romanticises the inhabitants of the region—especially the women of Fanø, whom she depicts as forming a matriarchal community alongside silent men. While this carries echoes of Blicher's exoticising gestures, Nors's emphasis is less ethnographic than symbolic: the landscape and its people serve as figures in a psychological tableau.

The book concludes with a childhood memory, where Nors sees herself from the outside standing on Grenen, which is the very tip of Jutland:

> The seas meet. A ship sails across the reef, and there's the girl. Her hand is sticky with ice cream. She's standing on a coastline with a foot in each sea. She will stay, she will go, she can do nothing else.
>
> (Nors 2022, 227)

Again, the seascape serves a symbolic function, illustrating psychological and existential questions.

The symbolic duality in Nors' fictional and essayistic coastal writing—where the coastal landscape and the inner life of the writer are inseparable—can be related to René Dietrich's theory of the coastline's "liminality". In an article on coastal poetry, Dietrich understands liminality as concerning not only the shore and its amphibious landscape between sea and land but also the poet who writes from within it:

> For a discussion of poets who in speaking from the shore practice a poetics of liminality, this brings into closer focus not only the liminality of the shore itself, but of the speakers as well as of their acts of poetic speech.
>
> (2009, 452)

In the same way, Nors' symbolism is closely linked to her coastal poetics: the journey along the coast is not merely a setting for psychological reflections, but the writing and the insights emerge precisely from this concrete landscape.

In Nors, then, the Wadden Sea is not primarily a historical, ethnographic, sublime, or phenomenological site, but a symbolic and emotional one. The coast, and especially the Wadden Sea region, becomes a medium for self-discovery, a place where memory, loss, and existential acceptance unfold. Her littoral writing is interiorised, making the tidal zone a mirror of psychic tides.

Conclusion

Blicher's *Vestlig Profil af den Cimbriske Halvøe* functions as a hypotext for later Danish coastal writing. Yet, although Seeberg and Nors retrace his geographical route (each in their own direction), they do not simply repeat his literary approach. Blicher's project combines topographical observation, political reflection, and Romantic fascination with sublime nature, exoticism, and cultural otherness. He does not write about himself, but mediates Jutland to a cultivated metropolitan audience, discovering in the Wadden Sea a theatre of deep geological time and human endurance.

Seeberg, by contrast, rejects the exoticising gaze. His interest lies in the unspectacular and in the small residues of everyday life: plastic boats, modest houses, and forgotten bakers. The Wadden Sea becomes for him not a sublime frontier but a quiet field of perception, where distance, haze, and atmosphere structure experience. His writing is less about the coast's dramatic history than about the phenomenology of looking, remembering, and dwelling in seemingly marginal places.

Nors takes the littoral turn inward. While her book also follows the coastline, its true trajectory is psychological. The Wadden Sea becomes a symbolic landscape in which grief, memory, and existential acceptance unfold through the description of concrete places. Rather than offering a purely external ethnography or a phenomenology of perception, Nors renders the tidal zone as a site where the boundaries between inner and outer landscapes dissolve.

All three writers, writing in each their century, engage with the Wadden Sea's distinctive amphibious character, yet they do so in strikingly different registers: the Romantic sublime, the unspectacular phenomenological, and the psychological-symbolic. Together, their works demonstrate how coastal writing is shaped not only by geography but by the literary and intellectual horizons of its moment and by the authors themselves. The Wadden Sea region, in their hands, becomes more than a landscape and a seascape: it is a shifting zone of encounter between nature and culture, perception and memory, and world and self, where each writer charts a different line between land and sea.

Notes

1 See also the forthcoming volume of readings of the Wadden Sea landscape edited by Anders Ehlers Dam (Dam 2026).
2 All translations are my own, unless otherwise indicated.
3 Cf. Søren Frank and his advocation for a blue humanities approach to literature. Frank calls for an "Amphibian Comparative Literature", cf. Frank (2022, 11). Also, Colin Williams use this expression. He calls the tidal marsh an "amphibian zone" or "amphibian land" (Williams and Brodde 2023). From an environmental-historical perspective, van Dam (2017) speaks of the Netherlands as an "amphibious culture".
4 For a reading of selected Wadden Sea works from an ecocritical and geological point of view, see Ritson and de Smalen (2021).

Works cited

Behschnitt, Wolfgang. 2006. *Wanderungen mit der Wünschelrute: Landesbeschreibende Literatur und die vorgestellte Geographie Deutschlands und Dänemarks im 19. Jahrhundert*. Ergon Verlag.
Behschnitt, Wolfgang. 2017. "Jutland and the West Coast as Liminal Spaces in Danish Literature." In *Nordic Literature: A Comparative History*, edited by Thomas A. Dubois and Dan Ringgaard, Vol. 1: *Spatial Nodes*, edited by Steven P. Sondrup and Mark B. Sandberg. John Benjamins Publishing.
Blicher, Steen Steensen. (1838) 2013. *Vestlig Profil af den Cimbriske Halvøe fra Hamborg til Skagen*. Gyldendal.
Braudel, Fernand. 1980. *On History*. Translated by Sarah Matthews. University of Chicago Press.
Collot, Michel. 2014. *Pour une géographie littéraire*. Édition Corti.
Corbin, Alain. 1994. *The Lure of the Sea: The Discovery of the Seaside in the Western World 1750–1840*. University of California Press.
Dam, Anders Ehlers, ed. 2026. *Vadehavet i litteraturen*. Forlaget Wunderbuch.
Dejgaard, Jonas Ullerup. 2026. "'En ganske anden Strandvej': Blichers vestkyst mellem betragtning og brug." *Passage: Tidsskrift for litteratur og kritik* 94.
Detering, Heinrich. 2026. "Det queere landskab: H.C. Andersen i Vadehavet." In *Vadehavet i litteraturen*, edited by Anders Ehlers Dam. Forlaget Wunderbuch.
Dietrich, René. 2009. "Towards a Poetics of Liminality in This 'Space between Spaces': The Shore Lines of Contemporary American Poetry." *Anglia: Journal of English Philology* 125 (3): 448–64.
Frandsen, Johs. Nørregaard. 2023. *Hvor ligger provinsen*. Gads Forlag.
Frandsen, Steen Bo. 1996. *Opdagelsen af Jylland: Den regionale dimension i Danmarkshistorien 1814–64*. Aarhus Universitetsforlag.
Frank, Søren. 2022. *A Poetic History of the Oceans: Literature and Maritime Modernity*. Brill.
Frederiksen, John, ed. 2022. *Vadehavet—grænseløs natur- og kulturarv*. Gyldendal.

Genette, Gérard. 1997. *Palimpsests: Literature in the Second Degree.* Translated by Channa Newman and Claude Doubinsky. University of Nebraska Press.
Heringman, Noah. 2023. *Deep Time: A Literary History*. Princeton University Press.
Kluwick, Ursula, and Virginia Richter. 2015 "Introduction. 'Twixt Land and Sea: Approaches to Littoral Studies." In *The Beach in Anglophone Literatures and Cultures: Reading Littoral Space*, edited by Ursula Kluwick and Virginia Richter. Routledge.
Nors, Dorthe. 2008. *Karate Chop: Stories*. Translated from Danish by Martin Aitken. Minneapolis: Graywolf Press.
Nors, Dorthe. 2022. *A Line in the World: A Year on the North Sea Coast*. Translated from Danish by Caroline Waight. Pushkin Press.
Nors, Dorthe, and Henrik Saxgren. 2021. *North Sea*. Translated from Danish by Caroline Waight. Gads Forlag.
Pearson, Michael N. 2006. "Littoral Society: The Concept and Its Problems." *Journal of World History* 17 (4): 353–73.
Ritson, Katie. 2019. *The Shifting Sands of the North Sea Lowlands: Literary and Historical Imaginaries*. Routledge.
Ritson, Katie, and Eveline R. de Smalen. 2021. "Imagining the Anthropocene with the Wadden Sea." *Maritime Studies* 20: 293–303.
Seeberg, Peter, and Katrine Ussing. 2000. *Vestlig profil af den jyske halvø*. Gyldendal.
Sloterdijk, Peter. 2025. *If You Have Never Thought Gray: A Theory of Color*. Translated by Corey Dansereau and Robert Hughes. Polity.
van Dam, Petra J. E. M. 2017. "An Amphibious Culture: Coping with Floods in the Netherlands." In *Local Places, Global Processes: Histories of Environmental Change in Britain and Beyond*, edited by Peter Coates, David Moon, and Paul Warde. Oxbow Books.
Williams, Colin, and Marco Brodde, eds. 2023. *Hjemsøgt af tidevandet: The Society of Wildlife Artists i Vadehavet*. Nationalpark Vadehavet.

6

THE INDUSTRIALISED NORTH SEA

Introducing tough oil fiction

Karl Emil Rosenbæk Reetz

Introduction

In this chapter, I explore the importance of a no-nonsense "Yankee" mentality in relation to offshore oil work in the pioneering years of Norwegian oil. Practical, corporal labour reign supreme, questions are ill heard, and unionisation is unthinkable. Introducing Touch Oil Fiction as an analytical frame, I study three literary texts portraying the early days of Norway's oil engagement in the North Sea: Gunnar Bull Gundersen's short story "Vi som fragter oljen" (1960; We Who Freight the Oil), Magnus Johnsgaard's novel *Svart regn* (1995; Black Rain)—referred to by the publisher as "the first Norwegian oil novel"—and Einar Grønlie Olsen's memoir of being an early oil worker, *Oljefeber* (1984; Oil Fever), written by Olsen together with journalist Gunnar Hagen Hartvedt.

In the beginning of Johnsgaard's *Svart regn*, for instance, we hear how the Norwegian protagonist Sjur Willy is introduced to a certain work mentality as he and his workmates Tengs and Bruin ask for a jacket to withstand the cold on the windy platform in the middle of the North Sea:

> The toolpusher [Peck] hadn't heard anything that foolish. There was just one thing that helped against the cold—hard work!
>
> - If that doesn't help, then?
> - Then you work even harder, said Peck.
> - And if that doesn't help?
> - Then I hope you freeze to death (1995, 23)[1]

DOI: 10.4324/9781003647638-9

A scene like this works to instantly inform the reader (and the newly hired Norwegian roustabouts) about the laws of the platform. "There was no job for people who liked to discuss who should do what, how, and in what order. There was just one order. There was just one speed, and that was full-on" (Johnsgaard 1995, 80–81). Of course, exchanges and proverbial doctrines like these are a matter of posturing and pulling rank as much as anything else, but that only seems to further emphasise the overall, masculinist, message—suck it up, stop whining. What is remarkable is not so much the presence of a machismo worker mentality in the oil business in the 1950s–1970s. That was to be expected. The interesting aspect is the degree as to which these three offshore oil fictions more than anything express admiration and reverence towards this imported macho-culture. In a recent book by the Norwegian historian Helge Ryggvik, we find a distinctly different description of the presence of Americans and the locals' feelings about it in these nascent years of the Norwegian offshore oil adventure:

> In the pioneering years on the Norwegian sea shelf, sure enough the Americans were respected for their technological competence. But at the same time, many Norwegian oil workers experienced that the American oil companies in particular were hostile towards unionisation and promoted a macho-based anti-regulation culture that undermined the security.
>
> (2021, 142)

Here we see a clear demarcation between the American macho way and the social-democratic welfare-state way of collective agreements and worker regulations. The Americans, Ryggvik writes, were technically skilled but their demeanour clashed with the Nordic habitus. In my reading, Ryggvik's book suggests a Norwegian oil worker class that disagreed with the imported American way; however, in the literary representation of the pioneering age, there is a rather curious valorisation of this behaviour that revolves around a longing for simpler—less bureaucratic, unregulated—times.

Practical reason at sea

In her seminal article from 2010, Margaret Cohen explains the lack of interest displayed by twentieth-century literary critics in maritime fiction by their underappreciation of what she calls "practical reason". This particular knowledge form of modernity—"the intelligence distinguishing people who excel in action"—Cohen argues, often finds its expression in maritime fiction, whereas terrestrial-based novels of the modern period are more affected by Enlightenment capacities of rationality, abstract thought, and "disinterested scientific knowledge" (660). This is not to say that all land-based novels are introspective and entirely without action, but the reasonings and explanations guiding the protagonist are more often than not based on a different set of knowledge forms than those taken place in maritime settings. At sea—or in settings related to ocean activities—practical reason expresses itself much more than it does on the civilisation's terra firma, as this knowledge form "thrives in dynamic zones at the edge of expansion, which are zones of great unpredictability but also discovery, profit, and innovation" (660). Where modernity has often been conceived as "a project of progress and perfectibility", a rationalist development that one can critique for its mistakes and/or celebrate for its advancements, Cohen introduces the "project of adventure" as a more "morally ambiguous" form that has played "as decisive a role in modernity as calculation and scientific reason" has (661). And the paradigmatic setting for the explorer—the morally ambiguous adventurer—is the ocean, Cohen concludes. Following Margaret Cohen's argument, I therefore venture seaward to the inception of a highly productive and adventurous site: The oil endeavour of the North Sea of the 1960s and 1970s, where great uncertainty, gusto, visceral labour, and a clear macho jargon coalesce.

Compared to other early offshore oil sites, exploring and extracting in the North Sea attracted particular awe for its harsh working conditions. Moreover, the prehistory to the first Norwegian oil discovery dates back centuries. As it is widely referenced in the Norwegian blogosphere and trade newspapers, the bishop of Bergen, Erik Pontoppidan (1698–1764), actually delivered the first written mention of possible North Sea oil back in 1752:

> Next to its saltness, the oil, or fatness, . . . of the North-Sea, is a remarkable property of it . . . . It is moreover not improbable, that small springs or currents of rock-oil, naphtha, sulphur, or pinguous effluvia of coals, and other slimy and oleaginous juices, may arise in the sea as well as the earth.
>
> (Pontoppidan 1755, 73)

Still, it would be another 217 years before oil extraction began. At this point in time—the 1960s—the primary offshore oil reference was the Gulf of Mexico, where the distance to shore was minuscule in comparison and the weather conditions much more pleasant (at least outside of the predictable hurricane season). Historically, the first tentative attempts at offshore drilling anywhere date back to the 1890s, off the coasts of California. Then, in 1926, "modern seismology was developed", and in 1947 the first "out-of-sight of land" platform was built "mark[ing] the beginning of the modern offshore industry as it is known today" (National Ocean Industries Association (NOIA) 2006). The nomenclature (offshore) originates from these early attempts at drilling oil from the seabed close to land. Yet, as the techniques developed, the practice has been extended further and further out into the sea. In the North Sea context, there is definitely nothing off*shore* about the water-based oil enterprise.

In 1969, the Norwegian oil enterprise in the North Sea really took off with the discovery of the massive oil field Ekofisk, some 300 kilometres from the Norwegian coast. Whereas other oil nations eventually developed offshore oil engagements as an appendix to their onshore extraction activities, Norway's oil history, as hinted earlier, is special as "[t]he Norwegian petroleumscape did not expand incrementally into the offshore realm; rather, it emerged directly at sea" (Couling 2021, 120).

In the modern era, up until the 1960s, the North Sea had primarily been exploited by fishers and as a highway for merchant shipping that connected the Nordics and the Baltics to the Atlantic world. Historically and culturally, the North Sea was inscribed with Viking lore and marine stories of excellent seamanship as well as numerous tales of famous naval battles. As such, the North Sea was in no way pristine waters when the oil business took off. Still, at the time of its inception as a possible oil site, the oceanic knowledge and marine equipment needed to construct and erect an oil platform in situ at sea levels of "120m–150m water depth of the North Sea" did not yet exist (Sandberg 2019, 4–5). To even begin the Norwegian industrial offshore oil production, then, the operators quickly discovered that an entirely new concrete casting technique had to be invented in order to produce the platforms on land, tow them out to sea through the Norwegian fjords, and fill the hollow legs with water, thereby lowering them into position. The technique was called Condeep, short for Concrete Deep Water Structure, and the local invention of this technique surely added to the adventurous frontier-trope of breaking new land (or sea in this particular case) (Sandberg 2019, 4–5. cf. Rosenbæk Reetz 2026).

What further makes the Norwegian case relevant is the period of time during which this initiative commenced: The early warning shots of the energy crises of the 1970s had been delivered in the form of the first closure of the Suez Canal (1956–57), the production peak of conventional oil in Western nations such as West Germany (1966) and the United States (1970), the half-hearted 1967 oil embargo by a league of Arab nations on the second day of the Six-Day War, and the subsequent forming of OAPEC in 1968. While oil independence had been on the agenda ever since it surpassed coal as the "undisputed King" following the Second World War, the particular world events of the 1950s and 1960s intensified the scramble for secure sources of oil, preferably domestic (Yergin 1991, 541, 654). Therefore, as conventional oil production on land began to reach its highest potential and M. King Hubbert's peak oil theory circulated in political and public spheres, unconventional methods for domestic oil extraction were intensified (Hubbert 1956; see also Smil 2017, 168–70). In the petrocritical vocabulary, "tough oil" designates the introduction of new and more demanding extraction methods as "the era of easy oil is likely gone forever" (Wenzel 2014, 157). And according to Stephanie LeMenager, the initial step into uncharted oil territory—what she calls "the first Tough Oil frontier"—is the oceans (2016, 175). In that sense, Norway's oil engagement in the North Sea presents itself as a key entry point to scrutinise what I term "Tough Oil Fiction".

This maritime industrial adventure has contributed to a larger complex of oceanic oil endeavours, increasingly turning the world oceans into "petroleumscapes"—with the North Sea becoming "the most industrialized sea in the world" (Couling 2021, 110). In that sense, the North Sea oil is a clear example of how an "adventure complex" of combined "emotion, discourse, narrative practice, and intelligence has played as decisive a role in modernity as calculation and scientific reason" (Cohen 2010, 661). Correspondingly, I show how these Norwegian Tough Oil Fictions draw on features from adventure fiction to stipulate the magnitude and importance of the industrialisation of the North Sea as they inscribe this new spatial and economic maritime sphere into Norwegian literature.

I swear allegiance . . .

In *Svart regn,* the sense of venturing into uncharted territory is made explicitly clear for the reader: "None of them had worked on a drill rig before. Sjur Willy was the only one with a certain experience with the oil business. He had worked at a Shell petrol station for a year" (Johnsgaard 1995, 13). This tongue-in-cheek line is clearly meant to illustrate the sheer amount of inexperience these Norwegian workers have when it comes to working on an oceanic oil platform. A couple of pages further on, we also hear how the entire oil vocabulary is intriguing, exotic, and completely unfamiliar to the workers—"drillpipes", "joints", "monkeyboard", "derrick", "drawwork", "doghouse", etc. (Johnsgaard 1995, 19). As a reader, you are (most likely) also in the dark as to what many of these words specify. Hereby, the book draws its readers into an adventurous and strange journey together with Sjur Willy. This conveys a strong sense of shared discovery and thus identification with the protagonist.

So, while the technological development still continues in the North Sea and further north (see Ryggvik 2021), *Svart regn* clearly suggests that a particular adventurous gung-ho frontier understanding of North Sea oil reached its apex in Norway in the budding years of extraction—also known as the "pioneering age" (*pionertiden*) in the Norwegian public sphere. Likewise, linguist Johan Myking delineates how the 1980s marks the end point of the pioneering age in his extensive linguistic scrutiny of the attempt to create a Norwegian "oil language" (*oljespråk*) to replace the common language of communication in use on the offshore oil platforms, American English (1998, 242). As the oil enterprise increasingly became a national endeavour with large Norwegian oil operators—primarily, of course, Statoil (est. 1972), together with Norsk Hydro (est. 1965) and Saga Petroleum (est. 1972)—slowly gaining the majority of the concessions, Myking writes that language policies and worker rights were implemented to confirm and solidify Norwegian primacy on the continental shelf. The Working Environment Act (*arbeidsmiljøloven*) was extended to cover oil industry workers in 1977 (Meland 2019), while Statoil in 1983 stipulated Norwegian as the official working language on the Gullfaks field. As a consequence of these initiatives, Myking writes, "[t]he old pioneer-type of an oil worker with a *practical* background and trained in an American work culture has to a large extend been replaced by Norwegians who has learned about the oil business through formal and technical training" (1998, 243. My emphasis). In my reading, Myking's account illuminates how the primacy of what Cohen calls practical reason in the pioneering age slowly became absorbed and institutionalised through the education system, giving way to a more pensive, contemplative approach towards oil work.

This transition from one work culture and identity to another—and especially the tension between these two—is a common trademark in Norwegian oil-worker fiction: the practical, stalwart, tough, oil-*cowboy* on the one side, and the formally trained, rule-abiding, soft, oil-*worker* (in the social democratic sense) on the other. These representations and clashes take place almost exclusively on the open sea oil platforms, the primary site of Norwegian oil work. As I have discussed elsewhere, later Norwegian oil-worker fictions from the 2000s—published in the wake of the Global North neoliberal turn towards policies of deregulation—express a "clash between the liberalistic, global fossil capitalism and the egalitarian ideal of the Nordic welfare state model" in the shape of cautionary tales displaying the dangers of an unconstrained "Yankee" mentality (Rosenbæk Reetz 2026, 94; see also Rosenbæk Reetz 2024). In these later works, it seems the maritime adventure trope eventually has washed onshore, where different rationalities reign.

In the earlier Norwegian oil fictions, we also find this tension, although expressed somewhat differently. In these tales, the state plays a minor role and it is never venerated or endowed with nostalgia in the way that it is in later texts. In general, the North Sea is understood and represented as ill-tempered, unhospitable, and treacherous. A discursive trifecta of "wide sand dunes, roaring storms [and] shipwrecks" is noticeable in early modern tales from the coastal zone of West Jutland, for instance (Behschnitt 2017, 80). This, Behschnitt stipulates, evoked an aura of adventure. The early North Sea oil workers clearly identify with this spirit of pioneering and are very sceptical towards governmental involvement in their endeavours. These early texts draw on the libertarian and adventurous tradition underpinned by a Grotian *mare liberum* understanding of the waterway and the seabed, whereas later representations of the open sea oil work speak more positively and longingly for regulations regarding safety, worker rights, and insurance policies.

Already in the short story by Gundersen from 1960—that is, before Norway became an oil-producing nation in 1969—we see a variant of these ideals play out in the shape of a dialogue between robust "petro-masculinity" (Daggett 2018) and sexual frivolity. The novel tells the story of two Norwegian sailors on a night's leave from an oil tanker docked in a small city in England. The two men—a young ordinary seaman (*lettmatros*) and an older able seaman (*matros*)—need to kill a couple of hours before a dance hall opens its doors in the town, and they decide to sit down on a grassy slope beside the road to share a bottle of whisky. The dialogue between the two firmly reinforces the hierarchy. The younger man ruminates on some of the sacrifices life on sea brings with it, but the older and more practice-oriented seaman will not hear of it: "That's some damn nonsense, all that. I would know—I have sailed in nearly –" (Gundersen 1960, 89). Moulded by an unaccountable number of years at sea, the older seaman is clearly unwilling to scrutinise his own life choices or the costs these might have incurred, such as the inability to keep hold of a family or onshore friendships. And when the young man mentions the cargo they transport for Shell—oil—the older man quickly questions whether *they* are really the ones transporting this cargo:

> "We?"
> "Yes, it is us who carries it here—and all other places."
> "We have never carried any oil except the grease under our nails. It's the upper crust who freight—the shipowners."
> "But without us they wouldn't get far."
> "Without us they'd get exactly as far, with someone else."
> "But those others are like us, and we are like those others. We are all we."
> "Now you're getting too deep for me, again."
> "What I mean is, that we are all in on it."
> "Hell, we're not in on anything. I have a job and so do you. That's it. . . . You need to learn where you belong. And that's sure as hell not on the oil market. It's on deck with a broom and a bucket in hand." (Gundersen 1960, 90–91)

The two seamen belong on deck, not on the oil market, the older man scolds his younger colleague. They are completely insulated from politics, he thinks. They sweep, wash, move, and get their hands dirty, but they definitely do not play any part in the wider societal changes caused by the exploding rise in the global transportation of liquid power. Where the younger seaman shows some sense of responsibility or complicity, however minor, in the said changes, the older man adheres to a strict codex of utter unaccountability. His line of thought begins and ends with the sea. As a reader, we agree that he is not the primary agent of the energy transition from coal to oil—of course not. But his refusal to show the slightest interest in what he participates in carrying around the world still speaks volumes about his working mentality and how this explicitly opposes any social responsibility.

As the story concludes, the older seaman seems, rather paradoxically, to oppose social change even as he transports a material that enacts a high degree of conversion: with oil and gasoline come a collection of veritable sea changes in transportation and mobility, interconnectedness, the medical industry, and the military. Plastics enable all sorts of new modes of consumption, conservation, and transportation of goods. These sorts of social, cultural, and economic changes are not questioned in the novel, but as two couples park their car nearby and begin to kiss, the older man becomes livid. The younger man tries to calm him down—"they haven't done anything to us"—but the older man replies:

> "They haven't? How the hell do you know that?" Down towards the car: "You hear—worthless devils—here we sit—We who ship the oil."

> His voice cracks.
> "But you said that . . ."
> "To hell with what I said.—Those there, they just drive around—drive around—and then they sit and begin to—where we sit—We who ship the oil—but they know nothing—damn devils—they just drive—drive around—and here we are—we who ship the oil—but I'll teach them—damn." (Gundersen 1960, 95–96)

Clearly frustrated about the sexual frivolity in front of him—and inarticulate to the point of spluttering profanities—he suddenly turns around rhetorically and uses the younger seaman's prior argument to assert superiority over the kissing couples: "It is *we* who ship the oil, allowing them to cruise around and enjoy themselves" is the essence of his outburst. All of a sudden, he emphasises his involvement in the material transformation that allows these two couples to drive around purposelessly. Eventually, the scene escalates as the elder seaman throws his empty bottle at the car. The two men of the couples are rightously upset, and a brawl breaks out. Eventually, the police show up and the seamen are arrested for disturbing the peace (it is not clear from the text what happens to the men from the car).

What this short story registers is an odd allegiance to the maritime oil business. Firstly, the older man pompously stipulates that he holds the cargo in no regard whatsoever, but eventually his chauvinistic outburst reveals that he takes pride in providing the fuel of modernity and he does not want his involvement to be sullied by what he sees as the obscene social and cultural—and sexual—changes of modernity. He who delivers the oil defines the way of the world, he more than suggests to believe. As outsiders infringe on what he sees as the way of life, then, he transforms from a "broom and bucket" deckhand to a crucial part of an omnipotent entity—the We, with a capital *W*. The sacrifices that he has made in the name of seafaring—something that he earlier seemed to disavow—are now nevertheless belittled, to his mind, and his ostensibly disinterested attitude towards the freighted material is abandoned as it is the oil material itself that now empowers him. In that moment, his self-identity as a non-associate cowboy of the seas, jumping from job to job under his own command, is overlaid with an identity as an employee with a strong allegiance towards the firm that pays for him. Arguably, this has been the case all along and his true sensibility thus seems to emerge on these final pages. The bragging attitude towards the younger and slightly more idealistically inclined seaman exposes itself as just that; an attitude. Identifying as an oil worker, he suddenly considers himself to be in a position that allows him to regulate other people's behaviour—a position that he paradoxically expresses with the same amount of confidence and conviction that he earlier expressed his libertarian, I-don't-give-a-shit-about-anything-but-myself-identity. In that sense,

Gundersen's text is satirical as it ends up undermining the self-image of the older seamen: the freewheeling "cowboy" is in fact "just" a worker. However, I argue, the text does not criticise the primacy of an adventurous, tough frontier attitude at sea. If anything, it suggests the opposite. For it is only when the seaman exposes himself *as a worker*—subject to all kinds of allegiances—that trouble breaks loose.

Move fast and break things

Gundersen's text is written prior to Norway's domestic production of oil, but similar trademarks of tough self-sufficiency, self-contradictory allegiance to the oil business, and hot-headedness towards external interference are discernible in texts from the pioneering age, such as *Oljefeber* (1984) and *Svart regn* (1995). The aspect of modern oil adventure at sea, however, is accentuated much more explicitly in these two texts. Also notably, both of these texts are written by former oil workers. While the two texts are different in their application of fictionality—one a work of non-fiction with clear uses of fictionality in terms of plot, suspense, and figures of speech; the other a novel with obvious reference to real-life events in the Norwegian section of the North Sea—they intersect in their registration of a particular development: the death of the gung-ho oil mentality. Curiously enough, quite a few crucial scenes in *Svart regn* are actually almost verbatim reiterations of scenes found in *Oljefeber*.

Much can be said of these texts, but at their heart, they describe—with lamentation—the transition from oil exploration to oil production as a passage from the practice-based, frontier mythology of sailing into uncharted waters, to the contemplative, boring, and repetitive toil of administrating production. Whereas the first period is described with passion as an exciting and adventurous time, the second, while it might have been the keystone to Norway's current wealth, for the people involved, it brought with it a sheer lack of thrill. These texts suggest that as tough oil became easy, working life in the North Sea became boring.

Nowhere in the texts is the gung-ho mentality more present than when it comes to the issues of safety and the handling of waste: "Everything was just dumped [in the ocean]", as it reads in *Oljefeber* (Olsen and Hartvedt 1984, 101; cf. Johnsgaard 1995, 185–86). In relation to this, *Oljefeber* celebrates the "American way" as it cut through all the difficult red tape:

> They are simple minded. And consequently, they will in many cases—in practice—also be able to do things in a simple manner. That means, without the involvement of desk-procedures that must be supervised by many and controlled by even more.
>
> (1984, 89)

Of course, accidents happened but where would be the thrill if everything went by the book, as the subtext of both literary texts has it? A perfect illustration of this mentality is displayed in *Svart regn* when the drilling crew hears that they are to be sent on an obligatory security course:

> What the fuck was the point in that? Sjur Willy shook his head. What he needed was a three-week drilling course! He didn't even need that, what he really needed was an eighteen-week drilling course! Tengs agreed. What he didn't know about saving lives and first-aid wasn't worth knowing, he had the method of mouth-to-mouth locked in. Standing at his knees, blowing air into a plastic doll and pounding at its chest was at best a waste of time. He hadn't much to learn about fire extinguishing either, it was just a matter of pouring what they had onto it. If that didn't help, nothing would! The only thing left to do was to get onboard a safety boat or jump into the ocean. They didn't need a course to jump into the ocean, all they needed to do was jump, simple that!
>
> (Johnsgaard 1995, 300)

Where Helge Ryggvik in his 2021 account depicts a clear dialectic between the mentality of American and Norwegian oil workers in the pioneering age, *Svart regn* and *Oljefeber* show a much greater degree of awe and fascination towards the American way from the new-bred Norwegian oil workers: "It was literally the Wild West. Security precautions and worker rights were practically non-existent. And this did not bother the Americans. Their philosophy sounded like a refrain repeated by a drilling boss whose motto was expressed in this creed: 'There ain't but two cain't: can't do it, can't stay'" (Olsen and Hartvedt 1984, 59).

"As a situation-specific capacity, practical reason has historically taken many forms", Margaret Cohen writes (2010, 660). In the Norwegian pioneer age of oil extraction in the 1960s and 1970s, it seems the form of preference is one of unyielding action: "It was Mother Earth, the seabed, and the insane ocean they hammered on. Pitting oneself against such forces wasn't for everyone, this was a task for daredevils who stopped at nothing" (Johnsgaard 1995, 129). As both Ryggvik and Myking relay, this gung-ho attitude of Tough Oil was eventually replaced with a more disinterested, bureaucratic, and rationalistic practice. Norwegian oil went from adventure to administration.

Looking back at this reformation of the oil endeavour, one cannot but hope that we are beginning to see a similar culture of constraint closing in on the Wild West attitude of today's innovative tech-industry out of Silicon Valley. After all, nothing seems to inspire practical reason more than the motto of Facebook in its early days: "Move fast and break things." The recent tech-age repitition of the gung-ho attitude of venturing out into murky waters is exactly why it makes sense to go back and study the early North Sea oil texts: We might actually be able to see some of the hidden effects of this alluring, accelerative approach. Here, on the threshold of entering an era dominated by artificial intelligence, it seems more pertinent than ever to study earlier examples of adventurous enterprises at moments of crucial importance. And for the nation of Norway, the North Sea oil operations were surely such a period.

Note

1 When no translator is mentioned, the translations from Norwegian to English are my own.

Works cited

Behschnitt, Wolfgang. 2017. "Jutland and the West Coast as Liminal Spaces in Danish Literature." In *Nordic Literature: A Comparative History*, edited by Steven P. Sondrup, Mark B. Sandberg, Thomas A. DuBois, and Dan Ringgaard, 80–94. John Benjamins Publishing Company.

Cohen, Margaret. 2010. "Literary Studies on the Terraqueous Globe." *PMLA* 125 (3): 657–62. https://www.jstor.org/stable/25704462.

Couling, Nancy. 2021. "The Offshore Petroleumscape: Grids, Gods, and Giants of the North Sea." In *Oil Spaces: Exploring the Global Petroleumscape*, edited by Carola Hein, 109–26. Routledge.

Daggett, Cara. 2018. "Petro-Masculinity: Fossil Fuels and Authoritarian Desire." *Millennium* 47 (1): 25–44. https://doi.org/10.1177/0305829818775817.

Gundersen, Gunnar Bull. 1960. "Vi som frakter oljen." In *Vi som frakter oljen*, 82–97. H. Aschehoug & Co.

Hubbert, M. King. 1956. *Nuclear Energy and the Fossil Fuels*. Shell Development Company (Houston, Texas). June.

Johnsgaard, Magnus. 1995. *Svart regn*. Cappelen.

LeMenager, Stephanie. 2016. *Living Oil: Petroleum Culture in the American Century*. Oxford University Press.

Meland, Trude. 2019. "Working Environment Act." In *Industrial Heritage Ekofisk*: Norwegian Petroleum Museum.

Myking, Johan. 1998. "Oljespråket som laboratorieprodukt." *Syn og sagn* 104 (3): 242–49.

National Ocean Industries Association (NOIA). 2006. "About NOIA." https://web.archive.org/web/20100806100254/http:/www.noia.org/website/article.asp?id=123.

Olsen, Einar Grønlie, and Gunnar Hagen Hartvedt. 1984. *Oljefeber*. Gyldendal Norsk Forlag.

Pontoppidan, Erik. 1755. *The Natural History of Norway*. Translated by Andreas Berthelson. A. Linde.

Rosenbæk Reetz, Karl Emil. 2024. "Waterwork: The Clash Between Social Rights and Fossil Capitalism in Norwegian Fiction." *Arcadiana*. https://arcadiana.easlce.eu/2024/11/21/waterwork-the-clash-between-social-rights-and-fossil-capitalism-in-norwegian-fiction/.

Rosenbæk Reetz, Karl Emil. 2026. *Impasse Poetics at the Nordic Oil Coast: Contemporary Petrofiction from Denmark and Norway*. West Virginia University Press.

Ryggvik, Helge. 2021. *På kanten: Norsk oljevirksomhet i Barentshavet og hvorfor det må settes en strek*. Cappelen Damm.

Sandberg, Finn. 2019. "Condeeps. The Dinosaurs of the North Sea." *Journal of Energy History* (2): 1–15. http://energyhistory.eu/node/136.

Smil, Vaclav. 2017. *Oil: A Beginner's Guide*. Revised ed. Oneworld.

Wenzel, Jennifer. 2014. "How to Read for Oil." *Resilience* 1 (3): 156–61. https://muse.jhu.edu/article/569981.

Yergin, Daniel. 1991. *The Prize: The Epic Quest for Oil, Money & Power*. Simon and Schuster.

7

RIMBEREID'S WATERSCAPES

The North Sea viewed from Stavanger

Brandon Kaaz

Towed out to sea

In Norway, Øyvind Rimbereid is known as a pioneer in poetry and, more specifically, long poems. One outstanding feature of many of his poems is their relation to specific places. In this chapter, I will examine Rimbereid's Stavanger trilogy, comprised of the long poems *Jimmen* (2011), *Lenis plassar* (2017), and *Solaris korrigert* (2004), and its relation to the North Sea. I will do so in light of the blue humanities, an approach that, according to Serpil Oppermann, promotes hybrid forms and new strategies to look at the relations between the human and the oceanic:

> My argument . . . is that the blue humanities contests the assumptions of this kind of cultural inclination [such as anthropocentrism] through new representational and discursive practices by promoting hybrid forms and strategies to confront the complex contemporary realities of seascapes.
>
> (2023, 8)

In the Stavanger trilogy, Rimbereid looks into such hybrid forms by exploring human–animal, human–plant, and human–robot kinship, among other forms of human existence—such as a digital posthuman existence in *Solaris korrigert*. Furthermore, looking at the poems' form, the long poem—a genre with a traditional standing in Scandinavia (Madsen 2018)—may be viewed as a hybrid form of literature, as it finds itself somewhere between prose and poetry. Especially Rimbereid's long poems, which are highly experimental and examine the way in which humans relate to their surroundings, fit into this thinking of finding new ways to delve into one's environment through literature.

DOI: 10.4324/9781003647638-10

In a presentation held in 2024, Thomas Fechner-Smarsly conceptualised the Stavanger trilogy based on place: The three poems take place in and around Stavanger, Norway and depict human working conditions in the oil-boom town (Fechner-Smarsly 2024; Kaaz 2024, 256). The Stavanger trilogy, however, does not only tell three stories of workers who are affected by Norway's relation to oil. It also forms a historical imaginary chronicling the use of the North Sea as an oil and energy resource. The depiction of individual human fates related to this fact is used to explore the entanglement of oil, the North Sea, and humans.

Solaris korrigert, Rimbereid's most well-known poem, was added to a Norwegian literary canon only four years after its publication in 2004 (Lindholm 2008). The poem is written in a language created by Rimbereid, which, as many scholars such as Hadle Oftedal Andersen, Audun Lindholm, and Katie Ritson have pointed out, is an amalgam of the Stavanger dialect, languages spoken around the North Sea today, and Old Norse (Andersen 2020, 157; Lindholm 2008, 356; Ritson 2019, 136). Therefore, it has been described as a sort of North Sea oil language. This makes the poem essentially untranslatable, which is why in this chapter I will cite the original text and include my own rough translation solely to give the gist of the original. Even just linguistically, Rimbereid ties his poem to a specific place: the North Sea as viewed from Stavanger.

Over the course of the narrative, we are taken along on a voyage under the sea's surface:

> OU fara avgarde / til regio London, i hoy hoy speed- / tunnel undr seaen / ovfr platfurmvrak / og olda, emti gassbrunnar / der umkring dark seagrass vexr, / vexr som umkring ein munn vidout ord.
>
> (Rimbereid 2004, 28–29)
>
> ///
>
> TO travel / to region London, in the high high speed / tunnel under the sea / over oil rig wreckages / and old, empty gas wells / around which dark seagrass grows, / grows as around a mouth without words.

The Stavanger trilogy paints a map of the North Sea not only on the surface but also in its depths. Due to its innovative and hybrid form as well as the fact that it reimagines the way we relate to the ocean, it offers much in terms of blue humanities. As Oppermann puts it:

> In this approach, art, culture, and literature not only propound new cognitive modes of connecting with the seas but also help bridge the biocultural separation between the land and the sea, situating our thinking within the crises-ridden and traumatized Anthropocene seas.
>
> (2023, 6)

The Stavanger trilogy builds this bridge by telling the story of a North Sea that is continuously industrialised with energy and transportation infrastructure in the form of oil rigs and tunnels. In a way, it performs one of the central ideas of critical ocean studies:

> If there is any agreement about climate change, it is that our planetary future is becoming more oceanic. Scientific discourse has positioned the ocean as evolutionary origin for life on earth and, given the imminent threat of sea level rise, our anticipated destiny.
>
> (DeLoughrey 2015, 353)

As we will see, over the course of the Stavanger trilogy, humanity makes its way deeper and deeper into and under that big blue wet thing we call the ocean.

Places in the sea

Before we dive deeper into the poems, I would like to think about the concepts of space and place and how these relate to the ocean and, more precisely, the North Sea. Geographer Yi-Fu Tuan relates place to a sense of security, and space, conversely, to freedom:

> "Space" and "place" are familiar words denoting common experiences. We live in space. There is no space for another building on the lot. The Great Plains look spacious. Place is security, space is freedom: we are attached to the one and long for the other. There is no place like home.
>
> (1987, 3–4)

In his account, space and place are not only closely related but mutually dependent. Places fill space, are connected to each other by it. In being in a certain space, getting to know it and endowing it with value, it can become a place:

> What begins as undifferentiated space becomes place as we get to know it better and endow it with value. . . . From the security and stability of place we are aware of the openness, freedom, and threat of space, and vice versa. Furthermore, if we think of space as that which allows movement, then place is pause; each pause in movement makes it possible for location to be transformed into place.
>
> (1987, 6)

Differentiating between space and place depends on experience. Experiencing a certain area as space or place is a subjective experience brought upon by knowledge of and intimacy with one's environment. Space is freedom—movement. Too much space, however, and one might be swept away by nothingness. Place is stability—pause. Place without enough space brings with it claustrophobia, the inability to breathe. Space and place are experienced differently for each person, as Tuan remarks while comparing how a taxi driver, for example, would experience a city differently than a geographer (1987, 6). In Rimbereid's Stavanger trilogy, we are presented with five such subjective experiences of the North Sea and Stavanger region—the three human, one horse (in *Jimmen*), and one plant-choir (in *Lenis plassar*) narrators highlight how differently a place can be experienced by distinct entities.

Rendering a more precise account of the relationship between space and place, Tuan goes more into the role of movement:

> Space, we have noted, is given by the ability to move. Movements are often directed toward, or repulsed by, objects and places. Hence space can be variously experienced as the relative location of objects or places, as the distances and expanses that separate or link places, and—more abstractly—as the area defined by a network of places.
>
> (1987, 12)

Space connects places and at the same time separates them from each other. It therefore comes as no surprise that the ocean is seen by many as paradigmatic space, such as by Tuan (1987, 3) or Deleuze and Guattari, who describe the ocean as "a smooth space par excellence" (Deleuze and Guattari 1988, 479). In human experience, the ocean is one of the great unknowables, a thing we can ever only visit and therefore never become fully acquainted with. In its materiality—its primarily liquid form—the ocean resists the concept of place as defined by Tuan: It is constant interconnected movement across the entire planet. However, this makes the existence or creation of place within the ocean all the more important from a human perspective. Tuan remarks that it is the pause in movement which makes it possible for a place to be endowed with value: "Place is a pause in movement. Animals, including human beings, pause at a locality because it satisfies certain biological needs. The pause makes it possible for a locality to become a centre of felt value" (1987, 138). Without place, there would be nothing to hold on to. The emergence of meaning and intimacy would be swept away by the current. Establishing a sense of place in maritime spaces relies on infrastructure—boats, oil rigs, and tunnels—and also on the structures of poetry that make such places accessible to the literary imagination.

Adrift in the North Sea

When I think about the North Sea, two things come to mind immediately: heavy storms and the offshore oil industry. On the topic of oil and gas, seen in the context of the North Sea, both the human need for these resources as well as the ingenuity with which they are pursued seem to know no bounds. In the case of the North Sea, the intricate and complex system with which oil is extracted has led to the creation of what Nancy Couling calls the offshore petroleumscape:

> Similar to the one on land, the offshore petroleumscape is complex and palimpsestic, with the critical difference being that it is not entangled with other fixed structures. It has established completely new sets of relations to the established petroleum nodes as well as to new geographic places on- and offshore. Offshore, the petroleumscape is made up of unprecedented, tailor-made, extended territorial frameworks—the sea needed to be prepared for such an extension of industrial activities as it had previously been a fluid space held open in the interests of trade. The establishment of the offshore petroleumscape gave companies and corporations temporary rights to inhabit sea space, install fixed structures, and carry out a new type of production at sea.
>
> (2021, 109)

The discovery of oil beneath the North Sea turned it into one of the most lucrative resources on the entire planet. While the North Sea is known for its rough conditions, this was only one of the many obstacles requiring creative solutions. Offshore oil is not simply underground—you have to get to the bottom of the sea itself to start drilling.

> Now the most industrialized sea in the world, the North Sea has produced a particular type of offshore urbanity in conjunction with the construction of its petroleumscape. The offshore petroleumscape leaves a formidable territorial legacy: it has set a precedent in transforming the North Sea into an energy seascape, which is being appropriated, renovated, and redirected toward renewables at the same time as processes of optimization are extracting the last hydrocarbons from existing fields and finding other substances to inject into their hollowed-out subsea spaces.
>
> (Couling 2021, 110)

This industrialisation of the North Sea is followed by Rimbereid's Stavanger trilogy. *Jimmen* takes place in the first steps of oil infrastructure development, while *Lenis plassar* is relatively contemporary. Finally, in Rimbereid's *Solaris korrigert*, the petroleumscape is taken into the extreme: As we will see, in the proposed future, the North Sea has become riddled with tunnels and passageways not only for oil but for other energy resources and even human transportation. Humanity and oil become so intertwined that even the way in which the sea was divided by the Norwegian Offshore Directorate to sell licenses for oil exploration and production makes its way into the division of human living areas, for example.

While the negative effects brought upon the planet by the burning of hydrocarbons have not led to a renunciation of oil, they have brought some attitude adjustments for humanity's search for energy resources. Today, humans continuously come up with new ways of harvesting (renewable) energy from the North Sea, such as with offshore wind farms. Couling posits that the North Sea is going from being a petroleumscape to becoming a more broadly oriented energy seascape. This development also plays a role in *Solaris korrigert*—one premise of the poem is that the planet's oil wells are dried up, leading to an energy crisis. One of the attempted solutions in the poem, as we will see, is geothermal energy.

To conclude, the North Sea may be seen as one of humanity's planetary batteries or energy-producing machines. This is only possible due to a network of places that has been and continues to be laid upon what is perceived as the blank space of the North Sea. It tells the story of one of humanity's many attempts to subjugate a resource and extract as much value from it as possible—and one perspective of this story is told by a poet from Norway's oil capital, Stavanger: Øyvind Rimbereid.

Stavanger, oil, and the North Sea: making places

Tuan emphasises the importance of literature for the exploration of "intimate experiences, including those of place" (1987, 162). Literature may be one of the most effective ways in which the feeling of freedom in a wide space or the intimacy of a familiar place can be conveyed.

At this point, it is important to highlight the key role of geographical locations and topography in Rimbereid's poetry. In an essay on the topographic poem, Rimbereid provides some insights into how he believes topography can best be portrayed:

> Perhaps this is also how actual topography is best depicted, through poetry that constantly searches for new elements, nuances, and constellations, both physical and human. Attentive, curious. And that poetry must always continue with this attentive and curious work, because the topographic can never become a finished material, but will always be underway, always changing. Just as our lives are, we may hope.
>
> (Rimbereid 2006, 144–45)[1]

Rimbereid highlights three things here: the topographic is always in a state of becoming, the importance of being curious about the topographic due to this fact, and that the topographic is not only physical but also human. Rimbereid speaks out in favour of embracing the entanglement of the human and non-human and searching for creative ways to portray this relationship (for more on the role of the topographic in Rimbereid, see Mønster 2014; Andersen 2020). Therefore, Rimbereid's poetry is often tied to the idea of a certain place, as in the Stavanger trilogy.

While it is not explicitly stated in *Jimmen* and *Lenis plassar* that the narrative takes place in Stavanger, this fact is nonetheless relatively clear. That they are mostly written in the Stavanger dialect is one of the most prevalent characteristics of the poems linking them to the oil town. In the case of *Jimmen*, it is also possible to pinpoint the location of the narrative with the help of street names found in Stavanger (Hakestad 2013, 1–2). However, the most significant indicators that tie these two poems to the Stavanger area have to do with the North Sea oil industry, and more precisely the construction of offshore oil rigs. The Condeep platforms, a certain type of offshore oil rig, were all built in Stavanger by Norwegian Contractors AS between 1973 and 1995. In *Jimmen*, one of the most important scenes takes place when the two narrators of the poem—the horse Jimmen and its human owner—watch the construction and deployment of a Condeep platform bound for the North Sea (Rimbereid 2011, 32–33, 51–53).

Due to its dream-like narration, the places through which the main narrator wanders in *Lenis plassar* while searching for his lost wife are somewhat more difficult to localise or even fully grasp as specific places. However, once again, the concrete constructions in the North Sea lend us the foothold we need: While looking out a window into the fjord, the narrator remarks on the fact that this location is exactly where a Condeep platform glided out into the open sea in the past: "det e lenge sidå, / ufatteligt lenge sida, / ein condeep-plattform har glidd ud / og forbi akkurat her" (Rimbereid 2017, 31) /// "it's been a long time, / an unbelievably long time, since / a Condeep platform glided out / and passed by right here". This passage gives us a hint not only as to *where* the narration is set but also *when*. Again, this is difficult to pinpoint precisely. However, seeing as it is "unbelievable" how long ago the Condeep was deployed, it is safe to say that *Lenis plassar* takes place around the time of its publication in 2017. This is supported by an allusion to the construction of the Ryfylke Tunnel (constructed 2013–2019), as we will see. By contrast, *Jimmen*, which depicts Norway's transformation into a society based on oil, is easier to localise at a certain point in time. This specification has been given by multiple literary scholars, all who set the poem into the mid 1970s (e.g. Andersen 2017, 185; Vassenden 2014, 263)—the beginning phase of Norway's oil fairy tale, "det norske oljeeventyret".

We can now summarise that information on both location and time of the narrations of *Jimmen* and *Lenis plassar* is given by references to the offshore oil industry. The fact that Condeep platforms were built in a certain location has led to a sense of place, tying location, the oil industry, and cultural imagination together—and cementing Stavanger's standing as Norway's oil capital. Furthermore, this shows just how tightly the Norwegian oil industry, Stavanger, and the North Sea are interlaced. Finally, the sense of place created by the fact that a Condeep platform was dragged out to sea at a certain location is almost paradoxical—the sense of place is transferred from sea to land (and not the other way around). At the same time, it is located not only in the water (relative to some fixed surface) but also where spectators could watch the spectacle from with their feet planted firmly on the ground, interlacing land and ocean together as a place.

Lost in place

Unlike *Jimmen* and *Lenis plassar*, *Solaris korrigert* is very precise about its location in both time and space. The poem is set in the year 2480 in a futuristic Stavgersand, a portmanteau of the city names Stavanger and Sandnes:

> SIDDY Stavgersand, siddy min, / exist nearli ne at all, mang / konclud naw. KAN henda den exist / onli som ein olda namn, / som symbol an ein siddy? / JA, som om places i 2480 forts sku wiktig vera! / STAVGERSAND, ein nearli / emti place / midt I flowen af so mang / cells, organics, connects, pow og del- / lovar.
> (Rimbereid 2004, 12)

> ///
> CITY Stavgersand, my city, / barely exists at all, many / now conclude. MAYBE it exists / only as an old name, / the symbol of a city? / YES, as if places were still important in 2480! / STAVGERSAND, a nearly / empty place / in the middle of the flow of so many / cells, organics, connections, power lines and by- / laws.

Depicted is a time in which the names of places are seemingly irrelevant. Stavgersand—the name is already indicative to the fact that everything has grown together to one big "place"—is called a nearly empty place by the narrator, which is paradoxical because this description is followed up by the fact that Stavgersand is in the middle of many things, such as cells, organics, connections, and so on. We can interpret this as meaning two things. The place of Stavgersand is fully packed with things: places, high-speed connections such as tunnels, pipes, and transmission lines—there is no space left. Everything is in constant movement and abuzz, but one does not have the freedom to move around. Furthermore, this passage is indicative of the fact that not only Stavgersand is one big, crammed place but also that the entire world has become one big, crammed place devoid of free space, freedom—"as if places were still important in 2480!" The narrator lives in a small room with their partner Shiri, the streets are chaotic and full of humans (Rimbereid 2004, 13)—the depicted future brings forth the feeling of claustrophobia. The only real places—in the text literally "*wirklig* place" (Rimbereid 2004, 13)—are nature parks, such as "WILD-BEAUTI-PARK-NORWG" (Rimbereid 2004, 12), which people from all around the world visit. What is referred to as a *real* place is the only place where one seems to be able to experience a sense of space, freedom—away from the buzz of the industrious and constricting cities.

Going under

Now that we have established a sense of place for the area in which the Stavanger trilogy is set—that is, its ties to the oil industry and the North Sea, and how these ties inform and constitute the poems—it is time to cast our gaze out to sea. Following the chronological order of *Jimmen* → *Lenis plassar* → *Solaris korrigert*, one can trace a movement from land to sea and, more precisely, into the water. In *Jimmen*, the narrators only ever experience the sea by looking out towards it, such as when they watch the Condeep platform being towed out to the open sea. As we progress from *Jimmen* to *Lenis plassar* and *Solaris korrigert*, the place from which we view the sea shifts from land—from where we look out over the sea—to the undersea: The latter two poems take us underwater via tunnels, subsea structures, and empty oil wells—enabling a new perspective of the sea from inside of and under it.

Geometrically, offshore oil platforms form a relatively straight line perpendicular to the seabed. These vertical lines are complemented by horizontal lines in forms of pipelines and tunnels. The Stavanger trilogy paints a picture of a North Sea that is criss-crossed by these horizontal and vertical humanmade lines. They are the places—the underwater and subsea passageways that separate the wet from the dry—which establish the necessary pause for humans to be able to not only traverse but ultimately exist in (spite of) the tumultuous space of the North Sea.

To better understand this picture, I would like to give some facts about oil and gas lines in the North Sea and relate them to this geometric North Sea concept. According to a 2024 article in *The Guardian*, there are roughly 27,000 kilometres of gas pipelines in the North Sea (Salvidge and Hosea 2024). The Norwegian Offshore Directorate states that the Norwegian gas pipeline network consists of roughly 8,800 kilometres, with the Norwegian oil pipeline network coming in at around 1,650 kilometres (Norsk Petroleum n.d.c.). For comparison, the distance from the southernmost part of mainland Norway to the northernmost is about 1,750 kilometres—the distance covered by the Norwegian gas pipelines is almost six times that. Finally, the North Sea is home to 184 offshore oil rigs.

> Since production started in 1971, oil and gas have been produced from a total of 125 fields on the Norwegian shelf. At the end of 2024, 94 fields were in production: 69 in the North Sea, 23 in the Norwegian Sea and two in the Barents Sea. Two new fields started production in 2024.
>
> (Norsk Petroleum n.d.a.)

As we can see, the North Sea is full of pipes extracting and transporting oil and gas—straight lines drawn through the North Sea to keep the water out and connect places to each other. The narrator of *Solaris korrigert* is a labourer who controls

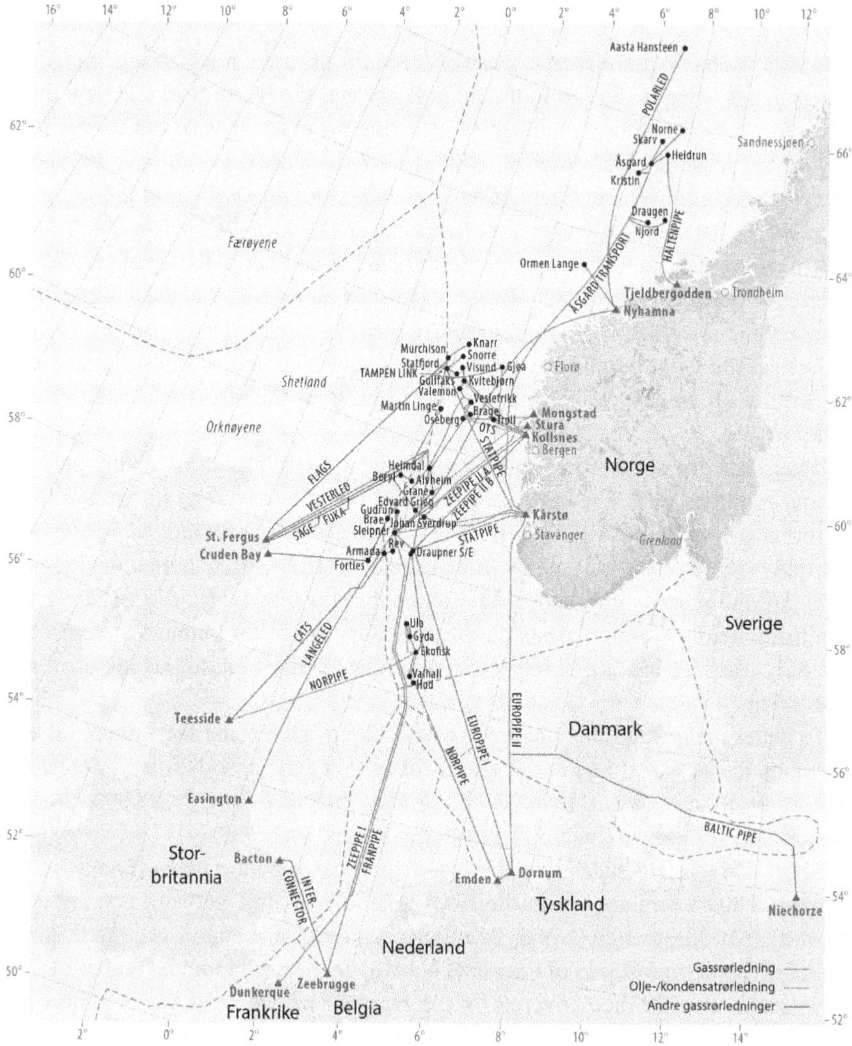

FIGURE 7.1 The Norwegian Gas and Oil Pipeline Network (Norsk Petroleum n. d.c.).
Source: www.norskpetroleum.no

robots that repair all sorts of pipes in the North Sea. This ultimately gives the narrator the expertise needed at *Seifa botten*, a sort of virtual world located in an empty oil well (Rimbereid 2004, 42). This virtual world becomes a new home for humanity over the course of the narrative.

The movement from land to sea carried out in Rimbereid's Stavanger trilogy reflects the urbanisation of the North Sea, such as its division into blocks remarked upon by Couling:

> In contrast to the earlier fluidity of extended commercial and cultural connections across the North Sea, the strategic instruments applied in the process of extended urbanization fundamentally changed the nature of the seascape into a fixed set of Cartesian coordinates designed to produce a seascape "operationalized" for the extraction and production of oil and gas.
>
> (2021, 113)

The fixed sets of Cartesian coordinates describe the division of the North Sea into quadrants by the Norwegian government, each with a number between 1 and 36. Each quadrant is divided into twelve smaller blocks. The division of living areas in Stavgersand, which are numbered in a similar fashion, parallels this—such as 14.6 where the narrator lives (Rimbereid 2004, 9), or 3.4 and 3.5, where more powerful social classes live (Rimbereid 2004, 22–23). Andersen points out that Norwegian readers will associate this with the way in which the continental shelf is divided into blocks (2020, 160). This demonstrates the entanglement of oil and society in the world of *Solaris korrigert*, and is indicative of how tightly this knot has been woven. However, this is a system created not only for oil extraction but for territorialising the sea—a system which then makes its way from water onto land. At any rate, this adds even more lines to our geometric structure of the North Sea, in addition to the material lines in the form of pipes.

The resource-transporting and human-drawn lines in the North Sea are complemented by human-transporting tunnels under the North Sea. In *Lenis plassar*, we have an allusion to such a tunnel being built:

> Leni, / som flerne ganger snakte / om denne nye tunnelen de bygge nere i byen, / den som ska kryssa fjorden, / ska visst verdens lengste ble, / langt der nere kryssa onna / før sakte opp igjen.
>
> (Rimbereid 2017, 42–43)
>
> ///
>
> Leni, / who many times talked / about the new tunnel they are building downtown, / the one that will cross the fjord, / will apparently be the world's longest / far down there crossing under / before slowly rising up again.

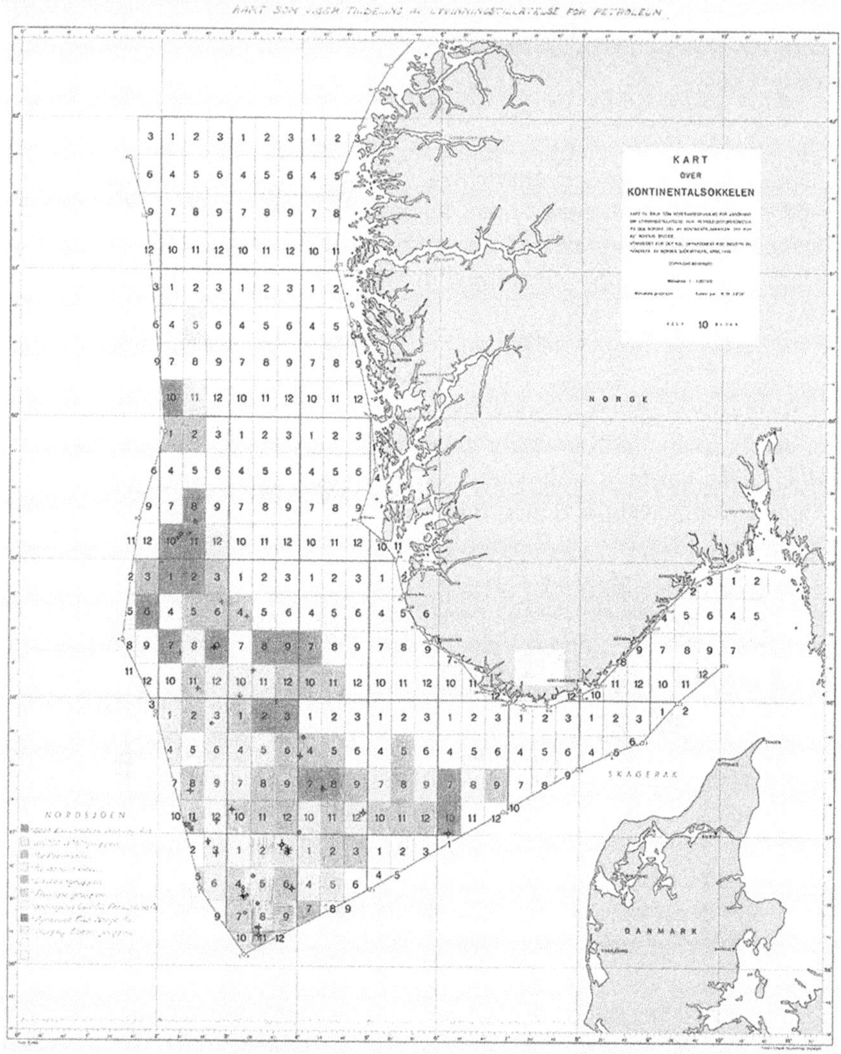

FIGURE 7.2 The first map dividing the Norwegian North Sea into quadrants and blocks (1965) (Norsk Petroleum n. d.b.).

Source: www.norskpetroleum.no

The tunnel mentioned here is most likely the Ryfylke Tunnel (NO: Ryfylketunnelen), a tunnel under the Horgefjord connecting Stavanger and the district of Ryfylke. The tunnel was built between 2013 and 2019, while *Lenis plassar* was published in 2017. As stated in the poem, the Ryfylke Tunnel is currently the world's longest subsea road tunnel, boasting a length of 14.4 kilometres. It is also the world's deepest subsea tunnel, reaching a maximum depth of 292 metres below sea level. An even longer, deeper tunnel—the Boknafjord Tunnel, also in Stavanger—is currently under construction. Much more interesting than these tunnel facts, however, is the way in which the narrator recalls Leni's thoughts about (being in) the tunnel:

> Leni sa flerne ganger at det va merkeligt / at alt det kolossala vatnet, heila havet / egentligt, ville då ligga øve oss / når me kjørte ner der, at alt der ville vera omvendt, / havet som himmel, / sa hu, flerne ganger, / ikkje skremt, men omvendt, likte å se / det for seg, någe eg ikkje heilt fatta. / Vatnet som himmel?
>
> (Rimbereid 2017, 43)
>
> ///
>
> Leni said several times that it was strange / that all that colossal water, the entire sea / really, would be above us / as we drove down there, that everything would be reversed, / the sea as sky, / she said, several times, / not afraid, but on the contrary, liked to see / it for herself, something I didn't quite understand. / The water as sky?

The tunnel is portrayed as a place in which everything is upside down—a place where the sea has become the sky. The mentioning of the tunnel can be understood as an intertextual reference to the earlier published *Solaris korrigert*, in which underwater and subsea tunnels seem to have become commonplace transportation routes.

The sea as sky

The Stavanger trilogy paints a picture of a North Sea that is becoming more and more industrialised, filled with pipes and tunnels, and metal and concrete. This picture in turn speaks to the relationship between humans and the North Sea, or even water in general. The main purpose of all of this construction is to displace water. Like water and oil, water and humans do not mix well. Yes, humans need to drink a lot of water to survive, but they cannot survive being deep inside of it for long times. Apart from the obvious danger of drowning, water in its liquid form is movement. Humans need pause in movement—places—to survive underwater. Therefore, we go to great lengths to circumnavigate the substance covering about 71 per cent of the planet. In the case of the North Sea, this is accomplished by creating a network of underwater and subsea places that suppress the movement of the sea, creating a sort of land and enabling human activities.

In the world ravaged by human expansion depicted in *Solaris korrigert*, a world mostly devoid of space and freedom, the sea (or rather the subsea) is presented as one of humanity's last refuges. As mentioned before, the narrator is one of the people sent to work on or live in the world computed by Breynmachin BK2884. It is described as a sort of virtual or holographic world in which all of humanity's needs are met (Rimbereid 2004, 35–36). Its placement is rather interesting—its home is an empty oil well one kilometre under the sea:

> MEN ogso dei emti brunnar ska naw fyllast. / EIN ny, seifa tiim er for lengst startat. / MANG so energic ofr detta. / WI ska ne emti left eftr uss! / DEPT i Nordsea, / i ein emti oilbrunn, 1 km undr sea, / fins allrede Breynmachin BK2884, *hidden og protectat.* / . . . OGSO aig der ned ska snart. / AIG vid min mekanical / novledg og miner robots / for ou arbeida i NEW-DEPT-SEA-VORLD. / I vorld undr slam, seagrass / og fish. / ARBEIDA vid geothermal energi / aig ska, piping den up.
>
> (Rimbereid 2004, 30–31)
>
> ///
>
> BUT also the empty wells are now to be filled. / A new, safe time has long begun. / MANY are energised over this. / WE shall not leave behind emptiness! / DEEP in the North Sea, / in an empty oil well, 1 km under the sea, / is already Brainmachine BK2884, *hidden and protected.* / . . . ALSO I shall go there soon. / I with my mechanical / knowledge and my robots / to work in the NEW-DEEP-SEA-WORLD. / IN the world under mud, seagrass / and fish. / WORK with geothermal energy / I shall, piping it up.

Humanity flees the surface, the real world, for a virtual copy of it in a subsea cave—a non-place hidden within the carcass of an exhausted oil well, that is, an empty place under the free space of the North Sea. The narrator's job is to pipe up geothermal energy from Earth's core. The network of lines and pipes in and around the North Sea gets bigger and delves deeper.

Before the narrator descends to *Seifa botten*, the sea is painted red by a light, and he questions whether this is because of light or poison: "SEAEN nearli red, skummande red, / syns den vera. AF ljus / or af poison? (Rimbereid 2004, 43) /// THE sea nearly red, foaming red, / it seems. BY light / or poison?" Finally, as the narrator reaches the bottom after an elevator ride, he is blinded by light—and most of the page is left blank. Katie Ritson reads the light metaphors as indicative to a sort of false Enlightenment:

> The proximity of light and poison suggests the possibility of a light of truth open to corruption, a false light. . . . The poem ends with the narrator's arrival in Solaris; the elevator stops with a click and the last line is "kvar vegg umkring meg fell" ("the walls surrounding me fell away") (45), the emptiness of the remaining page perhaps suggesting the blinding light into which the narrator has been dispatched. The light that beckons with truth and progress may in the end only signal humanity's next entanglement with the fate of the Earth.
>
> (Ritson 2019, 157–58)

In *Seifa botten*, the world consists of light, it is simulated by a supercomputer. Like King Oedipus in the Sophoclean tragedy, the narrator of *Solaris korrigert* sees too much, and the result is blindness—too much light resulting in darkness. In the end, Leni's words have become prophetic: The world is inverted and reduced to a digital shadow of its former self—humanity hides from a world it ravaged in a place where the sea has become the sky.

Conclusion

Jimmen, *Lenis plassar*, and *Solaris korrigert* are tied to a specific geographical location: the North Sea as viewed from Stavanger, Norway. This fact makes its way into the poems in multiple ways. One of the most evident is through language. While *Jimmen* and *Lenis plassar* utilise the Stavanger dialect to tie them to the region, *Solaris korrigert* is written in a language created by the author, which consists of languages spoken on and around the North Sea. Seen as a trilogy, these poems tell the story of the industrialisation of the North Sea. Beginning when the first oil rigs were towed out to sea, the poems take us through the present to an imaginary future in which the planet's oil has run out and humanity must seek refuge—and finds it in an empty oil well under the North Sea. In telling this story, the Stavanger trilogy creates a sense of place of the Stavanger area and the North Sea through their relation to the offshore oil industry, such as by marking locations as the places where people watched oil rigs being towed out to sea, or the use of numbers reminiscent of the way in which the Norwegian Offshore Directorate divided the North Sea to designate human living areas in the futuristic Stavgersand. Furthermore, the poems show how the industrialisation of the sea—the creation of places in the vast liquid space of the sea, separating the wet from the dry—is used by humans to be able to exist in (spite of) the North Sea, such as by creating tunnels and pipes for transportation. Finally, the Stavanger trilogy tells the tale of human hubris—the extractivist mentality leads to the end of oil and thereby a prerequisite for the human way of life, leading to a new, perhaps posthuman way of life and a shift to geothermal energy. Whether this ending is a positive development is up for the reader to decide.

Note

1 Unless stated otherwise, all translations have been rendered by the author of this chapter.

Works cited

Andersen, Hadle Oftedal. 2020. "Stavanger, Pre- and Postmodern: Øyvind Rimbereid's Poetry and the Tradition of Topographic Verse." In *Contemporary Nordic Literature and Spatiality*, edited by Kristina Malmio and Kaisa Kurikka. Palgrave Macmillan.

Andersen, Per Thomas. 2017. "'Der syddrå boble.' Om Øyvind Rimbereids dikt *Jimmen (2011).*" In *Åpne dører mot verden: Norske ungdommers møte med fortellinger om skyld og privilegier*, edited by Jonas Bakken and Elisabeth Oxfeldt. Oslo Universitetsforlaget.

Couling, Nancy. 2021. "The Offshore Petroleumscape: Grids, Gods, and Giants of the North Sea." In *Oil Spaces: Exploring the Global Petroleumscape*, edited by Carola Hein. Routledge.

Deleuze, Gilles, and Guattari, Félix. 1988. *A Thousand Plateaus: Capitalism and Schizophrenia.* Translated by Brian Massumi. University of Minneapolis Press.

DeLoughrey, Elizabeth. 2015. "Ordinary Futures: Interspecies Worldings in the Anthropocene." In *Global Ecologies and the Environmental Humanities: Postcolonial Approaches*, edited by Elizabeth DeLoughrey, Jill Didur, and Anthony Carrigan. Routledge.

Fechner-Smarsly, Thomas. 2024. "Syddr and Grease: Offshore-Oil-Technologies, Waste and the Perspective of Human Labour." Presentation held January 19, 2024 at The Sea in Scandinavian Literature Workshop, Frankfurt am Main.

Hakestad, Aud Jorunn Haugen. 2013. *Å vera til stades: Ei lesing av Jimmen (2011) av Øyvind Rimbereid.* University of Stavanger.

Kaaz, Brandon. 2024. "The Sea in Scandinavian Literature." *Nordeuropaforum—Zeitschrift für Kulturstudien*: 253–57.

Lindholm, Audun. 2008. " 'FOR aig vet existen af vorld.' Om Solaris korrigert." In *Norsk litterær kanon*, edited by Stig Sæterbakken and Janike Kampevold Larsen. Cappelen.

Madsen, Claus. 2018. *Det Udstraktes Poetik: Læsestrategier til det skandinaviske langdiktet.* Åbo Akademi.

Mønster, Louise. 2014 "På hestehove gennem himmelporte: Øyvind Rimbereid og den topografiske digtning, især *Jimmen* (2011)." In *Nordisk samtidspoesi: Særlig Øyvind Rimbereids forfatterskap*, edited by Ole Karlsen. Oplandske bokforlag.

Norsk Petroleum. n.d.a. "Fields." Accessed February 26, 2025. https://www.norskpetroleum.no/en/facts/field/.

Norsk Petroleum. n.d.b. "Norway's Petroleum History." Accessed March 3, 2025. https://www.norskpetroleum.no/en/framework/norways-petroleum-history/.

Norsk Petroleum. n.d.c. "The Oil and Gas Pipeline System." Accessed February 26, 2025. https://www.norskpetroleum.no/produksjon-og-eksport/rortransportsystemet/.

Oppermann, Serpil. 2023. *Blue Humanities: Storied Waterscapes in the Anthropocene.* Cambridge University Press.

Rimbereid, Øyvind. 2004. *Solaris korrigert.* Gyldendal.

Rimbereid, Øyvind. 2006. "Om det topografiske diktet." In *Øyvind Rimbereid: Hvorfor ensomt leve. Essays.* Gyldendal.

Rimbereid, Øyvind. 2011. *Jimmen.* Gyldendal.

Rimbereid, Øyvind. 2017. *Lenis plassar*. Gyldendal.

Ritson, Katie. 2019. *The Shifting Sands of the North Sea Lowlands: Literary and Historical Imaginaries*. Routledge.

Salvidge, Rachel, and Leana Hosea. 2024. "Abandoned Pipelines Could Release Poisons into North Sea, Scientists Warn." *The Guardian*, February 24. Accessed February 26, 2025. https://www.theguardian.com/environment/2024/feb/24/abandoned-oil-gas-pipelines-poison-pollution-risk-north-sea-scientists.

Tuan, Yi-Fu. 1987. *Space and Place: The Perspective of Experience*. Arnold.

Vassenden, Eirik. 2014. " 'Men en aent dyr.' Språk og erfaring i Øyvind Rimbereids *Jimmen*." In *Nordisk samtidspoesi: Særlig Øyvind Rimbereids forfatterskap*, edited by Ole Karlsen. Oplandske bokforlag.

8

LONGING FOR WHAT'S LOST

Evolution, ecology, and multispecies entanglements in Henrik Ibsen's *The Lady from the Sea*

Frederike Felcht

Losing ground

The Norwegian author Henrik Ibsen (1828–1906) is often considered the founding figure of modern drama and is one of the most staged playwrights worldwide. However, his work has hitherto not been very present in blue humanities scholarship—in contrast to, for example, Shakespeare. But water pervades Ibsen's plays: the sea is a metaphor for human relations and a symbol for the human psyche in *Brand* (*Brand*, 1866), a dramatic poem written before the so-called contemporary dramas Ibsen is most famous for. Of these latter plays, *En Folkefiende* (*An Enemy of the People*, 1882) is an outstanding example of Ibsen's engagement with water: The discovery of water pollution by the medical officer Dr. Stockmann that affects the recently opened town's spa and the question of how to deal with it are central to its plot (Detering 2013; Garrard 2007, 120–26). Another example is the later, more symbolic drama *Rosmersholm* (*Rosmersholm,* 1886) that includes numerous allusions to folk tales of *nøkker* (i.e. humanoid, often shape-shifting inland water spirits) and merfolk (Jacobsen and Leavy 1988, 121–30, 173–95). In *Fruen fra havet* (*The Lady from the Sea*, 1888), the sea plays a major role, as is indicated by the title.

Set in a small town on a Norwegian fjord, the drama refers to several different bodies of water. However, these bodies of water have been rather neglected in scholarly research that usually focuses on psychological or psychoanalytic aspects of the play (Heitmann 2003, 179). If analyses deal with the watery elements, they resort to abstract concepts like "the sea". Inspired by Cecilia Chen's emphasis on "the specificity of situated waters" (2013, 275), this chapter looks more closely at the concrete bodies of water represented in *The Lady from the Sea* and the relations

DOI: 10.4324/9781003647638-11

they entail. Based on Serpil Oppermann's observation of a "constant slippage between the geophysical properties and discursive construction of waterscapes" (2023, 11), it strives for a comprehensive perspective that highlights how the material and the metaphorical intertwine. For this purpose, I will show how the physical experience of bodies of water by humans relates to aquatic ecological properties, and how scientific discourse shapes the imagination of human–water relations in the play.

Henrik Ibsen wrote *The Lady from the Sea* in Munich, far away from the sea, where he lived for several years. His notes, however, show that the human–sea relation was central to his thoughts when he started writing the play—be it because of or despite his distance from the ocean:

> The sea's magnetic power. The longing for the sea. Human beings akin to the sea [Mennesker i slægt med havet.][1] Bound by the sea. Dependent on the sea. Must return to it. One species of fish is a vital link in the chain of evolution. Do rudiments of it still reside in the human mind? In the minds of certain people?
>
> Images of the teeming life of the sea and of "what is lost for ever"— .
>
> (Meyer 1971, 105; Ibsen 1909, 143)

These notes indicate that evolutionary ideas shaped Ibsen's understanding of the sea (Tjønneland 1998; Aarseth et al. 2009, 421–22; Shepherd-Barr 2015, 63–91, 2021) and that he considered the sea an important and threatened more-than-human habitat. According to American literary scholar Stacy Alaimo, Charles Darwin's idea of "physical relatedness [of humans and other animals] may provoke a rich ethical sense of kinship" (2012, 480). Ibsen's reflections on humans' descent from aquatic animals in *The Lady from the Sea* resonate with this idea and its ethical implications.

Reading *The Lady from the Sea* with a focus on the roles of the sea, I found my own situation mirrored in a scene of the play, when the teacher Arnholm and the physician Wangel try to decipher the relation between Wangel's wife, Ellida, and the Stranger, who has asked her to follow him to sea. Arnholm summarises the impossibility of their endeavour to understand what has happened due to the shifting memories of the spouses and the different stories they tell: "Altså tegn imod tegn" (Ibsen 2009, 638).[2] "Sign against sign" is a good description of this play, in which explanations become slippery and relations change constantly (Heitmann 2003, 184–87), and teaching the play brings this slipperiness to the fore. However, the physician Wangel is convinced that it is "reality" (Ibsen 1967, 69) that brings "recovery" (Ibsen 1967, 69). So, how should we understand the sea and its oscillation between imagination and reality in *The Lady from the Sea*? I would like to suggest that Ibsen depicts the (always problematic) longing for a passionate fusion with the more-than-human sea under the conditions of the Anthropocene by pointing to aspects such as the qualities of water, the development of fossil fuel sea traffic, and the history of humankind's un-becoming kin with marine beings.

> Mankind's growing influence on the environment was recognized as long ago as 1873, when the Italian geologist Antonio Stoppani spoke about a 'new telluric force which in power and universality may be compared to the greater forces of the earth,' referring to the 'anthropozoic era'.
>
> (Crutzen 2002)

However, it would be misleading to understand Ibsen as a prophet of today's environmental crises. But traces of Anthropocene change are present in the play and it reflects on the deteriorating relationships between humans and their environment.

Absent mermaids, swimming ladies, and the qualities of water

Initially, *The Lady from the Sea* was titled *The Mermaid* and numerous scholars (Meyer 1971, 106–7; Jacobsen and Leavy 1988, 131–36, 196–210; Wicklund 1997; Templeton 1997, 195–96; Heitmann 2003, 191; Moi 2006, 297; Aarseth et al. 2009, 421–22, 428, 444) have pointed out intertextual references to mermaids and mermen in, for example, Hans Christian Andersen's *Agnete og Havmanden* (*Agnete and the Merman*, 1834) and *Den lille Havfrue* (*The Little Mermaid*, 1837), and Paul Heyse's novel *Kinder der Welt* (*The Children of the World*, 1873). However, Ibsen's decision against the title *The Mermaid* emphasises the importance of the sea in the play. The title hints at the absence of mermaids; the mermaids that are never shown in *The Lady from the Sea*.

The relation between showing and telling has a self-reflective aspect in the play. The drama—usually the genre in which showing plays a major role—problematises artistic forms of visualisation, as Annegret Heitmann (2003, 180–97) has argued convincingly. Its plot is driven by what is told, and Ibsen scholar and translator Anne-Marie Stanton-Ife states that

> diegetic space and time offer Ellida infinitely more freedom than mimetic space and time . . . . Diegetic space . . . is unlimited, unfixed and multiple, and can unfold on all three time levels: past recollection, present elsewhere and future projection. It is thus only natural that Ellida should cling to her sea-narrative as it is the only one that does not bind her in any way.
>
> (1996, 46)

But the narratives are unreliable and thus emphasise the importance of the "real" body, place, and action. This reality is not fixed, but changes constantly in response to the often illusionary (self-)perceptions of the protagonists. The play thus constantly oscillates between physical and discursive constructions of reality.

In the first scene, the painter Ballested explains to Lyngstrand, a young spa guest and would-be sculptor, what he is painting: The picture—like the scenery in the theatre (Heitmann 2003, 180)—depicts the fjord they (and the audience) are seeing, as Ballested says,

> but it still needs a figure. There isn't a model anywhere here in town. . . . Here in the foreground beside this rock there is to be a mermaid lying half dead. . . . She has strayed in from the sea and can't find her way out again. So here she lies, dying in the brackish water, you see.
>
> (Ibsen 1967, 2)

Ballested, we learn, is a jack of all trades and master of none, which makes his artistic talent questionable and gives his artwork a comic touch. There is no mermaid in his painting, only a fjord, nevertheless, in the heads of the audience, the picture of the half-dead mermaid appears. Like the artist figures Ballested and later Lyngstrand, Ibsen plays with the gap between reality and imagination (Heitmann 2003, 180–82; Binding 2006, 47–80; Moi 2006, 301–5). Ocean is part of this play: The stage instructions clarify explicitly that the open sea of which the figures often speak is never seen (Ibsen 1967, 24, 2009, 551; Stanton-Ife 1996, 48–49). It is constantly evoked but absent—the setting thus emphasising the estrangement from the sea that the play describes. At the same time, most scenes take place outdoors, thus indicating the importance of more-than-human environments for its contents.

Ballested's mermaid is to die in brackish water. From an ecological perspective, brackish water can indeed become a difficult habitat at a certain salinity level (Schiewer 2008, 10). But it is not so much the salinity that disturbs the woman who gave Ballested the idea for his picture, Ellida Wangel; it is the unhealthy warmth of the water: "Fresh! Good Lord! The water here is never fresh! There's no zest, no sparkle. Uh! Here in the fjords, the water is sick" (Ibsen 1967, 11). It is not the mermaid who is sick here, but the water. The reply of Ellida's husband, "There's a splendid recommendation you are giving the place [badestedet]" (Ibsen 1967, 11, 2009, 526), reminds us of the earlier drama *An Enemy of the People* and parallels the physicians Wangel and Stockmann. The issue of water pollution thus becomes a subtext and an intertext of the play.[3]

The list of characters includes the remark that the play takes place "in a fjord town in Northern Norway" (Ibsen 1967, n.p.), most likely modelled after the town Molde (Binding 2006, 49–60; Aarseth et al. 2009, 513), thus indicating that it is the Norwegian Sea that Ellida comes from and longs for. Contrasting the water of the fjord and the open sea, Ibsen picks up the oceanographic difference based on the geological structures of the relatively deep Norwegian Sea basin and the flatter ground of the surrounding fjords (Skjoldal 2004, 15–17).

In the eyes of the others, Ellida has a special relationship to the sea and everything that belongs to it ("et særligt forhold både til havet og til alt, hvad havets er" [Ibsen 2009, 526]). Her husband calls her "havfruen" (Ibsen 2009, 525), the mermaid, at her first entrance, but Ellida's dress encodes her as a swimmer and thus a human being with a special relationship to water, without transcending the boundaries of what is generally considered a human body. Literary scholar Kirsten Shepherd-Barr emphasises that Ellida is no "freak of nature", and that Ibsen focuses on the "relationship between a mature woman and her environment instead" (Shepherd-Barr 2015, 75). Perceiving Ellida as a mermaid underlines how distanced the other characters have already become from the more intimate relation to the sea that Ellida (who comes from a place close to the open sea) embodies—and how distanced the male characters in the play are from the female (Wicklund 1997, 102–3).

Swimming is gendered in *The Lady from the Sea*: Ellida is a passionate swimmer, whereas her husband does not seem to swim with her. Arnholm states, "I am not much of a swimmer" (1967, 42). Lyngstrand almost dies in a shipwreck. Christine Hamm (2015) has shown how the literary depiction of the relation between women and the sea changed historically from women *at* the sea to women *on* the sea and, finally, women *in* the sea. The different positions correspond to an increasing degree of emancipation that is associated with the idea of the ocean as a place of freedom and of the healing power of seawater. Stanton-Ife emphasises that Ellida's relation to the sea rejects the traditional alliance of the woman and the home and states:

> The sea, with its associations of unboundedness, changeability and fluidity . . ., can be seen to represent the space in which Ellida can be 'inside herself', a container for-herself; a space in which her own pleasure and meanings can unfold.
> (1996, 34)

The sea reminds Ellida, furthermore, of her place of origin, a distant past that she longs for.

In contrast to this association of the sea with freedom and healing, for men like Lyngstrand, whose father sends him to sea for economic reasons, the sea rather resembles a dangerous prison. There are, furthermore, several moments in the play that indicate that the healing power of the water in the sea bath is perhaps not so effective. We have already seen that it is associated with sickness and pollution, and nobody except Lyngstrand himself believes that bathing in the fjord can heal him from his respiratory disease. The Wangel's son dies from unknown reasons at the age of "four or five months" (Ibsen 1967, 10), despite (or because of?) the fact that his mother swims every day. This child, like its mother, has a special relationship to the sea. Ellida tells her husband that the eyes of the child "changed colour with the sea. When the fjord was calm and sunny, his eyes were the same. The same when it was stormy. Oh, I saw it well enough, even though you didn't" (1967, 40). A genealogical end of the close human–sea relation is thus indicated with the child's death. The future brings an estrangement from the sea.

There are no concrete indicators for the pollution of the water, but one pollution factor plays a prominent role in the play. The fjord connects the town with the international tourist routes that go further north, and we learn that the visitors travel by steamer whose arrival and departure is an important event—the whole play is structured by it, from the beginning to the end. The fossil fuel economy has reached the town, and it has become a travel destination for bathing, breathing fresh air, and viewing beautiful scenery. The touristic consumption of the landscape threatens its status as an object of desire.

Fjord and pond: ambivalent acclimatisation

In Ibsen scholar Rolf Fjelde's topographical interpretation, the play is structured by the spatial division of land versus sea, and the analogy between Ellida and the figure of a mermaid is an emblem for her being torn between the land, embodied by Wangel, and the sea, embodied by the Stranger (Fjelde 1978). However, the play is neither set on the open sea nor inland, but at a fjord. In focusing our attention on bodies of water, we can see that fishpond, fjord, and open sea structure the play.

The third act starts with a stage instruction that includes a description of Wangel's garden: *The place is damp and marshy and overshadowed by large old trees. On the right can be seen the edge of a stagnant pond* [*skimlet dam*] (Ibsen 1967, 41, 2009, 587; italics in the originals). The fjord is not the open sea, as Ellida emphasises, but even less so is the fishpond, as doctor Wangel's daughter Bolette explains to Arnholm, who had also once been her teacher:

Bolette. Oh, it's true. I don't think life is so very different for us from what it is for those carp down there in the pond. They have the fjord close by where the great shoals of wild fish move in and out. But the poor tame local fish [de stakkers tamme husfiskene] know nothing of all this. They can never be part of that life.

Arnholm. But then I don't think it would particularly suit them if they did get out of here.

Bolette. Oh, I doubt if it would make much difference on that score.

Arnholm. Anyway, you can't say you're all cut off from life out here. Not in summer at any rate. These days this place is nearly like some great traffic junction, with the whole world in transit.

(Ibsen 1967, 43, 2009, 591)

This dialogue is firstly about living at a tourist destination without having the means to travel somewhere oneself; secondly about being a young woman, hungry for knowledge but restricted by the patriarchal norms of the nineteenth century; and thirdly about being a fish in a fishpond. It is important in this last context that Arnholm here sees only the difference between two different habitats, inland water and ocean, whereas Bolette also thinks of domestication—"de stakkers tamme husfiskene". She thus emphasises a dynamic aspect of ecological relations that can change over time. But as the "skimlet dam", literally "the mouldy pond", indicates, the domestication of (wo)men and fish has reached an unhealthy standstill, both are exploited (the carps are fished, Bolette works for a family that does not care about her own wishes) and need to find a way to break free. Bolette's focus on domestication resists a simple naturalisation of the existing state and thus opens rooms for interventions.

A further dynamic aspect of species-in-habitat is referenced through the concept of acclimatisation that the painter Ballested introduces in the play (a term that he, ironically, cannot pronounce):

Lyngstrand. Were you born here?
Ballested. No, I wasn't. But I've acc-acclimatized myself [akkla-akklimatiseret mig]. I feel bound to the place by ties of time and custom.

(Ibsen 1967, 3, 2009, 510)

Ibsen hereby refers to an idea that Darwin presented in *The Origin of Species* (1859), which Ibsen probably knew from the translation of the fifth edition from 1869 that the Danish author and biologist Jens Peter Jacobsen published in 1872 (Aarseth et al. 2009, 425). Darwin explains acclimatisation mostly with reference to domesticated plants and animals as a combination of habit and natural selection (1872, 183–87). In *The Lady from the Sea*, acclimatisation gets an emotional component ("I feel bound to the place") and is essentially a human capacity: Ballested explains later that "menneskene *kan* alki-a-klimatisere sig" (Ibsen 2009, 663), "humans *can* alci-a-climatise themselves" (James McFarlane's translation with "people" [Ibsen 1967, 79] does not imply the zoological connotations of "menneske" as a species); Ballested repeats this idea in the end of the play. Ballested himself acclimatises by combining as many niches as possible for earning his living, whereas Ellida accepts her roles as a wife and a mother when Wangel finally allows her to choose whether she wants to stay with him or not. Both Ballested and Ellida settle down and give up the more nomadic life form that is still embodied by the Stranger. Whereas Ballested and Ellida integrate into a place with strong human-built infrastructures, the Stranger remains an outsider with a special relationship to the sea. However, in contrast to the fishpond, Ballested's and Ellida's new home upholds a connection to the global marine circulation of people, goods, and non-human animals.

A reference to acclimatisation appears at the beginning and at the end of the play, it is marked as an outstanding human capacity and combined with the idea of evolution:

Ellida. [*smiling gravely*] Well, Mr. Arnholm, do you remember what we were talking about yesterday? Once a creature has settled on dry land [Når en nu engang er blevet en fastlandsskabning], there's no going back to the sea. Nor to a life that is linked to the sea [til havlivet] either.

Ballested. Why, that's just like the case of my mermaid!
Ellida. Yes, something like that.
Ballested. Except for one difference. On the other hand, men and women [menneskene] can acclam-acclimatize themselves. Yes, yes, I assure you, Mrs. Wangel. They can ac . . . cli . . . matize themselves.

(Ibsen 1967, 95–96, 2009, 696–97)

The idea that one becomes a creature of the solid land acknowledges the evolutionary impact of the environment on humans. This development has, Ellida ultimately decides, become irreversible. She links this new insight to the loss of relationships with the ocean and marine life. The first reply refers to an earlier dialogue:

Bolette [*with a sigh*]. Ah no, we have to be content with dry land [landjorden].
Arnholm. Well, that's where we really belong, I suppose.
Ellida. No, I don't think we do.
Arnholm. Not dry land [landjorden]?
Ellida. No, I don't think so. I think if only man had learnt to live on the sea from the very first . . . Perhaps even in the sea . . . We might have developed better than we have, and differently [så vilde vi nu ha' været ganske anderledes fuldkomne end vi er]. Better and happier.
Arnholm. Do you really believe that?
Ellida. Well, I sometimes wonder if we wouldn't. I have often talked to Wangel about this . . .
Arnholm. And what does he think?
Ellida. He thinks it could well be.
arnholm. [*jestingly*] Well, perhaps! But what's done is done. We've taken the wrong track and become creatures of the land [landdyr] instead of the sea [havdyr], and that's that! At all events, it's too late to put things right now.
Ellida. Yes, it's sad but true. And I believe that people [menneskene] suspect something of this themselves. And bear with it some secret sorrow. Believe me, here are the deepest springs of human melancholy [deri er det, at menneskenes tungsind har sin dybeste grund]. Yes, believe me.

(Ibsen 1967, 46–47, 2009, 598–99)

This passage refers to the Darwinian idea of "a kinship between humans and other animals" (Alaimo 2012, 480), humans have become "landdyr", literally "land animals"—James McFarlane's translation "creatures" still refers to the idea of creation that Darwin undermined. But humans have lost their connection with the sea animals and therein lies the deepest ground of their melancholia, their *tungsind*, literally their heavy mind. Estranged from their aquatic origins, human beings collapse to the ground, they do not float any longer in the water that lightens their weight.

Ejvind Tjønneland states that "[t]he melancholic variant of the theory of evolution in *The Lady from the Sea* doesn't have any counterpart in Jacobsen's Darwin articles" (1998, 190, my translation); he considers those articles to be the most likely source of Ibsen's knowledge about Darwin. Initially, *The Lady from the Sea* views evolution as a process of estrangement that induces a longing for the past rather than a process of increasing fitness. Even though this melancholic theory of evolution seems to change in the end of the play in favour of a more positive view, the outcome of Ellida's acclimatisation remains ambivalent, since we do not know whether Ellida and Wangel really manage to establish a better relationship.

Tjønneland and Shepherd-Barr emphasise that Ibsen was influenced not only by Darwin, but also by Ernst Haeckel. Shepherd-Barr explains that "Ibsen's engagement with evolution develops a clear openness to the Haeckelian emphasis on the influence of the environment on the organism", inspired by the "Haeckelian version of Darwin that downplayed natural selection" that Peter C. Asbjørnsen and J. P. Jacobsen popularised in Scandinavia (2015, 75). She sees in "Ellida's longing for her evolutionary past" an echo of "Haeckel's recapitulation theory" (2015, 73), according to which the ontogenetic development of an individual is based on the recapitulation of earlier evolutionary stages. Alaimo warns against origin stories that distract from environmental problems and "revel in a prelapsarian innocence" (2016, 126). However, Ellida's longing has an ecological dimension in *The Lady from the Sea*. With the term "ecological", I refer to Haeckel, too. The German marine biologist, physician, zoologist, eugenicist, and artist coined several definitions of "Ökologie" between 1866 and 1870. According to Haeckel, ecology is the science of the mutual relations ("Wechselbeziehungen") of organisms to each other and of their dependency on their organic and inorganic environments (Haeckel 1866; Toepfer 2016). It is this focus on mutual relations between organisms and their organic and inorganic environments that is prominent in *The Lady from the Sea*. Ibsen is more interested in the psychological impact of the environment than in its physical influences, but the longing for the sea is also embedded in a more-than-human concept of kinship.

Strange kinship

When Ellida first tells her husband about her relationship with the Stranger, she says that the Stranger changed his name from Friman to Albert Johnston and that he came from Finmark, was born in Finland, and immigrated with his father to Norway. Wangel states, "Altså en kvæn" (Ibsen 2009, 571[4]) and thus identifies him as a member of the Kveeni, a Finnish minority that migrated to Northern Norway in increasing numbers from the eighteenth century but suffered under the Norwegianisation policy of the late nineteenth century (Ki 2020, 419–20). Later, the Stranger became a sailor, Lyngstrand knows him as "the American" (Ibsen 1967, 20), and he writes letters from Russia, China, and Australia. The Stranger is thus depicted as a person who is always on the move and who does not even have a fixed name—his nomadic existence mirrors the movements of the inhabitants of the sea, to whom he has a special relationship and whose migratory existence is characteristic of the fauna of the Norwegian Sea. With Steve Mentz, we could read the attraction between Ellida and the Stranger as a symptom of "the erotic nature of oceanic desire" (Mentz 2020, 96). Before he leaves Ellida, "he said that we two were to be wedded to the sea" (Ibsen 1967, 35) and throws three rings into the sea. This strange wedding ceremony thus includes not only a human couple but also a third party, the sea. This is not as surprising as it could seem, given Ellida's description of their previous relationship:

Wangel. What did you speak about then?
Ellida. We spoke mostly about the sea.
Wangel. Ah! About the sea?
Ellida. About the storms and the calms. About dark nights at sea. And about the glitter of the sea on sunny days. But mostly we talked about whales and dolphins, and about the seals that lie out on the rocks in the warmth of the sun. And we talked about the gulls and the eagles and all the other sea-birds. And you know . . . isn't it strange?. . . when we talked about such things I used to feel that he was somehow of the same kith and kin as these sea-creatures [når vi talte om sådant noget, så stod det for mig, som om både sjødyrene og sjøfuglene var i slægt med ham.]
Wangel. And you?
Ellida. I too almost felt as though I were one of them. [Ja, jeg syntes næsten, at jeg også kom i slægt med dem alle sammen.]

(Ibsen 1967, 34, 2009, 571–72)

The kinship ("slægt") between humans and animals was widely discussed in the natural sciences after the popularisation of Darwin's evolutionary theory (Aarseth et al. 2009, 572). However, it is less a biological relationship that seems to be relevant in this passage and more an emotional attachment that is based on the experience of a shared habitat. This habitat consists of the ocean and the skerries that are depicted in sensory details that evoke the described experience.

For being able to integrate into her new surroundings, Ellida has to give up this emotional attachment and the idea of a shared habitat between human and more-than-human beings. The kinship with the sea and its creatures becomes "[d]et grufulde . . ., som skræmmer og drager" (Ibsen 2009, 653), "the gruesome . . ., that terrifies and attracts", an uncanny attraction that is located exclusively "in my own [i.e., Ellida's] mind" (Ibsen 1967, 74) and that she must overcome in order to establish closer attachments to her husband and stepdaughters. What initially had been a longing for the concrete ocean and its inhabitants is transformed into a dangerous abstract threat. Wangel warns: "This craving for the unattainable . . . for the limitless, for the infinite . . . will ultimately put your very mind in darkness" (1967, 92). The Stranger, who established the concrete relationship with the sea, "experiences three simultaneous symbolic deaths in the play", as literary scholar Magdalen Ki explains:

> In Norway, his legal status is that of a deceased, drowned man. His second symbolic death is to be embodied in artwork: he inspires Lyngstrand and his patron, but even in a work of art, he lives as only a ghostly avenger coming back for his unfaithful partner. In Lyngstrand's mind, the "American" is a victim of female infidelity, the symbol of male trauma when lovesickness converges with homesickness, the mutability of life, and the eternal return of the same love. The Stranger's third death refers to his emotional shipwreck as Ellida finally declares that she is free from his influence.
>
> (2020, 427)

The privileged relationship to the sea embodied by the Stranger disappears from the fjord town as it undergoes its development into a modern tourist destination.

Bolette reads botanical and geographical books (Ibsen 1967, 42, 2009, 590). Ibsen thus points to the popularisation of more scientific perspectives on more-than-human environments, but he does not refer directly to Darwin or Haeckel or to knowledge about the ocean. He wrote *The Lady from the Sea* in a period of an increasing interest in the sea: in the 1850s, influential Norwegian marine biologist Michael Sars discovered organisms in the deep sea that had been known only in fossil forms before (Sætre 2004, 38). The Norwegian herring that was of huge economic importance for the west coast of Norway was investigated by several Norwegian researchers. Herring fishery stopped abruptly in the 1870s when the regional fish stocks declined (Sætre 2004, 38–39), which caused economic distress that has left its traces in *The Lady from the Sea* (Ki 2020, 415) but the reasons are never explained in the play. The 1870s were furthermore the decade in which Norwegian deep-sea research started and the exploration of the Norwegian Sea began (Sætre 2004, 39–41). These new forms of knowledge do not substitute the intimate relationship to the sea that is being lost in *The Lady from the Sea.*

A happy ending?

Some researchers state that *The Lady from the Sea* has a happy ending (Mørkhagen 2019, 595; Moi 2006, 294). However, other readings are more critical. Ibsen scholar Joan Templeton points to the "sinister subplot" (1997, 197) of Bolette, who has to marry a man whom she does not love in return for the possibility of an education. As Toril Moi explains with reference to "the economic and sexual facts" presented in the dialogue between Arnholm and Bolette, the play does not simply push "a liberal ideology that wildly exaggerates the possibilities of free choice in modern society" (Moi 2006, 311–12). However, free choice is valued in *The Lady from the Sea*, and the freedom to make individual choices is combined with the integration into a modern consumer society. Free choice is thus not only restricted by economic and gendered power relations; its drawback is also the estrangement from the awareness that we share our habitat with non-human beings. And, if we take the death of the sea-child seriously, free choice betokens the end of our future.

Notes

1 The translation often cannot supply quite the same semantic nuances and associations as the original, and therefore, I have included the original wherever relevant for my argument.
2 In this chapter, I use the English translation of John McFarlane. However, McFarlane translates this sentence that is of central relevance for many interpretations with "Yes, it cuts both ways." (Ibsen 1967, 66) and thus obliterates its self-reflexive aspects.
3 *The Lady from the Sea* is embedded in a complex web of relations to other plays by Ibsen that all contribute to how one reads the text. For example, Ibsen scholars see in Ellida's decision to stay with her husband and family the counterpart to the famous end of *Et Dukkehjem* (*A Doll's House*, 1879) and Hilde reappears in *Bygmester Solness* (*The Master Builder*, 1892) (Moi 2006, 297; Templeton 1997, 199).
4 McFarlane translates with "I see, an immigrant" (Ibsen 1967, 33).

Works cited

Aarseth, Asbjørn, Hanne Sine Andresen, Tone Merete Bruvik, Erlend Lønnum, Vemund Olstad, Sybille Söring, Gaute M. Sortland, Hege Stensrud, and Vigdis Ystad. 2009. "Innledning. Ord- og sakkommentar." In *Innledninger og kommentarer*, edited by Vigdis Ystad. Vol. 8k of *Henrik Ibsens Skrifter*, edited by Vigdis Ystad et al., 413–562. Aschehoug.

Alaimo, Stacy. 2012. "Trans-corporality at Sea." *Interdisciplinary Studies in Literature and the Environment* 19 (3): 476–93. https://www.jstor.org/stable/44087131.

Alaimo, Stacy. 2016. *Exposed. Environmental Politics and Pleasures in Posthuman Times*. University of Minnesota Press.

Binding, Paul. 2006. *With Vine-Leaves in His Hair: The Role of the Artist in Ibsen's Plays*. Norvik Press.

Chen, Cecilia. 2013. "Mapping Waters: Thinking with Watery Places." In *Thinking with Water*, edited by Cecilia Chen, Janine McLeod, and Astrida Neimanis, 274–98. McGill-Queen's University Press.

Crutzen, Paul J. 2002. "Geology of Mankind." *Nature* 415: 23. https://doi.org/10.1038/415023a.

Darwin, Charles. 1872. *Om Arternes Oprindelse ved Kvalitetsvalg eller ved de heldigst stillede Formers Sejr i Kampen for Tilværelsen*. Translated by J. P. Jacobsen. Gyldendalske Boghandel. https://darwin-online.org.uk/converted/pdf/1872_OriginDanish_F643.pdf.

Detering, Heinrich. 2013. "Das Drama der Ökologie: Henrik Ibsens *En folkefiende* (1882)." In *Natur und Moderne um 1900: Räume—Repräsentationen—Medien*, edited by Adam Paulsen and Anna Sandberg, 121–41. Transcript Verlag.

Fjelde, Rolf. 1978. "*The Lady from the Sea*: Ibsen's Positive World-View in a Topographic Figure." *Modern Drama* 21 (4): 379–91. https://doi.org/10.1353/mdr.1978.0013.

Garrard, Greg. 2007. "Poodles and Curs: Eugenic Comedy in Ibsen's *An Enemy of the People*." In *Culture, Creativity and Environment: New Environmentalist Criticism*, edited by Fiona Beckett and Terry Gifford, 115–33. Rodopi.

Haeckel, Ernst. 1866. *Generelle Morphologie der Organismen*. 2 vols. Reimer.

Hamm, Christine. 2015. "Nordsjøens maskuline og feminine grammatikk." In *Nordsjøen i norsk litteratur*, edited by Christine Hamm, Jørgen Magnus Sejersted, and Eirik Vassenden, 225–39. Fagbokforlaget.

Heitmann, Annegret. 2003. *Intermedialität im Durchbruch: Bildkunstreferenzen in der skandinavischen Literatur der frühen Moderne*. Rombach.

Ibsen, Henrik. 1909. *Efterladte Skrifter*. Edited by Halvdan Koht and Julius Elias. Gyldendalske.

Ibsen, Henrik. 1967. *The Oxford Ibsen: The Lady from the Sea. Play in Five Acts.* Translated by John McFarlane. Oxford University Press.

Ibsen, Henrik. 2009. "Fruen fra havet." In *Vildanden/Hvide heste/Rosmersholm/Fruen fra havet*, edited by Christian Janss, Brenna Taugbøll, Tone Bruvik, Stine Merete, Hilde Bøe, Ellen Nessheim Viger, Sybille Söring, Tone Modalsli, and Vemund Olstad. Vol. 8t of *Henrik Ibsens Skrifter*, edited by Vigdis Ystad et al., 501–697. Aschehoug.

Jacobsen, Per Schelde, and Barbara Fass Leavy. 1988. *Ibsen's Forsaken Merman: Folklore in the Late Plays*. New York University Press.

Ki, Magdalen. 2020. "Migration and Ibsen's *Fruen fra Havet*." *Scandinavian Studies* 92 (4): 411–39. https://muse.jhu.edu/article/768718.

Mentz, Steve. 2020. *Ocean*. Bloomsbury Academic.

Meyer, Michael. 1971. *Henrik Ibsen: The Top of a Cold Mountain 1883–1906*. Rupert Hart-Davis.

Moi, Toril. 2006. *Henrik Ibsen and the Art of Modernism: Art, Theatre, Philosophy*. Oxford University Press.

Mørkhagen, Sverre. 2019. *Ibsen ". . . den mærkelige Mand"*. Gyldendal Norsk Forlag.

Oppermann, Serpil. 2023 *Blue Humanities: Storied Waterscapes in the Anthropocene*. Cambridge University Press.

Schiewer, Uwe. 2008. "Introduction." In *Ecology of Baltic Coastal Waters*, edited by Uwe Schiewer, 1–22. Springer.

Shepherd-Barr, Kirsten. 2015. *Theatre and Evolution from Ibsen to Beckett*. Columbia University Press. https://doi.org/10.7312/shep16470.

Shepherd-Barr, Kirsten. 2021. "Science." In *Ibsen in Context*, edited by Narve Fulsås, 82–90. Cambridge University Press.

Skjoldal, Hein Rune. 2004. "An Introduction to the Norwegian Sea Ecosystem." In *The Norwegian Sea Ecosystem*, edited by Hein Rune Skjoldal, 15–32. Tapir Academic Press.

Stanton-Ife, Anne-Marie V. 1996. "A Woman's Place/Female Space in Ibsen's *Fruen fra Havet*." *Scandinavica* 35 (1): 29–52.

Sætre, Roald. 2004. "Scientific Research in the Norwegian Sea: Background and History." In *The Norwegian Sea Ecosystem*, edited by Hein Rune Skjoldal, 33–64. Tapir Academic Press.

Templeton, Joan. 1997. *Ibsen's Women*. Cambridge University Press.

Tjønneland, Ejvind. 1998. "Darwin, J.P. Jacobsen og Ibsen." *Spring* 13: 178–99.

Toepfer, Georg. 2016. "Von der Naturgeschichte zur Ökologie (1750–1900): Entstehung und Geschichte der Ökologie bis zum Beginn des 20. Jahrhunderts." *Natur und Landschaft: Zeitschrift für Naturschutz und Landschaftspflege* 91 (9/10): 398–404.

Wicklund, Beret. 1997. "Ibsen's kvener og havfruer: Myter, samfunnskritikk og overføring i *Fruen fra havet*." *Edda* 1: 99–110.

9

MAPPING SEASCAPES WITH OLAV DUUN'S NOVEL *FLOODTIDE OF FATE*

Sissel Furuseth

The final work of Olav Duun (1876–1939)—*Menneske og maktene* (1938), translated into English as *Floodtide of Fate* (1960)—opens with an ominous prediction for the exposed archipelago where the frame narrative of the novel is set:

> It had been prophesied from earliest times that Øyvære would go under. There it lay, out in the sea; and the sea would take it. But it hadn't happened yet; wouldn't happen either, if it didn't happen this year. That had been prophesied too, by some prophet of misfortune.
>
> There were quite a few people out there, aboard Øyvære as they said; some were born there, others were newcomers. Recently, a family or two had moved away; each for its own reason.
>
> The name Øyvære sounded queer to other people, but it was an old name and good enough for its purpose. There were three islands in all, and a mass of skerries and holms; beyond were the reefs, with a nasty gap in them right to the west of the islands.
>
> (1960, 7)

Øyvære is a fictional place name, a compound word (*øy* + *vær* in the definite form), that literally means the fishing village on the island. There are few explicit geographical markers in the novel, but most Duun scholars acknowledge that Øyvære is inspired by the old fishing station Sør-Gjæslingan in the Vikna archipelago off the coast of Namdalen in Mid-Norway. As a teenager, Olav Duun joined his father and other fishermen-farmers on winter fishing in the waters around Sør-Gjæslingan, about fifteen nautical miles north-west from his home island Jøa, and he continued to do so throughout the 1890s. The natural disaster depicted in the closing

DOI: 10.4324/9781003647638-12

chapter, however, bears similarities with the 1901 storm surge that took place at Sandsundvær ninety nautical miles further north. On the western side, both archipelagos front towards the Norwegian Sea, a rough and unpredictable environment but also attractive because of the rich fishing grounds along the continental shelf edge. Nowadays, the region is also known for its offshore oil and gas fields.

In the 1920s and 1930s, Olav Duun was among the most celebrated novelists in Scandinavia. "He is a master of nature description, digs much deeper in his depiction of the soul than Knut Hamsun, and has a far greater immediate creative strength in his imagination than Sigrid Undset", historian Oskar Johnsen emphasised when he proposed Duun as a candidate for the Nobel Prize in Literature in 1924 (Vannebo 2004, 67). At that time, Duun had just finished the six-volume historical novel *Juvikfolke* (*The People of Juvik*), published between 1917 and 1923, a work ending with a dramatic scene where the protagonist Odin and his rival Lauris cling to the bottom of an overturned boat in a terrible storm. The wreck cannot take more than one man, and Odin lets Lauris have the boat. This act of self-sacrifice has been read as a striking example of Duun's heroic humanism (Solstad 2015, 57). Yet Duun never received the Nobel Prize, but on 11 November 1926, the Oslo newspaper *Aftenposten* announced him as the winner. This was an embarrassing mistake. The actual winner that year was the Irish playwright Georg Bernhard Shaw (Hageberg 1996, 83).

Overall, literary scholars have focused on the ethical and psychological depths of Duun's works. Political and rhetorical readings have also come to the fore. *Menneske og maktene* has been interpreted as an allegorical warning of the fascism that threatened Europe in the 1930s (Vannebo 2004, 72). "The sea", "the ship", and "the moon" are primarily rhetorical figures working on the symbolic level of the novel, Heming Gujord has claimed (1995, 315). Notwithstanding that Åsfrid Svensen (1970) early on warned against the temptation of symbolistic interpretations, emphasising that *Menneske og maktene* is a *realist* novel, it has become standard procedure to translate Duun's nature descriptions into something else. Even scholars with an interest in literary *place*, such as Hans H. Skei, make it clear how

> Duun's writing surpasses all geographical or sociological borders, and creates a fictional region in which man's injustice to man is the key question, and in which the metaphoric or figurative representation of landscape transforms hamlets and villages, islands and inlets, into places of the mind.
>
> (1996, 63)

Statements like this illustrate a widespread fear among academic readers of reducing Duun's work to *regionalism* and "heimstaddikting" ("literature from home areas"). The fact that Duun wrote his books in the dialect-based language variant of *nynorsk* (New Norwegian) might have reinforced this worry. That said, Skei acknowledges that typical for most of Duun's fictions—from his first collection of short stories (1907) to the final novel published three decades later—are their setting in landscapes where sea and land meet. The protagonists move around in small open boats, "fighting nature to eke out a living the way Duun himself had experienced" (1996, 69).

The interpretative strategy I will pursue in this chapter is partly inspired by literary critic Maria Olerud (2023), who recently raised the issue: perhaps *Menneske og maktene*, a more than 80-year-old novel where the sea is rising, can be read as a novel *about* the sea rising. Considering the reception history of Duun's work, Olerud's challenge is more radical than her seemingly innocent question may indicate. She is particularly critical of author Dag Solstad (2015) who in a much-discussed lecture concluded that Duun's oeuvre is outdated. On the contrary, Olerud parries; for a young literary critic born in the 1990s, *Menneske og maktene* feels more relevant than ever because it embodies how it is to live in the era of climate change. Duun's depictions of Øyvære make it easier to empathise with people living in other exposed island communities, such as Tuvalu or the Maldives, she contends. Although Duun's flood novel hardly can be categorised as climate change fiction, I agree with Olerud that a literal interpretation of the depicted storm surge may help readers today to take the problem of extreme weather and the associated hardships more seriously.

Another source of inspiration for my reading of Duun is the emerging Doggerland literature, which makes use of scientific surveys of the submerged North Sea land to construct imaginative zones of contact where different timescales and forms of agency are brought into play (Ritson 2025). Instead of submitting to the eschatological explanations offered by the religious characters in *Menneske og maktene*, or to the political-allegorical interpretation put forward by twentieth-century literary critics, I take the phrase "earliest times" in the opening line of the novel as a starting point, suggesting that a flooding has already taken place and continues to work as a cultural memory within the exposed coastal community. Rachel Carson set the stage for this approach in *The Sea Around Us* (1951), where she reflected on the folklore of lost islands. Although stories about Doggerland were not part of recorded history until the early twentieth century, when fishing trawlers began to drag up unexpected remains from the seafloor, Carson finds it likely that people who escaped a prehistoric flooding somehow must have "communicated this story to other men, who passed it down to others through the ages, until it became fixed in the memory of the race" (2021, 81).

Since Duun's and Carson's time, modern science and commercial businesses have contributed to an intensified production of knowledge about the Nordic seas. After more than fifty years of petroleum industry offshore, today's scientists have quite detailed information about the seafloor and the geology of the Norwegian continental margin. This knowledge may be activated in an ocean-centred re-reading of *Menneske og maktene*. At the same time, I am aware that Duun's 1938 novel and twenty-first-century geologists are mapping the sea in different ways. "How we know watery place", to quote Cecilia Chen, "is not separable from the material and discursive apparatuses that we use to articulate this knowledge"

(2013, 280). And as Stacy Alaimo has pointed out in an essay on the compressed temporalities of the Anthropocene oceans: "the paradox stings with the realization that much of the technology enabling discovery is due to the very industries—deep sea mining, deep sea fishing and trawling—that decimate species and habitats" (2017, 154). Duun wrote his novels at a time when marine sciences made huge progress, but they often did so in agreement with governmental authorities to ensure better profit for the fisheries (Kjesbu et al. 2021). In hindsight, we know that these developments were not only for the best. Before I delve into the depths of Duun's seascape, I will stay on the surface for a while and pay attention to how the sea affects everyday life in Øyvære.

Island dwelling

Menneske og maktene is a collective novel about a collective trauma. Literally, the title translates to "humans and the powers". Lisbeth P. Wærp is indeed right when she states that Duun's final novel "is about a home constantly under threat" (2017, 156). Structurally, the novel consists of seven chapters: I) "Aboard Øyvære", II) "The Saga of Helmer and Some Others", III) "Borghild and her Brother", IV) "Torger", V) "Roald the Wonder-Child", VI) "The Sea-troll", and VII) "When God Fled the Land". Whereas the opening chapter presents the prophecy and introduces Øyvære and its inhabitants, and the final chapter describes the catastrophe as it happens, the five middle chapters tell the individual stories of key characters. By using the term "saga" for these stories, Duun creates an atmosphere of ancient times, but other details give proof of a contemporary novel set in the hard 1930s. In *Menneske og maktene*, there are motor vehicles, cinema, sewing machines, and local fishermen ousted by foreign trawlers.

An important feature of *Menneske og maktene* is the constant shifts of voice and perspective. There is no authoritative narrator to guide the reader's evaluation of the events, which may explain why scholarly interpretations differ so much. As James Voss has emphasised, "each point of view, each fragment of revealed meaning, seems to bear some truth of their collective situation" (1980, 365). Nature phenomena play a crucial role in the shaping of these points of view. For instance, this is how teenager Roald perceives the indifferent look Aud gives him after he has rejected her: "it was bluish-grey like the sea when it's a long way away" (Duun 1960, 169). When the storm surge arrives, the sea has the agency of a greedy coloniser: "The sea spread itself out, it had to have more room; it took possession of a sizeable piece of land" (1960, 210). According to Voss, Duun's human forms emerge from a shapeless background, such as "the amorphous sea" (1980, 266), but the last example shows how the sea is assuming a distinct agency itself, becoming the central character of the novel. Skei has observed how Duun's "narrative techniques include an unusually frequent use of psychophysical parallelisms, which in simple language means that descriptions of nature invariably prepare, represent, become descriptions of human emotions and conflicts" (1996, 71–72).

This is certainly true, but parallelisms between inner and outer landscapes can just as well be framed the other way around: human perception mirrors the shifting nature of the sea.

It is revealing how Duun combines setting and character description. Helmer, originally a farmer coming from the mainland, is first overwhelmed by the poverty in Øyvære, but eventually he starts small-scale fishing in his old broken-down boat: "He rowed out for coal-fish along the shallows when the sea was calm. He didn't venture far out, a peasant was a peasant, but small coal-fish were fine fish too" (1960, 62). The more widely travelled Sea-troll despises this "fisher of snails and toads" (1960, 187), but Helmer's method is probably in line with long-established marine subsistence techniques inherited from Stone Age settlers. As archaeologist Håkon Glørstad has argued, the Mesolithic period was "characterized by small-scale movements by boats with modest seagoing abilities" (2013, 58). Helmer restricts himself to summer fishing and enjoys the calm evenings on the island. As a religious man, he also appreciates the spiritual qualities of nature:

> the grey rock sang in him, and with it the tangles of seaweed struggling in the waves, the bird sleeping out there on the rock, the sea and the endless expanse all around, the canopy of clouds above, the heavenly world up there over everything—all this truth which met his eyes when he woke up and saw.
>
> (1960, 66)

Torger, however, is a different kind of fisherman, born and bred on the island. Although Torger can appreciate a pretty song or a good story, poems about the sea are soon forgotten when he is working: "Out there was nothing but the sea, the grey naked sea, an eternal workaday for the fisherman. There the waves twisted and turned and the currents raced; there was no poetry in it, unless you made it up yourself" (1960, 103). Nevertheless, Torger is a man of the sea and content with his life as it is.

The east–west axis constitutes the geographical orientation in the novel: the east representing secure farmlands; the west the vulnerable fishing community. This does not mean, however, that farming is framed as the future and fisheries belong to the past. For Torger, who believes in better times to come, investing in a new boat was the right thing to do. He cannot see the point in holding on to technologies of the past.

> It was the same with the old methods of fishing; they saw that the sailing-boat was out-of-date, but they bought one themselves and patched it up and hoped that the good old days would come back. They didn't come back. And they oughtn't to come back either. What's out-of-date is out-of-date. Torger said a plague on the whole lot of them. He got himself a motor fishing-boat and tackle to match, although he was only a young lad and went against his father's will.
>
> (1960, 101)

Voss points out that "Torger is inseparably attached to his island dwelling" (1980, 378) but being an island dweller is not the same as being in constant harmony with the sea. As anthropologist Tim Ingold underlines, it is "a mistake to equate dwelling with rest or stasis. For being at home in the world entails action and perception, and to act and perceive one must move about" (2005, 507). This is literally the case in Duun's novel, in which the most far-sighted characters move higher and higher as the water rises. Øyvære is still their home. Very few choose to flee the archipelago before the disaster.

Remembering natural disasters

Living with the sea also means being prepared to die with the sea. This knowledge is a crucial part of islanders' habitus, partly self-experienced and partly transmitted through stories, oral and written. In 1627, pastor Michel Mogensen in the Nærøy Parish (which includes the islands in Outer Namdalen) published "Trenologia Numedalensis", a poem of twenty-five stanzas, to honour the memory of 210 fishermen who died in a terrible storm in February 1625. Poet-priest Petter Dass wrote about a similar disaster in 1660 (Gåsvær 1999, 14). As reported from several of these fatalities, multiple forces work together to cause the extreme situation: the sea, the winds, the moon, and sometimes seismic disturbances. In August 1819, one of the most powerful earthquakes in Northern Europe in recorded history hit Lurøy on the Helgeland coast, and a large number of earthquakes have occurred along the coast of Nordland over the years (Bungum and Olesen 2004, 251). Such quakes also happen offshore, as the so-called seaquakes, where they can cause extremely high waves.

Rolv Thesen, the first Duun biographer, claims that the Sandsundvær accident in 1901 must have made an indelible impression on the young writer because Olav Duun returns to this motif on several occasions. The first reference is an essay Duun wrote at the teacher training college in Levanger the same year as the accident happened, and the last is the author's very "testament", *Menneske og maktene* (Thesen 1942, 330). What happened on 22 January 1901 was that a spring tide in combination with storm and enormous breaking waves wiped out thirty-four human lives at the fishing station Sandsundvær on the outer Helgeland coast. In these low-lying islets, people had no say when confronted with the furious seas. Duun never visited Sandsundvær himself (Gåsvær 1999, 11), but he was clearly familiar with storm surges in similarly exposed environments. In *Menneske og maktene*, Helmer suggests that the men in Øyvære should build a more robust harbour on the sound by themselves (1960, 121). He does not trust "the State" in such matters. In real life, Statens Havnevesen (the national port authority) was responsible for building a breakwater to protect the harbour in Sandsundvær after the accident. The jetty was finished in 1907, the same year Duun made his debut as a writer.

In *Menneske og maktene*, memories of natural disasters seep out in various ways, often almost unnoticeably in passages where interpersonal relations are in the foreground. This is how Torger and his wife Borghild exchange their taciturn concerns when sitting on the top of a crag in Øyvære:

> He held her hand; so that she shouldn't disappear, he joked, for it wasn't the cosiest place in the world to sit up there. "I wouldn't like to sit here when there was an earthquake," she said. "Or on a stormy night," he said. Her hand got warm inside his. But the hand didn't say anything, neither did she. They climbed down in lighter spirits than they climbed up. Life was such a reasonable thing. (1960, 124)

In Duun's novel, the powers are many, and so are the interpretations of their causes and meanings. Torger is portrayed as a modern man with a secular worldview, and there is an evolutionary tone to it when single mother Kari brings her two children out to Øyvære "to save her offspring" (1960, 54). Others in the fishing village take refuge in religion. Members of the pietist meeting-house community believe that God will protect them. When they finally realise that the sea has become too powerful this time, the Apocalypse is hailed with joy: "The Lord will soon appear in the clouds. He is coming to fetch us!" (1960, 212). In a darker mood, the Sea-troll invokes a coming disaster in terms of the Deluge: "perhaps the storm from the west would come some time, the ocean would come and cleanse away all this human refuse" (1960, 179).

The novel also includes folkloristic elements. Helmer remembers "evenings when he had heard the bearded seal out amongst the skerries, such an unexpected and fearsome sound, as if it wanted to make one shiver" (1960, 203). In coastal communities, seals are often associated with drowned people. According to archaeologist Christer Westerdahl, who has described fishermen's superstitions and taboos, seals often "mean danger because of their position, not being of the land, and still as mammals resembling human beings, at sea" (2005, 9). The merman and the mermaid are other ambiguous creatures with disturbing meaning. In *Menneske og maktene*, the young couple Roald and Vebjørg are associated with these figures: "Roald was born on the island; a hundred per cent merman, as they said" (1960, 128). Roald for his part remembers Vebjørg as a little child saying after a sea bath: "I dreamt I became a mermaid. It was fun. I grew together all the way from up here and right down, and the soles of my feet became a bluish-green fishtail" (1960, 177). Roald and Vebjørg are among the people who are taken by the sea in the end, which indicates that the merman/maid references can be read as a foreshadowing.

Under the sea

Most of the events in *Menneske og maktene* take place on the surface but the novel also acknowledges the power of seismic energy. Borghild's worry about a potential earthquake refers to such underground forces. Roald's description of Aud as being "more sensitive than barometers and seismographs" (1960, 168) is another reminder of the vertical dimensions of existence and how a sea-oriented lifestyle is influenced by both weather systems and activities taking place beneath the earth's crust.

In Robert Macfarlane's *Underland* (2020), one of the chapters is titled "The Edge" and describes a trip to Andøya, north of Lofoten. Here, the British nature writer meets the outspoken fisherman and anti-oil activist Bjørnar Nicolaisen who educates his guest about the Edge ("Egga", in Norwegian), that is, the steep slope that marks the boundary between the continental shelf and the deeper parts of the Norwegian Sea:

> This drop-off, this *cliff*, it happens only a few nautical miles from the coast. This is why the fishing is so rich and easy here: the fish gather at the Edge, and we do not need to go far to harvest them. . . . It's the knowledge about what is *under* the surface that for all times has kept these coastal people and this coast *alive*.
>
> (Macfarlane 2020, 295)

Outside Andøya, Egga is located only seven kilometres from land. The same slope can be followed further northwards to the Arctic Ocean and southwards to Møre where the gigantic Storegga Slide hit about 8,200 years ago (Barthel 2024). Olav Duun's seascape is located in between, in a region where Egga bends westwards off the coast. Fish thrive there because of the strong currents that occur along the Edge. This underwater slope thus represents the very lifeblood of coastal dwelling but is also a reminder of the geohazards threatening the same dwellings.

Geologists have pointed out that "postglacial uplift seems to be playing a significant role for the contemporary seismicity" along the coast of mid-Norway (Bungum, Lindholm, and Faleide 2005, 142). Seaquakes can cause underwater landslides, which may in turn lead to tsunamis. It is well-established that the massive Storegga Slide "created a significant tsunami along the coast of Norway and surrounding land areas" (Solheim et al. 2005, 6). Even the submergence of Doggerland in the southern part of the North Sea has been traced back to this event. Since the discovery of oil offshore Norway in 1969, industry, academia, and private research foundations have cooperated to get better knowledge of "the seafloor morphology" and potential geohazards in the region. For the offshore industry, the main concern is risk assessment and cost-benefit analysis: to what extent is it safe and profitable to operate energy installations in this area? While the so-called "Ormen Lange Project" a couple of decades ago concluded that it probably "will take another interglacial-glacial cycle to form a situation which could lead to another major tsunami-generating slide" at the Ormen Lange gas field (Solheim et al. 2005, 9), a more recent study has estimated "that large-scale slope failures occur more frequently than previously assumed" (Karstens et al. 2023, 8). With the updated estimate, it follows that risk assessments and tsunami models for the region should be revised.

The geologists work with timescales stretching back to the Mesolithic and Neolithic eras, times hardly comprehended by the islanders in Duun's novel. Nevertheless, I maintain that the very awareness that an earthquake and a flooding might happen any time situates the people at Øyvære within a geological understanding of time. True, *Menneske og maktene* is not diving deep into the ocean to the extent recent works about related seascapes do, such as Macfarlane's *Underland*, Morten Strøksnes' *Havboka* (2015; *Shark Drunk*, 2017), or Sara Sølberg's *Seismiske smell* (2016, Seismic Shocks). Overall, Duun's work portrays the sea as perceived by ordinary fishermen-farmers moving around on the surface of both sea and land. Yet, read in the context of new nature writing and twenty-first-century ocean-oriented literature, *Menneske og maktene* may prove to be more up to date than Dag Solstad and the modernists appreciated. The way Duun portrays the storm surge as a bodily experience has the potential to engage readers across eras and regions. As rising seas are everywhere nowadays, Olerud is right in stating that *Menneske og maktene* has survived the ravages of time.

Dynamic mapping

What about Duun's fictional archipelago? Did Øyvære survive? Undoubtedly, land and seascape undergo dramatic transformations during the storm surge. The closing chapter renders in detail the rising tide, a strong vertical force controlled by the moon:

> They hauled up their boats time and time again. But after a little the sea came floating in with them. It reshaped the land, made new creeks and sounds amongst the rocks from one moment to another; soon one would be able to row over the meadow down by the sea instead of walking. Now it moved over the boundary it had reached last night; it moved like a clock, like time itself. The moon threw a glimmer on the sound now and again so they saw how it was there *now*; a strange sea [. . .].
>
> (1960, 199)

The islander's home gradually becomes unrecognisable as the moon and the sea work together to transform their dwellings. The passage above thus illustrates Cecilia Chen's point that shores morph with seasonal and tidal fluctuations (2013, 284). Watery places can be mapped in many ways because "the spatial imagination of place is tied to the duration and movements of time" (2013, 287). In *Menneske og maktene*, the moon is the central figure that makes the reader attentive to this interplay of time and space in the workings of the sea, and thereby also turning the spotlight on Olav Duun's technique as a "mapper-narrator" (Chen's term). Thinking with water, as Duun obviously does, means to engage with this form of changing maps.

After the storm surge, the archipelago returns to a recognisable shape, but houses are destroyed and people drowned. In the closing paragraphs, we follow the survivors bringing eleven dead bodies to the church in a creek farther east. Even more deceased are lost to the sea. The survivors purchase coffins and mourning bands, and "equip themselves for life again" (Duun 1960, 255). Despite the fatal disaster, the survivors have decided to return west to Øyvære and rebuild what has been broken. Duun's novel thus sends a strong message about ecological connectedness and human resilience in an exposed island community.

I regard *Menneske og maktene* to be a crucial resource text for the blue humanities. This unsentimental story about a godforsaken archipelago in the hard 1930s can help counterbalance "the detached awe of the spectator" that Stacy Alaimo finds problematic in many scientific works that take "the Anthropocene to sea" (2017, 154). Instead of portraying the seas as "alien", Alaimo supports "positions where wonder, epistemological humility, and material interconnection coexist" (2017, 155). Such a coexistence is generic for Duun's work in which the characters approach the sea kinaesthetically as both homey and troublesome.

Works cited

Alaimo, Stacy. 2017. "The Anthropocene at Sea: Temporality, Paradox, Compression." In *The Routledge Companion to the Environmental Humanities*, edited by Ursula Heise, Jon Christensen, and Michelle Niemann, 153–61. Routledge.

Barthel, Knut. 2024. "Egga." In *Store norske leksikon*. Accessed July 13, 2025. https://snl.no/Egga.

Bungum, Hilmar, Conrad Lindholm, and Jan Inge Faleide. 2005. "Postglacial Seismicity Offshore Mid-Norway with Emphasis on Spatio-temporal–magnitudal Variations." *Marine and Petroleum Geology* 22: 137–48.

Bungum, Hilmar, and Odleiv Olesen. 2004. "The 31st of August 1819 Lurøy Earthquake Revisited." *Norwegian Journal of Geology* 85: 245–52.

Carson, Rachel. (1951) 2021. *The Sea Around Us*. Edinburgh: Canongate.

Chen, Cecilia. 2013. "Mapping Waters: Thinking with Watery Places." In *Thinking with Waters*, edited by Cecilia Chen, Janine McLeod, and Astrida Neimanis, 274–98. McGill-Queen's University Press.

Duun, Olav. 1938. *Menneske og maktene*. Oslo: Olaf Norlis Forlag.

Duun, Olav. 1960. *Floodtide of Fate*. Translated by Ronald G. Popperwell. Sidgwick and Jackson.

Glørstad, Håkon. 2013. "Where are the Missing Boats? The Pioneer Settlement of Norway as Long-Term History." *Norwegian Archaeological Review* 46 (1): 57–80.

Gujord, Heming. 1995. "Å massakrere et øyvære—betraktninger om og omkring Olav Duuns roman *Menneske og maktene*." *Edda* (4): 315–33.

Gåsvær, Kolbjørn. 1999. *Hilderøya og Øyværet: Olav Duun og Gjæslingan*. Jøa.

Hageberg, Otto. 1996. *Olav Duun: Biografiske og litteraturhistoriske streiflys*. Oslo: Det Norske Samlaget.

Ingold, Tim. 2005. "Epilogue: Towards a Politics of Dwelling." *Conservation and Society* 3 (2): 501–8.

Karstens, Jens, Haflidi Haflidason, Christian Berndt, and Gareth J. Crutchley. 2023. "Revised Storegga Slide Reconstruction Reveals Two Major Submarine Landslides

12,000 Years Apart." *Communications Earth & Environment* 4 (1): 1–9. https://doi.org/10.1038/s43247-023-00710-y

Kjesbu, Olav Sigurd, Jennifer Hubbard, Iain Suthers, and Vera Schwach. 2021. "The Legacy of Johan Hjort: Challenges, and Critical Periods—Past, Present, and Future." *ICES Journal of Marine Science* 78 (2): 621–30.

Macfarlane, Robert. 2020. *Underland: A Deep Time Journey*. Dublin: Penguin.

Olerud, Maria. 2023. "Olav Duun og den første norske klimaromanen." *Vinduet*, April 14.

Ritson, Katie. 2025. "Doggerland Rising: Oil, Literature, and Resilience." In *Narratives of Resilience*, edited by Kate Rigby and Evi Zemanek, 41–59. Metzler.

Skei, Hans H. 1996. "Regional, Marginal, Universal: Questions of Literary Regionalism, with a Note on the Sense of Place in the Fiction of Olav Duun." In *Places Within, Places Beyond: Norwegian Regionalism in Literature*, edited by Wendy Griswold and Fredrik Engelstad, 63–76. Institute for Social Research.

Solheim, Anders, P. Bryn, H. P. Sejrup, J. Mienert, and K. Berg. 2005. "Ormen Lange—an Integrated Study for the Safe Development of a Deep-water Gas Field within the Storegga Slide Complex, NE Atlantic Continental Margin." *Marine and Petroleum Geology* 22: 1–9.

Solstad, Dag. 2015. *Dag Solstad leser Olav Duun*. Nasjonalbiblioteket.

Svensen, Åsfrid. 1970. "Sammenbrudd og livsfortolkning i Duuns *Menneske og maktene*." *Edda* 6: 345–62.

Thesen, Rolv. 1942. *Mennesket og maktene: Olav Duuns dikting i vokster og fullending*. Olaf Norlis Forlag.

Vannebo, Einar. 2004. "Olav Duun (1876–1939)." In *Dictionary of Literary Biography: Volume 297: Twentieth-Century Norwegian Writers*, edited by Tanya Thresher, 66–72. Gale.

Voss, James. 1980. "Olav Duun's *Menneske og maktene*: Form, Vision, Contemporary Significance." *Scandinavian Studies* 52 (4): 361–80.

Westerdahl, Christer. 2005. "Seal on Land, Elk at Sea: Notes on and Applications of the Ritual Landscape at the Seaboard." *The International Journal of Nautical Archaeology* 34 (1): 2–23.

Wærp, Lisbeth Pettersen. 2017. "The Island in Nordic Literature." In *Nordic Literature: A Comparative History: Volume I: Spatial Nodes*, edited by Thomas A. DuBois and Dan Ringgaard, 146–62. John Benjamins.

PART III

The Iceland Sea, the Greenland Sea, and the Arctic Sea

10

"NARRATING THE SEA"

The journey to Iceland in travel narratives of the modern period

Alessia Bauer

"Narrating the sea"[1]

Is it possible to grasp the nature of the sea—this strange, otherworldly space that offers no fixed points of reference? This question prompted me to revisit travel accounts to Iceland, which I have been studying for some time.[2] These accounts usually start by mentioning the sea crossing, yet since readers are generally more interested in what the travellers encounter on land, the initial chapters often receive insufficient scholarly attention. The objection that the sea in travel narratives merely serves as a route to the destination—and is therefore inherently uninteresting—does not hold, as the narratives would then only begin upon arrival. However, this is not the case: the authors do mention the crossing, but they develop various strategies to give meaning to and fill the time spent at sea. By shifting the perspective, I aim to systematically (re)examine these passages to analyse how the sea is perceived and described.

The representative—though necessarily selective—collection highlights non-Scandinavian authors who engage with the Nordic region, particularly through their writing on the Atlantic, which becomes the unifying element of these heterogenic texts. This is, therefore, not Nordic literature in the strict sense (as it is written by non-Scandinavians in a variety of languages); rather, it belongs to what Ritson (2019, 5–6) defines as the "cultural landscape of the Nordic region".[3] Applying Ritson's concept of a "literary *sealandscape*", this corpus—even though it offers an exogenous perspective—should not be seen as standing in contrast to the other contributions in this anthology. Its very purpose is to transcend national and linguistic boundaries and to adopt the sea as a shared point of reference that links all the narratives. The exogenous perspective differs from the others in the sense that it presents foreigners seeking to appropriate a foreign place; on the other hand, it

DOI: 10.4324/9781003647638-14

is also similar to the indigenous perspective in the sense that it depicts individuals attempting (though with limited success) to appropriate the sea—a space that is, by nature, uninhabitable.

The northern space emerges as a *terraqué*—a hybrid world where land, numerous islands, and the sea are inextricably linked. From the eighteenth century onwards, adventurers from continental Europe and the United Kingdom embarked on voyages to explore the North Atlantic region, including Iceland. Their narratives frequently depict the Atlantic crossing as the initial stage of their journey in which the forces of nature take centre stage. The selection of these specific narratives proposed in this chapter represents only a small sample of the vast number of travelogues written by authors from various countries and with differing expectations and levels of knowledge. The aim here has been to provide an overview of the diverse narrative strategies employed, showing that some of them recur across multiple accounts and others are rather specific for single authors.

When examining the characteristics of the sea, the Italian philosopher Roberto Casati (2022) attempts to explain why authors and scholars encounter difficulties in accurately describing the sea as a natural space (a *space* is a natural, geographic location on Earth that is measurable and, in itself, not further connoted. Such a space becomes a *place*—that is, a mental construct—when associated with particular meanings).[4] According to Casati, this challenge arises from the sea's inherent resistance to definition: it lacks a fixed colour, instead reflecting the sky and shifting with weather conditions; it provides no stable points of reference, as the open sea consists solely of water; and it cannot be described using the spatial categories typically applied to land. Right at the beginning of *Le territoire du vide*, Alain Corbin does not merely describe the sea as elusive to the human mind, as Casati does; he defines it as an "unfathomable place, an image of infinity, a quivering expanse" (1988, 1), which in itself is terrifying because Man is unable to fully comprehend it. In a more recent book, *Terra incognita* (2020), Corbin explains furthermore that the feeling of fear is generated by ignorance, and that our lack of deeper knowledge of the sea leads to this state. Given these properties of the sea, Casati argues for the necessity of developing a specialised vocabulary dedicated exclusively to the sea. In the absence of such conceptual tools, authors have employed various strategies to interpret and represent the sea, often filling this perceived "void"[5] and often transforming it into a constructed *place* filled with mental imagery.[6]

A similar issue arises in cartography. Beginning with Olaus Magnus' *Carta marina* (1539) (see Knauer 1981), early modern cartographers were reluctant to depict an empty sea—unlike in portolan maps—and instead filled the vast oceanic space with sea monsters and fantastical creatures.[7] This approach served as a strategy to represent the sea as essential fearsome because of its otherness. This visual convention persisted beyond Olaus Magnus, influencing the representations of the Atlantic Ocean in the works of cartographers such as Abraham Ortelius and

Sebastian Münster (see Grétarsson 2017, and also Spaans and Willumsen in this volume).

In the following, I will examine various narrative strategies, as each author employs distinct approaches to depicting their experience of the sea crossing. Some opt for brevity, reducing the account of the journey to a few striking details or nearly omitting it altogether. Others deviate from direct narration, incorporating digressions that evoke mythical stories of the sea and its deities. Still others seek to convey the emotional dimensions of the crossing, emphasising personal experience and perception. In line with Ritson (2019, 161), a comparative reading across various literary periods, languages, and traditions can more effectively uncover shared discourses embedded in this cultural land- and seascape.

All these strategies prove relevant when attempting to represent the vast, unstructured expanse of maritime space, as Casati emphasises. However, what authors consistently struggle with is portraying this natural space in its components (colour, smell, form, etc.). According to Casati, this difficulty stems from the fundamental fact that "man does not inhabit the sea, and the sea remains foreign to him" (2022, 49).

The sea in saga literature

Before addressing modern travel accounts, it is useful to take a brief retrospective look at Old Norse sagas to outline their treatment of the sea (see Annette Lassen in this volume). Although the sea is an inseparable element of the Viking world, it is striking that in Old Norse sagas, this space rarely occupies a central place in the narrative and is almost never described at all. Instead, the sea is typically mentioned only when it plays a significant role in the story. This tendency, however, is not unexpected, as it aligns with the sagas' characteristically laconic style, which prioritises essential elements and plot progression over elaborate descriptions (this is true for all kinds of natural surroundings, as long as the landscape does not serve to advance the plot).

While protagonists frequently engage in the preparation of ships (Old Norse *buá skip*), actual scenes of sailing are relatively rare. The sagas appear to be primarily land-centred, although the liminal space of the coast plays a prominent role. Ships arrive and depart, yet the events that take place at sea seem to generate little narrative interest. One typical example of this pattern can be found in *Gunnlaugs saga ormstungu* (Chapter 5), where the protagonist departs from Iceland and arrives on the other side of the Atlantic without any mention of the journey itself: "When they got a favourable wind, they took to the sea and arrived with their ship in Norway. They sailed up to Trondheim, entered the Niðaróss estuary, and pulled their ships ashore" (my translation).

In saga literature, we often observe that, at the moment of the crossing, the narrative time is highly condensed. This narrative economy reinforces the idea that the sea, though ever present, remains largely an unarticulated background element in saga storytelling. In this regard, Bintley rightly affirms:

> No matter how accomplished medieval Icelanders may have been as seafarers, like the rest of our species they were terrestrial: reliant and dependent upon the solidity of the earth, and engaged on a daily basis with the same struggle that confronts humans today in our global engagement with our environmental commons.
>
> (2014, 247)

The sea in modern travel accounts

The question now arises as to whether modern travel accounts from the eighteenth century onwards similarly neglect the sea or, instead, consciously perceive and describe this space in greater detail. The answer is both affirmative and negative: unlike the saga authors, most modern writers accept the challenge of narrating this elusive environment. Given that travel narratives—unlike works of fiction—do not aim to describe the sea metaphorically for artistic purposes (as is often the case in pirate or adventure novels), the narrative strategies I will discuss here appear to result primarily from the inherent difficulty of describing this natural space in objective terms.

In most travelogues, the sea crossing constitutes an integral part of the narrative. The typical itinerary involved departing from Copenhagen, passing through Scotland, the northern British Isles, and the Faroe Islands, before finally reaching the southern coast of Iceland. Consequently, both the journey itself and its narration begin with the Atlantic crossing, a phase during which the protagonists must overcome their first trials before they can attain their goal.[8] Casati characterises this phase as "the transition from a known place to the elsewhere that we want to reach" (2022, 89), a necessary step towards encountering the unknown.

When the sea crossing is explicitly mentioned, it generally follows a tripartite structure: firstly, the optimistic anticipation of departure, culminating in the act of setting sail; secondly, the voyage itself, frequently marked by physical discomfort, loss of hope, and moments of despair; and, finally, the sighting of land, which rekindles hope and signals the journey's progression. Just as perceptions of the destination vary, so too do accounts of the sea crossing, ranging from expressions of admiration to profound fear. In the following, some selected passages from various travelogues will illustrate these three different steps and the variety of meanings ascribed to them.

The anticipation of departure

Before venturing into the unknown, authors frequently describe the lands they pass along their route. Ida Pfeiffer (1797–1858),[9] Konrad Maurer (1823–1902),[10] and Josef C. Poestion (1853–1922),[11] each sailing from Copenhagen towards the North

Sea, are quite loquacious and provide detailed observations of their surroundings, with coastal landscapes and islands between Denmark and Sweden serving as their primary points of reference. However, once they move beyond the land's "zone of influence", they mark the moment when the shores of the Skagerrak—the strait between Denmark and Sweden—disappear from view. From this point onwards, their narratives suggest a sense of surrender to higher forces. At sea, they then invoke either pagan deities (Poestion) or the Christian God (Pfeiffer), placing their trust in fortune and fate to ensure a safe passage across the ocean.

These accounts emphasise the fundamental notion that human beings are not naturally adapted to life at sea; they rather exist in a state of symbiosis with the land. This innate orientation towards terrestrial spaces is reflected in a remark by J. C. Poestion, who observes that as long as the coastline remains visible, it is nearly impossible to turn one's gaze and focus on anything else:

> The next day, when I went up on deck, *there was no land in sight*. We had left the Skagerrak behind and were already in the North Sea proper . . . . I now had the opportunity to pay more attention to the ship and my first-class traveling companions than I had the day before, when *the sight of the enchanting coastal landscapes* and the awareness of finally being on my way to Iceland had captivated my mind.
>
> (2025, 36, my italics)[12]

Only when all land—that is, all that is known and familiar—has disappeared can the gaze turn forward, not towards the sea, however, but towards the fellow travellers.

The crossing of the Atlantic Ocean

Despite the incapability to find words apt of describing this unfamiliar landscape, authors often exhibit a strong desire to fill the narrative void. As a result, the experience of crossing the sea sometimes stimulates the imagination, leading to the emergence of pagan water deities, as observed by the French travel writer Louis-Frédérique Rouquette (1884–1926)[13]:

> Leaning on the bridge,. . . it seems to me that I will see the deities of the Polar North rise from the sea. The terrifying gods riding on formidable icebergs. The Elves of legend, hieratic and white, guardians of the Land of Ice, and the Trolls prowling anxiously in the depths of the basalt trenches where the waters plunge with a howl.
>
> (1925, 25)

It is remarkable that, instead of depicting the creatures that actually inhabit the sea, modern authors draw on the creatures of old cartography (cf. *Carta marina* by Olaus Magnus) or mythological figures.

Where intellect falters, authors frequently turn to myth to fill the space, as also seen in J. C. Poestion's account:

> We passed the mostly silted-up but still inhabited island of Anholt and the much larger island of Læsø (Old Norse: Hlésey), known from the Edda as the home of *the mythical sea giant or merman Ægir* (also Hlér, from which the island gets its name), where his hall, illuminated by gold, once stood, where the gods once held a feast. Since Ægir is considered the personification of calm seas suitable for sailing, I wished for him as my patron for the rest of my journey, which had so far taken place in calm waters. The beacons of Læsø reminded me involuntarily of Ægir's servants, Eldir (Fire) and Fúnafengr (Spark-Catcher).
>
> (2025, 35)

Poestion refers to the myths concerning cosmology as presented in the *Poetic Edda*. Similarly, Konrad Maurer's imagination is stirred by the sight of a particular swell, leading him to believe he perceives "in the wide and rounded crests of the waves; *as if a terrifying sea monster were attempting to rise from the depths*" (2017, 23).[14]

During the crossing, a noticeable shift in mood occurs: initially marked by confidence and excitement—fuelled by anticipation and the hope for favourable winds—the journey soon gives way to anxiety, seasickness, and disturbing dreams once the reassuring sight of familiar shores disappears. At times, the sea announces itself in a menacing tone, as highlighted by Rouquette:

> The Irish Sea is a smooth mirror, undisturbed by anything. But after rounding the Isle of Man, the dance begins. The current sweeps us into the North Channel, between the northern tip of Ireland and Scotland. The winds and currents pass through the corridors formed by the long rocky and insular trail of Scotland and the Hebrides.
>
> (1925, 23–24)

The subchapter ends with a kind of looming threat, stating: "We will not see land again until Iceland" (1925, 24). This negative premonition soon becomes reality, when travellers find themselves at the mercy of the stormy sea for several days ("After several days of very rough seas"). This transformation in perception underscores the psychological and emotional dimensions of maritime travel, where the unknown expanse of the sea evokes both awe and apprehension.

Ida Pfeiffer conveys the loss of confidence in a particularly evocative manner, as if she were literally losing the ground beneath her feet. As the other ships remain behind and bid farewell, her vessel is left alone to its fate, sailing into the cold:

> The entrance once passed, we entered the Cattegat, which already looked more like the great ocean: the coasts retired on each side, and most of the ships and

> barques, which till now had hovered around us on all sides, bade us "farewell!" . . . we alone, on the broad desert ocean, set sail for the icy north.
> (1852, 56)

Detaching from land is experienced as a moment of peril, perceived as a form of loss. The inability to fully comprehend the sea is underscored by the fact that, whenever possible, authors maintain their reference points on land.

By contrast, Scottish geologist George Stuart MacKenzie (1780–1848) provides only a brief account of the sea crossing, limiting his description to just two pages (1811, 77–78). The French theologian, writer, and lawyer Isaac de la Peyrère (1596–1676) takes an even more radical approach by entirely omitting the crossing from his narrative: his *Relation d'Islande* begins *in media res*, directly in Iceland, where he asserts that he has already been residing for some time (1663, 2–3). No mention is made of how he arrived there, effectively erasing the sea voyage from his travel experience.

Unable to describe the sea itself—its water, colour, or the movement of the waves—his compatriot Pierre Martin de la Martinière[15] (1634–1676) employs a substitution strategy, shifting the focus to another central element: the wind.[16] In his account of the crossing, the wind emerges as the primary force shaping the journey—sometimes favourable, sometimes excessively strong or perilous. It remains the principal agent of the narrative, its fluctuations dictating the voyage's progression, with the narrative emphasising deviations from expected conditions, such as sudden storms or prolonged periods of dead calm that impede progress (notably, calm weather is framed not as a relief but as an obstacle to navigation and a source of monotony for the sailors).

During his passage through the North Sea along the Norwegian coast, La Martinière also references the *Maelstrom*, which he describes as the "largest water vortex in the Norwegian Sea" (1671, 30). Here, hyperbole functions as a rhetorical device to heighten the perception of maritime dangers—an image reinforced by its inclusion in Olaus Magnus' *Carta marina*, from which it was likely derived. In this way, realistic description is interwoven with traditional mental imagery. Conversely, in chapter 43, describing his return from Iceland to Norway, the text employs a "missing" description reminiscent of the Old Norse sagas, where narrative time is drastically condensed, and almost nothing is recounted apart from the wind: "With a favorable north wind, we lifted anchor and departed, setting our course to the South-Southeast. After sailing for a few days, a strong breeze from the North-Northwest caught us and carried us to the coasts of Norway" (1671, 175–76).

The impossibility of describing the sea leads the French explorer Victor-Edmond Meignan (1846–1938) to adopt a different strategy—namely personification—in his travel account *Pauvre Islande!* (1889), thereby presenting an animated vision of the sea as a stylistic device. Like a human being, the sea sometimes becomes enraged (1889, 8), while at other times, it appears merciful (1889, 9).

The unifying feature of all these narratives is the omnipresence of weather—particularly stormy winds—and the feelings of fear and apprehension they evoke in travellers. The sea is often depicted as a harbinger of death and, almost always at least, of illness. A recurring motif in these accounts is seasickness, which serves as a symbolic trial. The narrator who endures it unscathed attains a certain heroic status, elevating themselves above their fellow travellers—such as Poestion, who observes the affliction of all his companions while remaining unaffected. Yet, the challenge of the crossing is evident in Poestion's account too, whose experience underscores the sea's hostility towards all humans. A rough sea (*unruhige See*) and ominous fog (*unheimlicher Nebel*) then become the constant companions of his voyage, stimulating his imagination and reminding him of Norse cosmogony: "This was the true *Niflheimr*, the cold and misty world of the Eddic cosmogony, which one could vividly picture here" (2025, 117). It is as if the sea cannot be compared to anything from "this world" and can only be associated with a primordial space or the elements of the primordial world. This once again leads us to the concept of the sea's complete otherness, as postulated by Casati.

Ida Pfeiffer shares a similar experience, falling ill like the others on the journey and even fearing she might not survive:

> During the night we had left behind us the "Cattegat" and the "Skagerrack," and were driving through the stormy German Ocean. . . . The naiads of this stormy region seized me, and bandied me to and from, until they threw me into the arms of what was, according to my experience, if not exactly after Schiller's interpretation, "the horrible of horrors,"—sea-sickness. . . . I became worse and worse, till I was at length obliged to remain in my berth with but one consoling thought, namely, that we were to-day on the open sea, where there was nothing worthy of notice.
>
> (1852, 56–57)

Like other authors (such as Poestion and Maurer), Pfeiffer feels justified in invoking mythological figures in her depiction of her encounter with the ocean—specifically, the Naiads of Greek mythology, a sort of female spirits, presiding over all sources of water. It appears symptomatic that this experience, lying beyond the confines of everyday life, cannot be adequately conveyed through ordinary imagery.

Apart from the references to mythological figures, a noteworthy point is Pfeiffer's concluding remark in the passage: the open sea holds no appeal and, therefore, is deemed unworthy of description. Maurer explicitly states that the sight of land on the horizon offers the hope and comfort that are otherwise absent in the vastness of the open sea (2017, 26). In several instances, the German scholar complains about the desolate nature of the open sea, which he repeatedly describes as an *empty space* (2017, 15).

However, it is precisely the awareness of being lost in an empty space that generates the contrasting sensation of the sublime. Paradoxically, at the moment when the narrator sees nothing but water surrounding him, he suddenly perceives the ocean as majestic: "Now, for the first time, I saw the sea in all its majesty" (2017, 23).

Interestingly, of the five senses, it is almost exclusively sight (or its absence, in the case of fog) that is referenced across all accounts. The olfactory experience is entirely absent in the descriptions; the sea seemingly has no smell. Furthermore, from the deck of the ship, it cannot be touched, rendering the sense of touch equally absent. And according to the accounts, the sea only reveals its "voice" during moments of storm; otherwise, it remains silent. Surprising, then, is the complete lack of any synesthetic experience of the crossing; that at no point are all senses engaged in a holistic manner with the sea.

Approaching the coast

Once the crossing is completed, the narration shifts to an inverted perspective, depicting the land and coastline as seen from the sea. However, even this view does not always provide relief to the travellers. Rouquette perceives an additional danger: after surviving the perils of the sea voyage, the land that awaits offers no reassurance: "Lagunes, marshes, shifting lands, torrents rushing down from glaciers, Skeiðarársandur and Brunasandur—do you know how many carcasses rot within your domain of death?" (1925, 26). In many travel accounts, the moment of arrival provokes conflicting emotions, which can be interpreted through the lens of the sublime: when viewed from the water, the land appears both majestic and inhospitable. Confronted with the coastline, the observer struggles to find suitable words to describe the landscape; instead, it evokes a blend of impressions and imagination. Rouquette, once again, captures this in evocative terms, with a learned reference to Dante's *Inferno*:

> Dawn on June 14. In the mist, tattered by the wind, the massive bulk of Öræfajökull rises on the horizon—the giant glacier of Iceland, standing upon Vatnajökull, one of the largest glacial formations in the world. Under the clear sunlight, it gleams as if within reach, yet it lies dozens of miles away, an advanced sentinel that seems to say: "Do not come here, turn away, or if you enter, like the other before you, abandon all hope".
>
> (1925, 26)[17]

Ida Pfeiffer, for her part, experiences great disappointment. Everything she believes she sees from the ship and describes—grassy hills, bushes, and even groups of trees—turns out to be false and illusory. The small forests, the green hills, etc. are merely an illusion, and the reality is entirely different (1852, 62). Similar to this

Austrian adventurer, Meignan was also mistaken when, from a distance, viewing from the water, he mistook the grass-covered roofs of Thorshavn on the Faroe Islands for a meadow (1889, 15).

Coming to Iceland, MacKenzie does not seem particularly enthusiastic either about what he sees from the ship, but he feels very relieved to have survived the crossing:

> Moderate weather now succeeded, and as soon as it was clear day-light, we had the pleasure of being summoned upon deck to enjoy the first view of Iceland. . . . and though we felt considerable joy on finding ourselves so near the end of our voyage, we could not help being impressed with the very *uninviting* appearance of the country.
>
> (1811, 77)

Fortunately, his initial negative impression was not confirmed once they had landed; otherwise, the dangerous undertaking would have been in vain.

Conclusion

We should now return to the initial question—"Is it possible to narrate the sea?"—and think about Roberto Casati's statement once again, "The sea is the most radical elsewhere" (2022, 49), which helps explain the difficulty, or even the impossibility, of describing it with adequate expressions and words.

The travel accounts presented in this chapter—of which many others could be cited—attest to this challenge. The authors demonstrate great creativity, employing an extensive range of possible strategies: some authors opt for concision, compressing narrative time to a minimum; others deviate from direct narration through digressions, invoking mythological images; still others employ rhetorical devices such as personification, synecdochic substitution, comparison, and other stylistic strategies. Only a very few entirely forgo describing the crossing.

Regardless of the different strategies employed, the unifying element across all these accounts is the pervasive sense of danger and the feeling of helplessness that accompanies the crossing, revealing the otherness of this natural space. Despite the varying perspectives and personal experiences, there is a fundamental similarity in tone throughout these narratives: crossing the Atlantic emerges as a physically and mentally demanding endeavour, one that pushes travellers to their limits and forces authors to sharpen their intellectual and emotional faculties. All attempts to fill the void of the maritime surface with diverse narrative strategies are of considerable literary interest. Whether they truly bring us closer to understanding the essence of the sea is another question.

Notes

1 Through the reading of Laurence Devillairs' *Petite philosophie de la mer* and Roberto Casati's *Philosophie de l'océan*—both published in 2022—I have come to realise that the distinction between "sea" and "ocean" is largely artificial and lacks a strong conceptual foundation. For this reason, I will employ the term "sea" in an inclusive sense, encompassing the ocean as well.

2 Travel accounts constitute a distinct literary genre situated between fact and fiction. While typically rooted in an actual journey, this is not a necessary condition. The author, often identical with the protagonist—as in autobiographical writing—should be regarded as a literary construct. For a definition of travel literature, see Brenner 1990.

3 See Katie Ritson's concept of a literary landscape, as outlined in the introduction to *The Shifting Sands of the North Sea Lowlands* (2019), where she discusses the creation of an imagined and literary landscape around the North Sea basin which is transnational and multilingual.

4 See Bachelard 1957; Tuan 1977; Westphal 2007. Ritson (2019, 5–6) uses the term *landscape* to refer to the notion of *place*, viewing it as a combination of real and imagined space.

5 In the travel account *Pauvre Islande!* (Poor Iceland), Victor Meignan directly refers to the emptiness of the sea, characterising it negatively: "It cost her a lot to have been drawn to *this void*, we will have the opportunity to observe this later" (1889, 7).

6 Ritson speaks of "the human imagination of the physical world" (2019, 13), rather than of the physical world itself. With regard to the mental images associated with Iceland and the North Atlantic region, see Sumarliði R. Ísleifsson 1996 and, more recently, 2015. In this regard, see also Oslund 2011.

7 This graphical representation finds a parallel in the depiction of the sea in Eddic poetry, as preserved in the *Codex Regius*. In this context, Bintley observes that the sea (meaning the Atlantic Ocean) is "for the most part a barrier to be overcome" (2014, 240). He emphasises that maritime space represents a constant threat and that, in *Hávamál*, it repeatedly emerges as an unreliable and treacherous force.

8 In this regard, one of the travelogues highlights this aspect through its title, namely "Reisen *nach* und auf Island" (Travel *to* and in Iceland) by J. C. Poestion.

9 Ida Pfeiffer was an Austrian female traveller and explorer without formal academic education. She began travelling at the age of 45, following the death of her husband. Pfeiffer journeyed across nearly all continents, including Asia, the Americas, Northern Europe, and Africa. Her travel accounts were translated into several languages.

10 Konrad Maurer was a German professor of legal history and a renowned expert in Old Norse literature. He spoke Icelandic and travelled to Iceland in 1858 with the explicit aim of tracing the remnants of Old Norse culture and to get in contact with the Icelandic society of his days.

11 Josef C. Poestion was an Austrian academic and translator of Icelandic literature as well as an excellent connoisseur of Old Norse literature. He worked at the Library of the Ministry of the Interior in Vienna, which he headed from 1888 until his death, serving as its director from 1896. He published several studies on Icelandic and Scandinavian literature.

12 Unless otherwise indicated, the translations into English are my own. The page of the original texts will be put in brackets after the translation.

13 Louis-Frédéric Rouquette, who died at the age of 41, was the author of around twenty publications. After a brief period of fame, this author has now largely fallen into oblivion.

14 The author was likely referring to the *miðgarðsormr* (Midgard Serpent), which is said to encircle the earth. This motif is attested in both the *Poetic Edda* and the *Snorra Edda*.

15 La Martinière was a physician and surgeon. In 1670, he had the opportunity to be enlisted as a surgeon on board a Danish ship outfitted for a five-month expedition to the Arctic Ocean, following the coasts of Norway and Lapland, and continuing on to Greenland and Iceland. He later published an account of the expedition, which became the first published travel report written by a Frenchman describing the Arctic coasts.

16 I would like to thank Katie Ritson for highlighting a very curious point: In a personal communication, she observes that the sea, though visible, is harder to describe, while it is often replaced by the wind, which is materially invisible. Quite often in the travelogues, it is the wind rather than the water that accelerates or slows down the pace of a sailing voyage. It is difficult to find a clear explanation, but my guess is that this is linked to human experience: All of us have felt the wind on land, but not everyone has experienced being at sea. The wind is part of the shared horizon of knowledge between the author and his readership.

17 Another direct reference to Dante's *Inferno* is to be found in Meignan's travelogue (1889, 90–91) concerning a barren, dramatic, terrifying, and hideous region of Iceland.

Works cited

Primary sources

Gunnlaugs saga ormstungu. In *Borgfirðinga sögur*, edited by Sigurður Nordal and Guðni Jónsson, 49–107. Reykjavík: Hið íslenzka fornritafélag (Íslenzk fornrit 3), 1938.

MacKenzie, George Stuart. *Travels in the Island of Iceland during the Summer of the Year MCCCCX*. Cambridge: Cambridge University Press, 1811.

de la Martinière, Pierre Martin. *Voyages des pais septentrionaux*. Paris: Louis Vendosme, 1671.

Maurer, Konrad. *Reise nach Island (im Sommer 1858)*. Edited by Alessia Bauer and Kurt Schier. München: Utz Verlag (Münchner Nordische Studien 31), 2017.

Meignan, Victor. *Pauvre Islande!* Paris: Ernest Kolb Editeur, 1889.

Pfeiffer, Ida. *Reise nach dem skandinavischen Norden und der Insel Island im Jahre 1845*. Pest: Heckenast, 1845.

Pfeiffer, Ida. *Visit to Iceland and the Scandinavian North*. Translated from the German of Madame Ida Pfeiffer and with Numerous Explanatory Notes and eight tinted Engravings. London: Ingram, Cooke and Co., 1852.

de la Peyrere, Isaac. *Relation d'Islande*. Paris: Thomas Jolly, 1663.

Poestion, Joseph C. *Kulturgeschichtliche Wanderungen auf Island*. Edited by Marion Lerner and Alessia Bauer. München: Utz Verlag (Münchner Nordistische Studien 53), 2025.

Rouquette, Louis-Frédéric. *L'île d'enfer*. Paris: J. Ferenczi, 1925.

Secondary sources

Bachelard, Gaston. 1957. *La poétique de l'espace*. Paris: PUF.

Bintley, Michael. 2014. "Fluid Dynamics: Aquatic Agency in the *Poetic Edda* of the Codex Regius." In *The Elements in the Medieval World: Interdisciplinary Perspectives: Water* (Elements, Nature, Environment 1), edited by Mirilina Cesario, Hugh Magennis, Elisa Ramazzina, 227–52. Leiden and Boston: Brill.

Brenner, P. J. 1990. *Der Reisebericht in der deutschen Literatur: Ein Forschungsüberblick als Vorstudie zu einer Gattungsgeschichte*. Tübingen: Niemeyer.

Casati, Roberto. 2022. *Philosophie de l'océan*. Paris: Presses universitaires de France/Humensis.

Corbin, Alain. 1988. *Le territoire du vide: L'Occident et le désir du rivage 1750–1840*. Paris: Champs Flammarion.

Corbin, Alain. 2020. *Terra incognita: Une histoire de l'ignorance*. Paris: Albin Michel.
Devillairs, Laurence. 2022. *Petite philosophie de la mer*. Paris: Éditions de la Martinière.
Grétarsson, Reynir Finndal. 2017. *Maps of Iceland: Antique Maps of Iceland 1482–1850*. Reykjavík: Crymogea.
Ísleifsson, Sumarliði R. 1996. *Ísland framandi land*. Reykjavík: Mál og menning.
Ísleifsson, Sumarliði R. 2015. *Tværeyjar á jaðrinum: Ímyndir Íslands og Grænlands frá miðöldum til miðrar 19. aldar.* Reykjavík: Háskóla útgáfan Sagnfræði stofnun Háskóla Íslands.
Knauer, E. Regina. 1981. *Die Carta Marina des Olaus Magnus von 1539: Ein kartographisches Meisterwerk uns seine Wirkung*. Göttingen: Gratia-Verlag.
Oslund, Karen. 2011. *Iceland Imagined: Nature, Culture, and Storytelling in the North Atlantic*. Seattle: University of Washington Press.
Ritson, Katie. 2019. *The Shifting Sands of the North Sea Lowlands: Literary and Historical Imaginaries*. London: Routledge.
Tuan, Yi-Fu 1977. *Space and Place. The Perspective of Experience*. Minneapolis and London: University of Minnesota Press.
Westphal, Bertrand. 2007. *La géocritique: réel, fiction, espace*. Paris: Les éditions de Minuit.

11

PLENTY AND PERIL

The sea in the sagas of Icelanders

Annette Lassen

Sailing plays a crucial role in the sagas of Icelanders, reflecting the fundamental importance of the sea in Norse society. The Nordic languages, both in the Middle Ages and in the period from around 800 to 1050, attest to the fact that the sea was not perceived as a boundary in the way forests and mountains were. In this Nordic world, the sea was not understood as an obstacle but as a conduit for communication, trade, and exploration. It functioned as a link, an oceanic highway that facilitated the spread of Norse culture to Iceland, the Faroe Islands, Greenland, Shetland, the Orkney Islands, Ireland, even Vínland (an area in present-day Canada), as well as to the British Isles and Russia. At a fundamental level, one could argue that seafaring is a precondition for the Icelandic sagas—for without Norse voyages, Iceland would not have been settled, and the sagas would never have been written. Thus, to speak of an "oceanic turn" in the sixteenth and seventeenth centuries, as coined by Carl Schmitt (1942/1981, 40), makes little sense in a Nordic context, as Søren Frank has pointed out (2022, 8–9, 60–61, 420).

Given the centrality of the sea, it is hardly surprising that it is rare to find a saga of Icelanders in which the sea, sailing, and ships are not mentioned, or in which no one departs from or returns to Iceland. Yet despite its persistent presence, the saga writers show no concern for the sea as nature, as an autonomous ecosystem independent of human perception or use. Rather, the sea is present primarily in relation to human activity and experience. For the saga writers, the sea is of interest insofar as it serves as a traversable or treacherous route between lands, a source of wealth through fisheries and whales, or a domain of lethal magic and dangerous sea-monsters, beings which, in the context of the saga age and medieval worldview, were regarded as real and potentially encounterable entities. The sea *per se*, as an abstract or natural phenomenon disconnected from human meaning-making,

DOI: 10.4324/9781003647638-15

appears to have held no narrative or conceptual interest (see Bauer's chapter in the present volume). Instead, the sea emerges as a site of negotiation, between safety and danger, wealth and loss, and identity and alterity.

To grasp the significance of the sea in the sagas, this chapter approaches the North Atlantic and the Arctic Ocean in the sagas from a *mentalité*-historical perspective. This approach is in accordance with Steve Mentz who, within the field of blue humanities, places cultural history "in an oceanic rather than terrestrial context" (2018, 69). The chapter will approach the sea around Greenland and Iceland through two themes: firstly, the wealth of the sea, as depicted in accounts of abundant fish and whales during the Age of Settlement and the subsequent environmental changes noted by saga writers; and secondly, the perils of the sea, as ocean voyages are frequently portrayed in the sagas as hazardous, often due to supernatural forces and paranormal phenomena.

The sagas of Icelanders: between history, fiction, and cultural memory

Written in the Middle Ages, the sagas of the Icelanders are, with few exceptions, set during the period from 930 to 1030, the so-called Saga Age.[1] Although it is impossible to date the sagas with certainty, most scholars agree that their composition began in the early thirteenth century and continued through the Middle Ages. The narrative style of the sagas is renowned for its concise and scenic style and can easily mislead readers into perceiving them as purely oral in origin. For a long time, scholars believed that the sagas were historical compositions preserving an "authentic" memory of past events and that they were products of traditional oral storytelling. Today, however, it is widely recognised that both oral and written narratives are shaped by culture and context. But even though the sagas are best understood as written works influenced by literary trends, they may still reflect elements of oral tradition and collective memory of the past.

There is currently a growing tendency in scholarship to view the sagas as expressions of cultural memory: texts that construct continuity between past and present by embedding historical imagination within a shared cultural framework (see, e.g. Glauser 2009 and Hermann 2014). Memory is inherently flexible and shaped by ideological, historical, and social conditions—as all fiction (see Lassen 2022b). Many of the sagas are today considered fictional, yet they served a distinct function within the contexts in which they were composed and transmitted. Although sagas recount the past, they inevitably reflect the ideals and values of the groups involved in their production. Saga writers often anchor the memory of bygone eras in their own contemporary landscape. Through their descriptions of past events, we can gain insight into their relationship with the sea, with which they still lived in close connection.

Plenty at sea: abundance, memory, and marine resources

The sagas of Icelanders are, however, all set in a distant past, in a world that looked very different from the writers' own time. Life was different, and at times the contemporaneous landscape bears traces of the past, traces that saga writers do not neglect to mention. They knew that the settlers who first arrived around 870 had left their mark on the land, altering it, and that the landscape of Iceland had originally been different.[2] According to the saga writers, the riches of the sea were also of another kind at the time of settlement. The *Saga of the People of Eyri* (*Eyrbyggja saga*) states: "Back then, there was an abundance of food, both from the islands and from other forms of sea fishing" (chapter 4).[3] Similarly, in *Egil's saga* (*Egils saga Skalla-Grímssonar*), we read about the settlement of Skalla-Grímr:

> He built a farm in Álftanes . . . There, he fished from a boat, hunted seals, and gathered eggs, all of which were abundant *at that time*. Driftwood was also brought home. Many whales came by, and one could take as many as desired. The hunting ground was undisturbed, and the animals were not accustomed to humans.
>
> (chapter 29)

When Skalla-Grímr explores the land, he discovers that "every lake is full of fish" (chapter 28). This theme of past abundance recurs in other sagas. According to *The Saga of Grettir the Strong* (*Grettis saga*), the fjords were full of fish (chapter 10). Furthermore, *Bard's Saga* (*Bárðar saga Snæfellsáss*; chapter 8) and *Viglund's Saga* (*Víglundar saga*; chapter 10) describe rich fishing grounds off Snæfellsnes. Unsurprisingly, rumours of this wealth reach Norway, prompting Norwegian chieftains to consider emigration to Iceland. In *The Saga of the People of Kjalarnes* (*Kjalnesinga saga*; chapter 10), Ingimundr hears from a prophetess in Norway that in Iceland "the cattle can stay outside in winter and find their own food [and] that every lake is full of fish". Similarly, in *The Saga of the People of Laxardal* (*Laxdœla saga*; chapter 3), two brothers discuss emigration:

> Björn and Helgi wanted to go to Iceland because they found many things about the land appealing. . . . There were said to be plenty of whales washing ashore, an abundance of salmon to catch, and fishing spots available all year round.

To set the appeal of these riches in context, it should be noted that their father preferred to go to the British Isles, which he knew from raiding. Their sister and her son also chose that destination; her son went to become king of the Scots before the Scots killed him, prompting her to set sail for Iceland instead. These examples show that the abundance of the sea, lakes, and rivers was remembered as a driving force behind the settlement of Iceland.

A recurring motif in the sagas is that killings and conflicts often arise over the flensing of stranded whales. While this took place on land, Icelanders also hunted whales at sea, as noted in *The Saga of Grettir the Strong* (chapter 25) and *The Saga of the Sworn Brothers* (*Fóstbræðra saga*; chapter 9), which describe whale hunting off Strandir in the Westfjords. This custom is reflected in the byname "reyðarsíða" (*Whale-Side*) borne by a character in *Njal's Saga* (*Brennu-Njáls saga*, chapter 114), and *The Tale of Thorstein Bull's-Leg* (*Þorsteins þáttr Uxafóts*).

Some sagas suggest a retrospective awareness of environmental decline, not only on land but also at sea. *The Saga of Grettir the Strong* emphasises that, back then, the fjords were full of fish. *Bard's Saga* and *Viglund's Saga* similarly note that, at the time of settlement, the fishing was good off Snæfellsnes. This awareness of decline aligns with historical evidence. Recent mitochondrial DNA analysis has revealed that Iceland once had its own walrus population, with unique genetic lineages. The Icelandic walrus population became extinct following the settlement. These findings suggest that the settlement may have been more motivated by the exploitation of marine resources, perhaps particularly walrus ivory, meat, and fat, than previously recognised (Keighley et al. 2019).

Place names also attest to the former presence of walruses in Iceland, for instance, Rosmhvalanes (Ness of the Walruses), which likely preserves the memory of past hunting grounds.[4] Some saga characters are described as carrying riches from walruses; they wear ropes of walrus hide or carry swords with walrus tusk hilts (*Bard's Saga*, chapter 10; *The Saga of the People of Laxardal*, chapter 29). However, walrus hunting is mentioned in only one saga: *Kormak's Saga* (*Kormáks saga*, chapter 18), and the walrus in question is suspected to be a shape-shifting witch: "a walrus surfaced near the ship. Kormak threw a heavy spear at it, striking the whale, and it sank. People thought they recognized Þorveig's eyes in the creature. The whale did not resurface." One reason for the lack of reference to walrus hunting may be the varied terminology used for the animal. In *Kormak's Saga*, the walrus is referred to both as a whale and as a walrus. In the *King's Mirror* (*Konungs skuggsjá*, p. 29), an Old Norwegian didactic work from c. 1250, the walrus is categorised as a *fiskr*, "fish". This term in Old Norse/Icelandic encompasses all sea creatures, not just fish in the modern sense.[5] The *King's Mirror* debates whether the walrus belongs among whales or seals: "Now one species is left which Greenlanders enumerate among the whales but it appears to me that it should rather be counted with the seals; it is called walrus (*hrosshvalr*)" (p. 16). Given this ambiguity, it is possible that saga references to an abundance of fish, seals, and whales may also have included walrus. The *King's Mirror* was written in the same period as many of the sagas and shared terminology with the saga writers. Walruses are typically hunted on land, like seals.

Walrus tusks and hide were highly prized from the tenth to the thirteenth century. Tusks were used for, among other things, book ornaments, reliquaries, and chess pieces (Perdikaris 2007, 208), while hide was used for exceptionally strong ship ropes. The settlement of Greenland was probably driven by the economic potential of walrus hunting. Nevertheless, *Eirik the Red's Saga* (*Eiríks saga rauða*) and *The Saga of the Greenlanders* (*Grœnlendinga saga*), which recount the Icelandic settlement of Greenland, do not mention the riches of nature—unlike the descriptions of Iceland and Vínland. The latter is described as abundant in self-sown wheat, with vineyards wherever there were rocky slopes, and: "Every stream was full of fish. Where the sea and land met at high tide, they made some pits, and when the tide receded, halibut lay in the pits" (*Eirik the Red's Saga*, chapter 10).

In late medieval sagas, as these texts increasingly begin to resemble what we would now term fiction, we find descriptions of Greenland's marine wealth. For example, *The Saga of Ref the Sly* (*Króka-Refs saga*) describes Refr's settlement in Greenland in the following way: "when they came to the bottom of the fjord, there was a good harbor. . . . There was an abundance of animals. Driftwood lay on all the beaches, and there were plenty of fish" (chapter 6). In the same saga, the Norwegian king sends a man to Greenland, instructing him to bring back "walrus tusks and walrus ropes" (chapter 10). The following summer, he returns with three treasures:

> The first was a polar bear, fully grown and very well tamed. The second treasure was a board game made of walrus tusk, very skilfully crafted. The third treasure was a walrus skull with all its teeth intact. It was engraved all over, and a large part of it was covered in gold. The teeth were firmly set in the skull.
>
> (chapter 11)

When Refr himself returns to Scandinavia, "his ship [is] loaded with Greenlandic goods, walrus ropes, walrus tusks, and furs" (chapter 14). He travels to Denmark, where the king rewards him with fourteen farms in exchange for the gifts he has brought. The most prized gift is the walrus-hide ropes (chapter 18). Beyond the ropes, Refr and his sons present an abundance of wealth, including "walrus ivory, and furs, and . . . many kinds of Greenland wares which were seldom seen in Denmark. They had five polar bears, and fifty falcons including fifteen white ones" (chapter 18). While these numbers are exaggerated, they nonetheless reflect historical realities: for example, we know that in 1252 the Norwegian king Haakon presented the English king Henry III with a polar bear. In *The Tale of the Greenlanders* (*Grœnlendinga þáttr*), set in the twelfth century, a Greenlandic chieftain sends his son to Norway with "a great number of walrus tusks and hides so that he could make his way among princes and noblemen" (chapter 1). Recent DNA research has revealed that no less than 80 per cent of the walrus tusks found in medieval Europe originated from Greenland (Perdikaris and McGovern 2007). The riches of the Greenlandic sea were thus a crucial source of income for the

Norse Greenlanders as well as for the merchants who braved the journey across the North Atlantic.

Perils at sea: monsters, magic, and maritime danger

But who dared to cross the North Atlantic a millennium ago? According to the sagas, the voyage between Iceland and Greenland was fraught with danger. Both *Eirik the Red's Saga* and *The Saga of the Greenlanders* recount the perils of sailing these waters. When Eiríkr the Red and his followers left Iceland in 982, twenty-five ships set sail for Greenland, yet only fourteen reached their destination. Upon arriving in Greenland, a man from the Hebrides, who had been aboard one of the first ships, is said to have composed a poem, *The Drapa of the Sea Fences*, though only a fragment has survived. The "sea fence" in the title refers to an immense wave that suddenly rises in the open ocean, steep as a wall, a rogue wave. In the poem, the poet prays to God to protect him during the journey (*The Saga of the Greenlanders*, chapter 1).

Another passage, in *Eirik the Red's Saga*, describes a ship attacked by maggots or shipworms while sailing from Vínland to Greenland (chapter 13). As the vessel begins to sink, the crew must decide who will be allowed into the lifeboat and who must go down with the ship, for the small boat can only hold half of them. The lifeboat, however, is safe, having been coated with seal tar, which the maggots cannot penetrate. Their commander, Bjarni, declares that lots should be drawn; rank and status should not determine one's fate at sea. He himself is among the fortunate ones. But as he prepares to step into the boat, a young man, one of the unlucky ones, reminds him that before setting sail for Vínland, Bjarni had made him other promises. Bjarni steps back onto the ship, allowing the young man to take his place in the lifeboat. The saga concludes: "People say that Bjarni died there in the sea of worms, along with the others aboard his ship. The ship 's boat and those on board it went on their way and made land, after which they told this tale" (chapter 13). The aforementioned perils are not paranormal, but the Greenland Sea was also believed to be inhabited by sea monsters, one of which is described in the chapter about the whales of Greenland in *King's Mirror*:

> There is yet another fish . . . which I hesitate somewhat to speak about because of its size; for most people will think it incredible; and very few can say anything about it with certainty; for it almost never comes near the shore or to places where fishermen go, and I do not believe there are many of its kind in the seas. In our language, we most often call it *hafgufa* ('sea steam').
>
> (p. 17)

Despite doubts about their existence, sea monsters are a learned phenomenon; they are described in medieval bestiaries, from where they found their way into the sagas.

In *Physiologus* (preserved fragmentarily in a manuscript from c. 1200, AM 673 a I 4to), another sea monster is described: *aspedo*, a shortened form of *aspidochelone*. In Old Norse/Icelandic, *aspedo* is known as *lyngbakr* (literally "heather-back"), a whale with a forest (*skógr*) growing on its back. When sailors see it, they mistake it for an island and go ashore. Then it dives, and the sailors drown.

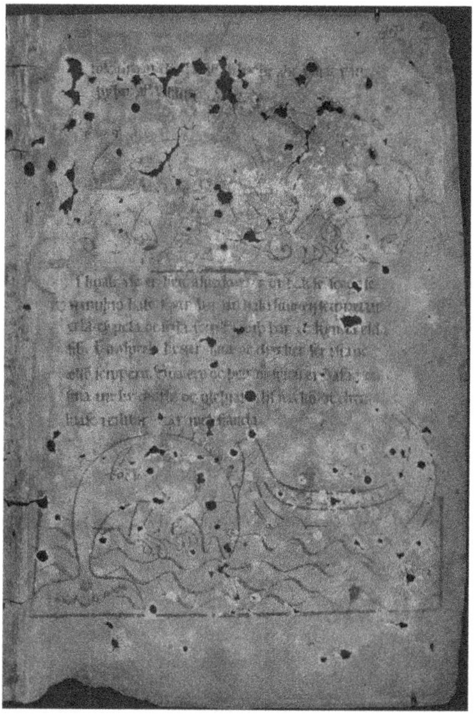

FIGURE 11.1 A drawing of the sea monster *aspedo*. The text, which is in Old Icelandic, reads: "There is a whale in the sea called *aspedo*, and on its back it is as though there were a forest. But out in the middle of the sea it raises its back, and sailors believe it to be an island and fasten their ship to it and then light a fire. But *aspedo* feels the fire and dives into the sea with all the sailors. Thus, it is also with those who place their hope in the devil and delight in his deeds—they drown in eternal torment with the devil. It also has another nature. When it is hungry, it opens its mouth, and it is as though it lets a fragrance come out of its mouth. But the small fish smell the fragrance and gather in its mouth. And when its mouth is full, it closes its mouth and swallows them. Thus, are faithless people overcome in the devil's manifold temptations like small fish; but the greater ones defend themselves and do not come near him." AM 673 a II 4to, 3r (Reykjavík, The Árni Magnússon Institute for Icelandic Studies). Photo: Handrit.is.

The sea monsters of medieval learning, however, do not appear in the sagas of Icelanders, but in one of the sagas of ancient times, *Arrow-Odd's Saga* (*Ǫrvar-Odds saga*), a *lyngbakr* surfaces in the Greenland Sea:

> Around midday, they spotted a large island. . . . When Oddr's men reached the island, they had only been there for a short while, when the island sank and all of them drowned.
>
> Heather grew on top of the island, but they never saw it rise again. The cliffs were also gone when they looked around. Oddr was greatly puzzled by this and asked Vigni what had caused it.
>
> Vigni said: ". . . there are two sea monsters. One is called *hafgufa*, the other *lyngbakr*. *Lyngbakr* is the largest of all whales in the world, while *hafgufa* is the greatest monster that lives in the sea. Its nature is to swallow men and ships and whales and everything it can catch. It stays under the water for days at a time, and when it sticks its head and snout above water, it never does so for less time than a full tide. When we sailed through the sound, we were actually between its jaws, and its nose and lower jaw were the cliffs you thought you saw in the sea, while *lyngbakr* was the island that sank".

In *Physiologus*, the theological significance of the *aspedo* is explicated as a symbol of the devil's schemes. In the medieval worldview, reality is perceived as having (at least) two dimensions. One cannot trust the surface of things; all appearances may be deceptive. This principle applies universally: everything is understood to possess a deeper moral or Christian significance. Yet everything might also be a deception orchestrated by the devil, no matter how innocent or familiar it may seem.

While no one saves the men in the sea of worms, the Christian God, according to the survivor from the Hebrides, rescues the men from the sea fences. In this world, where the devil tricks men and God can intervene and decide their destinies, it is unsurprising that evil forces are often believed to cause maritime hazards. In the sagas, storms and monstrous waves are frequently attributed to malevolent people with magical abilities. With their dark magic, these antagonists of the saga heroes can cause harm and death.

Sorcery and storms: magical danger in the sagas

In *Bard's Saga,* the scene is set at the fishing banks off the coast of Snæfellsnes (chapter 8), where a magical storm nearly claims the life of a man named Ingjaldr. The witch Hetta, a malevolent sorceress, has previously harmed him by killing his livestock but offers compensation by revealing the location of a fishing bank teeming with fish. Ingjaldr ventures out to sea alone. The morning begins with fine weather and an abundance of fish, but soon the wind picks up, bringing drifting snow and frost.

During the storm, Ingjaldr encounters a mysterious, red-bearded fisherman named Grímr (The Masked One), who insists that Ingjaldr remain at sea until he finishes fishing. As the weather worsens, Ingjaldr loses his equipment, and his boat begins to take on water. Meanwhile, back home, an ominous verse is recited, foreshadowing his fate. Just as Ingjaldr nears death, he is rescued by Bárðr, the saga's eponymous hero, who invites him aboard his boat. Upon Bárðr's arrival, Grímr vanishes, and the saga comments that some believed Grímr to be the Norse god Thor.

One of the theological methods for interpreting the nature of pagan gods and the disbelief of pre-Christian peoples is demonology.[6] Medieval Icelanders likely understood Thor in this passage as a demon, even though the saga does not explicitly state this. A similar association appears in *The Saga of the People of Flói* (*Flóamanna saga*), where the dangers of the Greenland Sea are likewise attributed to Thor. Here, Thor attacks the newly baptised Þorgils in a dream. Following their encounter, Þorgils recognises the dangers of the journey to Greenland but, as a devout Christian, refuses to be deterred. Without revealing the dream to his companions, he sets sail.

Once on the open sea, the wind dies down, and they drift for months. As autumn approaches, some aboard suggest invoking Thor, but Þorgils forbids it. Thor then appears again, tempting Þorgils with a promise that if he worships him, they will reach a harbour within seven days. Þorgils again refuses. After three months at sea, just before winter, the men are shipwrecked in Greenland's wilderness. Only a few survive.

The saga does not explicitly state that Thor caused the shipwreck, but he is clearly the source of the hardship at sea.[7] The sea is dangerous and deceptive, but in the world of the sagas, it does not possess agency in itself; rather, it is employed by evil humans and devilish beings to bring about shipwreck, dead calm, disorientation, and death. Demonisation of the sea along with sea monsters continues into the early modern period (see Spaans and Willumsen's chapter in the present volume).

Sometimes, individuals with magical abilities are believed to survive shipwrecks in mysterious ways—especially when their bodies do not wash ashore. In *Njal's Saga*, a man who is described as a male witch and a problematic character is lost at sea while fishing. According to the saga, this is a paranormal event: some believed that he had "entered the mountain Kaldbakshorn, where he was warmly received. But others contradicted this and said that there was no truth to it" (chapter 15).

Humans with magical powers could also summon storms at sea to control where their enemies would make landfall. A well-known example appears in *Egil's Saga*, as part of the famous conflict between the Norwegian queen Gunnhildr (who, according to other sagas, was educated in magic) and Egill, who had capsized and killed her son. Seeking revenge, Gunnhildr performs a magical ritual to ensure that Egill would never find peace in Iceland until he sees her again. As a result, Egill sets sail and encounters a violent storm on the open sea, leading to a shipwreck that delivers him directly into Gunnhildr's hands—precisely as she had intended (chapter 60).

Also in the Icelandic fjords, storms could be perceived as the work of sorcery. In *The Saga of the People of Laxardal* (chapter 35), Þórðr accuses a family from the Hebrides of sorcery, summons them to the Alþingi, and demands their outlawry. He then returns to his ship and sets sail. When the family hears of the accusations, they chant magic songs, which conjure a violent storm targeting Þórðr. He loses control of his ship; people on shore watch him throwing everything overboard to lighten the vessel—everything except the crew. Just as it appears that Þórðr might be safe, a wave crashes against the ship and capsizes it. Þórðr and everyone on board perish.

The prevailing sentiment is that such sorcery, as practised by this family, warrants death. Consequently, they are hunted down and killed. In the aftermath, the ship's keel washes ashore on an island later named Keel Island (Kjalarey), and Þórðr's shield drifts to Shield Island (Skjaldarey). Soon after, the bodies of Þórðr and his men also wash ashore and are buried at a site that becomes known as Gravemound Ness (Haugsnes).

As noted earlier, past events are often inscribed into the landscape of the saga writers' present. The memory of the magic storm and the shipwreck has shaped the perception of the landscape, as evidenced by the place names: Keel Island, Shield Island, and Gravemound Ness. The names testify to a cultural memory—and a present awareness—of the dangers of life with the sea, regardless of the saga's supposed lack of historical accuracy.

Magic and paranormal events were integral to the medieval worldview, shaping how people interpreted the perils of the sea. From a modern psychological perspective, some of these beliefs may have functioned as coping mechanisms for people living in close proximity to maritime dangers. At the same time, these narratives reflect broader social dynamics, particularly in their tendency to attribute magical storms to marginalised groups, whether non-Norse settlers, such as the sorcerous family from the Hebrides, or women. Written largely from a male perspective (with only a few exceptions), the sagas often reinforce patterns of othering and blame.

Conclusion: the inescapable sea

In the Viking Age society depicted in the sagas, sailing was not a choice but a necessity, essential for survival, trade, and exploration. Dependence on the sea's resources was likewise a matter of both subsistence and economic need. The authors of the sagas of Icelanders offer no romanticised depictions of seafaring or life aboard ships; instead, they portray a constant struggle against storms, fog, disorientation, water surging into the boats, and quarrels erupting as men attempt to bail out the water.

Nevertheless, the wealth and resources offered by the sea rendered remaining on land an impossible alternative. This mentality was not restrained by Christian moderation or Stoic principles of non-transgression, and it persisted after the conversion to Christianity. The pursuit of riches and new hunting grounds, particularly in Greenland, likely in search of walrus, drove Icelanders even further north. The sea, in the sagas of the Icelanders, emerges as an inescapable and integral part of human existence, reflecting the lived realities of the saga authors. Their connections to other countries—through trade, travel, and education—lay across the sea. The sea beckoned with its many opportunities.

Though the people of the sagas never set sail for pleasure, they set sail all the same.

Notes

1 The sagas of Icelanders (*Íslendingasögur*) are just one genre of saga literature. There are also sagas of ancient times in the Northern lands (*fornaldarsögur Norðurlanda*), kings' sagas (*konungasögur*), courtly romances (*riddarasögur*), contemporary sagas, bishops' sagas, and apostles' sagas. Besides the sagas, a wide range of prose texts are preserved from medieval Iceland. The poetic corpus of medieval Iceland includes Eddic and skaldic poetry along with ballads (*rímur*). For a general introduction to the sagas of Icelanders, see Lassen 2025.

2 In a study on common pool ressources in the sagas of Icelanders, Reinhard Hennig argued that the sagas served "to promote behavior that could lead to long-term sustainability in the managing of natural resources as commons" (2020, 58). This ecological awareness is, however, anachronistic; the sagas are more concerned with conflicts between people and their resolution. Even in the eighteenth century, it was not yet understood that the destruction of Icelandic nature was man-made, as Páll Vídalín's *Deo, Regi, Patriae* (1699) attests.

3 In Lassen (2025), datings and summaries of the single sagas can be found. Unless otherwise mentioned, the translations are my own.

4 Cf. Glauser about the mnemonic function of place names (2009, 209).

5 Language use generally varied with respect to whale/fish far into the modern period. Early modern authors, such as Hans Christian Andersen, used the term *whale-fish* for whales, which zoologists until the eighteenth century classified among the fishes. See also Brito (2024, 400).

6 For the medieval strategies for explaining paganism, see Lassen (2022a).

7 For a discussion of the theology of the saga, see Lassen (2019).

Works cited

Primary sources

AM 673 a II 4to. The Árni Magnússon Institute for Icelandic Studies, Reykjavík.

Bárðar saga Snæfellsáss. Edited by Guðbrandur Vigfússon. In *Bárðarsaga Snæfellsáss, Víglundarsaga, Þórðarsaga, Draumavitranir, Vǫlsaþáttr*, 1–46. Copenhagen: Berlings bogtrykkeri. 1860.

Brennu-Njáls saga. Edited by Einar Ólafur Sveinsson. In *Íslenzk fornrit* 12. Reykjavík: Hið íslenzka bókmenntafélag. 1954.

Egils saga Skallagrímssonar. Edited by Finnur Jónsson. In *STUAGNL* 17. Copenhagen: S.L. Möllers bogtrykkeri. 1886–1888.

Eiríks saga rauða. Edited by Sven B. F. Janson. "Handskrifterna till Erik den rödes saga." In *Sagorna om Vinland*, 26–81. *Kungl. Vitterhets Historie och Antikvitets Akademiens Handlingar* 60:1. Stockholm: Wahlström & Widstrand, 1945.
Eyrbyggja saga. Edited by Einar Ólafur Sveinsson and Matthías Þórðarson. In *Íslenzk fornrit* 4. Reykjavík: Hið íslenzka bókmenntafélag. 1935.
Flóamanna saga. Edited by Finnur Jónsson. In *STUAGNL* 56. Copenhagen, S.L. Möllers bogtrykkeri, 1932.
Fóstbrœðra saga. Edited by Björn K. Þórólfsson. In *STUAGNL* 49. Copenhagen, S.L. Möllers bogtrykkeri, 1925–1927.
Grettis saga. Edited by Guðni Jónsson. In *Grettis saga Ásmundarsonar, Bandamanna saga, Odds þáttr Ófeigssonar, Íslenzk fornrit* 7, 3–290. Reykjavík: Hið íslenzka bókmenntafélag, 1936.
Grœnlendinga saga. Edited by Guðbrandr Vigfússon and C. R. Unger. In *Flateyjarbók*, Vol. 1, 430–32, 538–49. Oslo: Malling, 1860.
Grœnlendinga þáttr. Edited by Guðbrandr Vigfússon and C. R. Unger. In *Flateyjarbók*, vol. 3, 445–54. Oslo: Malling, 1868.
Kjalnesinga saga. Edited by Jóhannes Halldórsson. In *Kjalnesinga saga, Jǫkuls þáttr Búasonar, Víglundar saga, Króka-Refs saga, Þórðar saga hreðu, Finnboga saga, Gunnars saga Keldugnúpsfífls, Íslenzk fornrit* 14. Reykjavík: Hið íslenzka bókmenntafélag, 1959.
Konungs skuggsjá. Edited by Ludvig Holm-Olsen. In *Norrøne Tekster* 1. Oslo: Kjeldeskriftfondet, 1945.
Kormáks saga. Edited by Einar Ólafur Sveinsson. In *Vatnsdœla saga, Hallfreðar saga, Kormáks saga, Hrómundar þáttr halta, Hrafns þáttr Guðrúnarsonar, Íslenzk fornrit* 8, 203–302. Reykjavík: Hið íslenzka bókmenntafélag, 1939.
Króka-Refs saga. Edited by Jóhannes Halldórsson. In *Kjalnesinga saga, Jǫkuls þáttr Búasonar, Víglundar saga, Króka-Refs saga, Þórðar saga hreðu, Finnboga saga, Gunnars saga Keldugnúpsfífls, Íslenzk fornrit* 14. Reykjavík: Hið íslenzka bókmenntafélag, 1959.
Laxdœla saga. Edited by Kr. Kålund. In *STUAGNL* 19. Copenhagen: S.L. Möllers bogtrykkeri. 1889–1891.
Páll Vídalín. *Deo, Regi, Patriae.* In Um viðreisn Íslands: Deo, regi, patriae. Landsbókasafn / Háskólabókasafn. Reykjavík: Örn og Örlygur, 1985.
Víglundar saga. Edited by Jóhannes Halldórsson. In *Kjalnesinga saga, Jǫkuls þáttr Búasonar, Víglundar saga, Króka-Refs saga, Þórðar saga hreðu, Finnboga saga, Gunnars saga Keldugnúpsfífls*. Íslenzk fornrit 14. Reykjavík: Hið íslenzka bókmenntafélag, 63–116. Þorsteins þáttr uxafóts. Edited by Guðbrandr Vigfússon and C. R. Unger. In *Flateyjarbók*, Vol. 1. Oslo: Malling, 1860, 249–63.
Ǫrvar-Odds saga. In *Fornaldarsögur Nordrlanda eptir gömlum handritum*, edited by C. C. Rafn, Vol. 2, 161–322. Copenhagen: Popp, 1829.

Secondary sources

Brito, Christina. 2024. "'Um Grande Peixe, Dona Baleia da Costa': The Whale in Portuguese Early Modern Natural History." In *Ichthyology in Context (1500–1880)*, edited by Paul J. Smith and Florike Egmond, 389–414. *Intersections: Interdisciplinary Studies in Early Modern Culture*, Vol. 28. Brill.
Frank, Søren. 2022. *A Poetic History of the Oceans: Literature and Maritime Modernity*. Brill.
Glauser, Jürg. 2009. "Saga of Icelanders (Íslendinga sögur) and þættir as the Literary Representation of a New Social Space." In *Old Icelandic Literature and Society*, edited by Margaret Clunies Ross, 203–20. Cambridge University Press.
Hennig, Reinhard. 2020. "Sagas for Sustainability? Commons, Conflict and Cooperation in the Sagas of Icelanders." *Scandia: Journal of Medieval Norse Studies* 3: 36–63.
Hermann, Pernille. 2014. "Islændingesagaer og erindring." *Kultur & Klasse* 118: 245–60.

Keighley, Xénia, Snæbjörn Pálsson, Bjarni F. Einarsson, Ævar Petersen, Meritxell Fernández-Coll, Peter Jordan, Morten Tange Olsen, and Hilmar J. Malmquist. 2019. "Disappearance of Icelandic Walruses Coincided with Norse Settlement." *Molecular Biology and Evolution* 36 (12): 2656–67.

Lassen, Annette. 2019. "Perseverance and Purity in *Flóamanna saga*." *Journal of English and Germanic Philology* 118 (3): 313–28.

Lassen, Annette. 2022a. *Odin's Ways: A Guide to the Pagan God in Medieval Literature*. Routledge.

Lassen, Annette. 2022b. "Moulding Myth and Memory: The Plasticity of Old Norse/Icelandic Traditional Tales." In *Crossing Disciplinary Boundaries in Studies of the Viking Age*, edited by Daniel Sävborg, 277–94. *The North Atlantic World*, 4. Brepols.

Lassen, Annette. 2025. *The Sagas of Icelanders*. Routledge.

Mentz, Steve. 2018. "Blue Humanities." In *Posthuman Glossary*, edited by Rosi Braidotti and Maria Hlavajova, 69–72. Bloomsbury.

Perdikaris, S., and Thomas H. McGovern. 2007. "Cod Fish, Walrus, and Chieftains: Economic Intensification in the Norse North Atlantic." In *Seeking a Richer Harvest: The Archaeology of Subsistence Intensification, Innovation, and Change*, edited by Christopher T. Fisher, and Tina Thurston. Vol. 3, 193–216. Springer.

Schmitt, Carl. (1942) 1981. *Land und Meer: Eine weltgeschichtliche Betrachtung*. Reclam.

12

INVESTIGATING THE ARCTIC ECOTONE

The Arctic Ocean in Peter Høeg's *Miss Smilla's Feeling for Snow*

Hanna Eglinger

The irritating *Miss Smilla*

Peter Høeg's novel *Miss Smilla's Feeling for Snow* (1993)[1] was a Danish and international bestseller at the time and can now be considered a classic of Nordic Noir literature. Who doesn't know Smilla, the recalcitrant half-Danish, half-Greenlandic protagonist with her rough edges, her identity crises plaguing here from luxury-clad Danish high society to Greenlandic social welfare; who doesn't remember the tireless energy with which she, as a self-appointed private detective and "female James Bond" (Heede 2020), first explores the dark criminal underworld of Copenhagen to solve the murder of a Greenlandic boy, and later embarks on a sea voyage that takes her into the Arctic ice and science fiction eco-crime scenarios in order to get to the bottom of her ruthless opponents?

Høeg's novel was a milestone in Nordic Noir literature *avant la lettre* and also in terms of its postcolonial depiction of Arctic land- and seascapes, which was decisive both for contemporary imaginations of the Arctic and for the Western conception of Greenland's colonial heritage and the connection between Greenlandic and Danish culture. While Western ideas of the Arctic stemmed from the heroic polar expeditions of the nineteenth century and their representation in literature, consisting primarily of the Arctic as a remote threatening death zone and the Arctic Ocean as incomprehensibility aestheticised (Ortlieb 2010), *Miss Smilla's Feeling for Snow* manifested a more nuanced picture of the Arctic, characterised by unresolved colonial entanglements and ecological complexity, but also a material richness expressed in the different states of frozen water. In this way, in the basic argument of my chapter, Høeg's novel places a stronger literary focus on the Arctic Ocean as a habitat and site of multispecies encounters, the metaphorical potential

DOI: 10.4324/9781003647638-16

of which is used to represent the (postmodern) complexity of the Arctic Ocean as an ecotone at different levels of meaning. I will examine here in more detail how aesthetic and ethical aspects intertwine to reveal the ecotonic structures of the novel.

The enduring fascination with Høeg's thriller lies in its potential to irritate, which stems primarily from the unresolved contradictions and points of friction that are reflected on the level of character, the level of content, and—in the skilful play across various registers—on the level of genre. The book has been criticised for limiting its colonial critique to a largely clichéd juxtaposition of different cultures that fails to open up a postcolonial "third space" (Thisted 2002, 334), and for Smilla's contradictory character, which offers little to satisfy a psychological reading. And research has emphasised that Smilla's recalcitrance should be understood in the light of a Nordic Noir agenda of criticising social dysfunctions (Stougaard-Nielsen 2017, 140) and the postmodern deconstruction of binary oppositions (Rühling 2002). The mixing of genres, which leads to a dissolution of boundaries between a classic, realistic whodunnit and a dystopian science fiction thriller (Hamm 2025, 42), has also often disappointed one-dimensional expectations. In this chapter, I focus on the literary deployment of the Arctic Ocean in order to offer an interpretation that accords this potential for irritation a central aesthetic and ecological significance.

Arctic literacy

Right from the beginning, we learn how to read traces in the snow. The first-person narrator Smilla explains to investigators (and readers) why the traces left in the snow by the murdered boy Isaiah are conspicuous and why she is sceptical that his death was merely an accident. Throughout the novel, a close connection is developed between reading and detection (cf. Hamm 2025, 39), linking Smilla's semiotic skills with those of the reader. The multilayered reading and interpretation skills that the novel demonstrates in the character of Smilla are therefore also demanded of its readership. In a criminalistic context, Smilla's detective literacy increases the tension and enables her to escape the traps set for her by her adversaries at the last moment.

> It's dark by the time I take the train back. The wind carries a light snow with it, *pirhuk*. . . .
>
> In a big city you adopt a particular way of regarding the world. A focused, sporadically selective view. When you scan a desert of an ice floe, you see with different eyes. You let the details slip out of focus in favour of the whole. This way of seeing reveals a different reality. . . .
>
> With this way of seeing, a person's breath in the cold—that veil of cooled drops that forms in the air in temperatures under 8°C—is not merely a phenomenon

> fifty centimeters from his mouth. It's something all-encompassing, a structural transformation of the space surrounding a warm-blooded creature, an aura of minimal but definite thermal displacement. . . . I sense him when I'm fifty metres away, before he hears me.
>
> (Høeg 1993, 220–21)

Quotes like this clearly illustrate Smilla's Arctic literacy—her ability to read and interpret Arctic natural phenomena ("thermal displacement") and to adapt her way of life and movement accordingly. As a glaciologist, Smilla knows how to name and describe different ice phenomena scientifically and in detail. As a former Arctic resident of Indigenous origin, she also conveys to her readers how she uses the eco-semiotic signs of her surroundings to live within the Arctic environment and move around it with confidence: she can recognise animal tracks down to the rhythm of their movement (1993, 150), smell the proximity of Greenland (1993, 275), and estimate distances in *sinik*, an embodied "concept of space-time" (1993, 278).

We can explain Smilla's Arctic combination of Indigenous environmental and scientific literacy in the light of ecosemiotics: If we understand the way of using local literacy in an ecosemiotic sense of "reading" the land- or seascape (finding one's way around it by interpreting its signs), the literal "reading" (reading a text) is the guiding metaphorical image. Ecosemiotics conceives landscapes or environments as "texts". The "textuality of nature" can thus be understood as an ecosemiotic extension of the material of cultural studies, which understands all products of culture as "texts": as "a basic unit . . . which transports, as a whole, some meaning and function" (Maran 2010, 80). Conversely, we need local anchoring and "embedded and embodied" (DeLoughrey 2019, 163) knowledge in order to open up fictional worlds. Literary story-worlds as "conceived space" refer to "lived and perceived space" (Lefebvre 1991, 38–40), which means that "meaning arises between the sensory and the cognitive in interaction with the external environment" (Brudin Borg 2024, 252). Reading *Miss Smilla's Feeling for Snow* can thus be an ecosemiotic project where reading, detection, and ecoliteracy are strongly connected.

Smilla's Arctic literacy is coined by a multifaceted worldview, which is both rational-scientific and intuitive and embodied. As a first step, therefore, we have to focus on the materiality of the Arctic environment, which, as explained, represents the semiotic prerequisite for Smilla's actions and perceptions.

The material Arctic: resources and aggregates

In Høeg's novel, the Arctic Ocean, with its material treasures, forms the basis and setting for Western greed for profit. The points of contact between capitalist civilisation and the Arctic Ocean are exploitative, extractivist companies such as *Cryolite Corporation Danmark* or the floating oil platform *Greenland Star*, which have an exclusive interest in the material resources of the Arctic. Their hidden presence on the periphery is significant: As Karl Emil Rosenbæk Reetz puts it, "concealment is to be considered one of the main instruments that make the capitalist world-system function" (2026, 84). Smilla's detective journey into the Arctic Ocean thus also serves as an eco-critical project to uncover hidden power structures: As the ship's course towards the Greenlandic Arctic progresses, the Danish-Greenlandic colonial relationship turns into a criminal relationship of capitalist environmental exploitation. The resources range from cryolite and oil to a mysterious meteorite that serves as a particularly obvious example of the capitalist unscrupulous handling of risk. The criminals' greed for profit ignores the fact that the heat of the meteorite in the ice has produced a parasitoid mutation of the Arctic worm *Dracunculus borealis* that poses an uncontrollable danger to humanity.

Closely linked to the material resources of the Arctic Ocean is its capacity to resist, its threatening agency, and its potential to "strike back"—its dimension of the uncontrollable. This dimension is most visibly reflected at the end of the novel, in the still unforeseeable consequences of the deadly parasitoid Arctic worm, whose spread, as a response to human hubris, embodies the ultimate deadly threat. However, the dimension of the uncontrollable is also evident along the way in the various states of aggregation of the Arctic Ocean. In contrast to ice, Smilla informs us,

> I've always been afraid of the sea. . . . One of the reasons I'm fond of ice is that it covers the water and makes it solid, safe, negotiable, classifiable. . . . On the open sea there are no landmarks, there is only an amorphous, chaotic shifting of directionless masses of water that loom up and break and roll, and their surface is, in turn, broken by subsystems that interfere and form whirlpools and appear and disappear and finally vanish without a trace. Slowly this confusion will work its way into the labyrinth of my ear and dissolve my sense of orientation; it will fight its way into my cells and displace their salt concentrations and the conductive power of my nervous system as well, leaving me deaf, blind, and helpless. I'm not afraid of the sea merely because it wants to strangle me. I'm afraid of it because it will take away from me my orientation, the inner gyroscope of my life, my awareness of what is up and down, my connection to Absolute Space.
>
> (Høeg 1993, 231–32)

In contrast to the solid ice surface, the sea represents a sphere of disorientation, which—as a power imbalance—is also the prevailing principle on the ship: The frustration of the disoriented captain, who does not know his destination and, in view of the obscure machinations of the villains, can only ever "navigate by sight", illustrates his powerlessness due to a lack of knowledge and orientation. Thus, the uncertainty of the journey is paralleled by the detection process of the crime story.

Even the ice is uncontrollable. Smilla emphasises the invincibility of the ice in connection with the offshore company's efforts to overcome its power with floating drilling platforms:

> This magnificent harbour is nothing more than a tin tray. If the ice came by, it would carry it off as though it had never been there. . . . . They haven't beaten the ice . . . . The battle hasn't even begun.
>
> (1993, 322)

The agency of the ice becomes particularly visible when it overtakes the ship and comes dangerously close to its crew: The "bad kind of ice . . . comes from the sea fog. It doesn't need any waves, it simply covers everything" (1993, 300) and transforms the deck into "a grey, solidified sea" (1993, 310). But unlike the mystification of this force of nature as an incomprehensible death zone, as has often been the case in Western literary polar tradition, Smilla is more concerned with pragmatically adapting to the circumstances and coming to terms with them: "You can try to live with the ice. You can't fight it or change it or live instead of it" (1993, 365).

Ecosemiotics (mapping vs. navigating)

As explained in the second section of this chapter, certain semiotic skills are essential for Smilla to be able to "live with the ice" (1993, 365). Smilla's Arctic literacy manifests itself in various knowledge systems.

As a glaciologist, Smilla has a petrocritical-activist past that proves her ecoliteracy. Among other things, she has written an article: "Ice Research and the Profit Motive in Denmark in Connection with the Exploitation of Oil Resources in the Arctic Ocean" (1993, 89), which resulted in her expulsion from the Danish Glaciology Society. The novel's ethically motivated cultural criticism is thus also essentially directed at ecological conditions (at a time when the Anthropocene had not yet been labelled as such, i.e. before the ecological wave in research had really begun). Smilla looks back critically on the beginning of her scientific socialisation as a glaciologist and her first successes.

> We . . . proved that the vibrations from the drills wouldn't affect the structure or elasticity of the ice. We were full of pride and scientific pioneer enthusiasm. It wasn't until we were writing the final report, in which we advised A.P. Møller, Shell, and Gospetrol to commence exploiting the Greenland oil reserves from platforms built on the ice, that we realized what we were doing. By then it was too late.
>
> (1993, 323)

Closely linked to this self-admitted colonial and capitalist complicity, to which Smilla has committed herself through her participation in Western scientific culture, is the practice of cartography. As Cecilia Chen emphasises, "[a]mong the technologies employed to support often violent colonial relations were the imported European cartographic practices that imposed an instrumental understanding of waters, lands, and peoples" (2013, 275). There is no doubt that Smilla has mastered this technique of map reading with confidence:

> I only have to look at a map once and the landscape rises up from the paper. It's not something that I learned. Although, of course, I had to acquire a nomenclature, a system of symbols. The ridged elevation peaks on the topographical maps of the Geodesic Institute. The red and green parabolas on the military maps of the ice pack. The discus-shaped, greyish-white photographs of X-band radar. The multi-spectrum scans of LANDSAT 3. The candy-coloured sediment maps of the geologists. The red-and-blue thermal photographs. But in the truest sense it has been like learning a new alphabet. Which you then forget about as soon as you start reading. The text about ice.
>
> (Høeg 1993, 71)

However, this implicit discussion about "learning" and "forgetting" makes it clear that Smilla's Indigenous embodied knowledge not only complements her geographical literacy but also confidently surpasses it. The inadequacy of (scientific) language is repeatedly emphasised: "Reading snow is like listening to music. To describe what you've read is to try to explain music in writing" (1993, 37). When it comes to embodied knowledge, language reaches its limits, as does the European scientific system. With regard to the representation of the Arctic Ocean in *Miss Smilla's Feeling for Snow*, Smilla's embodied combination of knowledge systems moves "away from narrow representations of water as a seemingly tractable resource towards an understanding that waters are situated, lively, and shared" (Chen 2013, 275).

> In North Greenland distances are measured in *sinik*, in "sleeps", the number of nights that a journey requires. It's not a fixed distance. Depending on the weather and the time of year, the number of *sinik* can vary. It's not a measurement of time, either. . . . *Sinik* is not a distance, not a number of days or hours. It is both a spatial and a temporal phenomenon, a concept of space-time, it describes a union of space and motion and time that is taken for granted by the Inuit but cannot be captured by any European everyday language.
>
> (Høeg 1993, 278)

In contrast to mapping the Arctic Ocean as a treasured resource, there is a particular focus on Smilla's orientation in the icescape as an embodied or "situated, lively, and shared" movement. As Chen puts it with reference to Whitridge (2004, 219–26), Indigenous

> "[e]laborate navigational concepts, skills, and technologies for moving through space along tens of thousands of kilometres of traditional travel routes," include the reading of snow and snow drifts, the deciphering of distant land configurations through their luminous reflections in low-lying clouds, orientation by sun, moon, and stars, and various other mnemonic mapping practices.
>
> (2013, 290)

Smilla's innate and instinctive ability to find her way on the ice even in the thickest fog (Høeg 1993, 37–39) is described as "a wild, bright feeling of elation, because I know with absolute certainty which way we should go" (1993, 38). However, the fact that Smilla equates this state with Newton's concept of Absolute Space suggests the proximity and intermingling of the various practices of knowledge. In this way, ecosemiotics appears to be a transitional field between Smilla's Western and Indigenous heritage, between scientific and embodied knowledge, between material and literal signs, and equally important for literary and scientific eco-stories. It is worth noting the role that metaphors (and allegories)[2] play "as a bridge or link between the world and language" (Brudin Borg 2024, 252), but also between literary and scientific narratives. Since allegories or metaphors represent abstract facts through images, they are essential narrative elements that are needed to make scientific facts understandable or vivid.

Material metaphors—metaphorical aggregates

Smilla's Arctic literacy means that her native environment—the Arctic icescape and its climatic characteristics—also provides the metaphorical potential with which she describes everyday phenomena, cultural injustices, and her philosophy of life. Thus, her knowledge of what is, from a scientific point of view, a paradoxical aggregate provides the recipe for "facing adversity in life":

> With instruments sent up in meteorological balloons we measured supercooled drops of water in Thule in the seventies. They live for a short time in very high clouds. The area surrounding them is cold but completely still. In a pocket of motionlessness their temperature will drop to –40°C. They ought to freeze, but they don't; they remain stationary and stable and fluid. That's the way I try to face adversity.
>
> (Høeg 1993, 332)

On the one hand, this paradoxical state between stable and fluid is also characteristic of Smilla's intercultural in-between position: "Smilla's most 'solid' sense of identity is, ironically, a 'fluid' and exclusive performance of her hybrid Danish-Greenlandic heritage" (Stougaard-Nielsen 2017, 141). On the other hand, a connection can also be made in gender-related terms between paradoxical or changeable states of aggregation and Smilla's resistance to conventional attempts at gender classification. In this respect, ice is an ambivalent metaphor that can express Smilla's fluidity between masculine-heroic agency, misbehaviour, and rough feminine charm (e.g. Høeg 1993, 300) as well as her attraction to her adversary Tørk Hviid, who can be seen as the personification of ice (e.g. 1993, 355).[3]

States of aggregation and metaphors are always closely intertwined in *Miss Smilla's Feeling for Snow*. In the opening pages, Smilla describes the crystallisation of ice in its various material properties and with metaphorical associations as "soapy mash called porridge ice", as "free-floating plates, pancake ice", and finally as it solidifies into "a single solid sheet":

> [N]ow the ice will stay, now the crystals have formed bridges and enclosed the salt water in pockets that have a structure like the veins of a tree through which the liquid slowly seeps; which . . . is one reason for believing that ice and life are related in several ways.
>
> (1993, 6)

This quote is notable not only for its metaphors (porridge, pancakes, bridges, and trees) but also for its precise description of the material qualities of ice, which appear as a "soapy mash", "free-floating plates", or "a single solid sheet". The transformation of water into frozen form involves many different material states and forms of aggregation. The changeability, transience, and transformations of water evoke a multifaceted diversity on a metaphorical level as well. The "snow" in the title is significant as an intermediate form between water and ice, but it too takes on many different appearances and forms of expression. One could say that Høeg is perpetuating a colonial stereotype with the idea of the countless Inuit words for snow and ice, but Smilla uses these words in concrete situations and to differentiate between various aggregates and qualitites (*qanik* (1993, 3), *hikuaq, putaaq, maniilaq, apuhiniq, agiuppiniq, killaq, sikussaq* (1993, 364–65)). This creates an impression of change, diversity, and dynamism, which forms an essential basis for Smilla's ecosemiotic as well as our literary reading experience.

In the above quote, the ice is significantly placed in an organic context: "ice and life are related in several ways." It is not only on a metaphorical level that the comparison with the veins of a tree allows ice to emerge as a subject or agent. The material mutability of snow also makes it appear as a living organism: "Even when there's no heat, no new snow, no wind, even then the snow changes. As if it were breathing, as if it condenses and rises and sinks and disintegrates" (1993, 7). The combination of material description, acknowledgement of the impact of time and change, and metaphorical meaning is also found later in the text, where the metaphorical theme of the tree is taken up again:

> In some ways ice is so transparent. It carries its history on its surface. Ice packs, knolls, and slush form when the ice melts and then freezes again. The blend of various ice ages in mosaic ice: the black fragments of *sikussaq*, ancient ice formed in protected fjords, released over time and forced out to sea. Now, in the last rays of the sun, a fine veil of *qanik*, snow flurries, falls from the clouds that the sun has ducked under.
>
> A reed stretches from the white surface straight into my heart. Like an extension of the saltwater tree within the ice.
>
> (1993, 365)

The connection between ice and life, established through the comparison with the saltwater tree, plays a role on various levels in Høeg's novel. Not only is it, as the title suggests, a special characteristic of the half-Greenlandic protagonist, whose "feeling for snow" can be traced back to her socialisation and childhood in the Greenlandic ice, but it also becomes relevant in the ambivalent portrayal of the Arctic Sea as a death zone and a lively habitat. The Arctic is not only a death zone and a threatening crime scene backdrop but also the home and hunting ground of the Inuit, the habitat of the Arctic worm, and a place where different cultures and species meet. In this way, Smilla's investigative journey across the Arctic Ocean and the associated depiction of water, snow, and ice in all their material complexity result in a tense and aesthetically condensed spatial order that can be understood—on a material and metaphorical level—through the concept of the ecotone.

The Arctic ecotone

Various ecocritical research approaches find value in the "idea of the ecotone". Chen refers to hydrological and ecological definitions of the ecotone

> as a "gradient across ecosystems" . . .—a three-dimensional zone of varying shape and thickness where diverse biological communities coexist. Where "eco" is derived from the Greek *oikos*, or dwelling, and "tone" means tension, identifying an ecotone requires that we first recognize the correlation of two or more overlapping systems. Moreover, the vibrant tension of this threshold requires an inclusive approach to mapping that encompasses the conflicts, negotiations, and unpredictable encounters within the tangled overlap—a robust approach that differentiates between layered systems while also recognizing the energies generated by their coexistence. Whereas "ecotone" is a term used more commonly in the physical sciences, it offers, here, a way to understand shores and coasts as

> lively places where land and water meet; it allows us to consider the particularities of encounter between diverse systems, populations, and even practices—of which some may be more land-based, others largely aqueous, and yet others existing only in the complex mingling of both.
>
> (2013, 286)

In Høeg's novel, the ecotone incorporates various systems of knowledge and classification, populations of diverse biological, ethnic, and ethical heritage, and different practices of detection, navigation, and orientation. As an area of "complex mingling" and unexpected twists, the Arctic Ocean in all its different shapes of water, snow, fog, and ice provides not only the ambivalent setting for Smilla's investigations as well as her search for identity but also, as Astrida Neimanis writes about ecotones, "a zone of fecundity, creativity, transformation; of becoming, assembling, multiplying; of diverging, differentiating, relinquishing" (2012, 93).

In purely material terms, the habitat of the parasitoid Arctic worm serves as an exemplary ecotone. It arises from the interplay of different temperatures of Greenland's icy coast, the mysterious meteorite and the seawater, and has biological consequences in the exchange of different species (see especially Høeg 1993, 395–400). If ecotones are described as "places and times that are pregnant with change" (Mortimer-Sandilands 2004, 46), the deadly mutation of the Arctic worm, whose catastrophic dimension manifests itself in contact with the human species, is such a change that has unforeseen consequences—both biologically and ethically. As Neimanis puts it, "an ecotone has a material fecundity that rejects an ontological separation between 'thing' and 'transition', between 'body' and 'vector'" (2012, 93). In this respect, Høeg's novel can be understood as an ethical lesson addressing vulnerability and responsibility in that "we must learn to be at home in the quivering tension of the in-between":

> In-between nature and culture, in-between biology and philosophy, in-between the human and everything we ram ourselves up against, everything we desperately shield ourselves from, everything we throw ourselves into, wrecked and recklessly, watching, amazed, as our skins become thinner.
>
> (2012, 94)

As such an embodiment of the in-between, Smilla follows the trail of criminal machinations into the Arctic Ocean, where thin-skinnedness and vulnerability become increasingly palpable. The ecotone is both the setting and the spatial order of the world depicted: material conditions and cultural knowledge systems intertwine with metaphorical attributions of meaning, all of which are inflected through individual experiences and perceptions. With its various layers of material and metaphorical meaning, the ecotone of the Arctic Ocean is not only a topographical space but a structure permeated by discursive, symbolic, and semantically coded implications.

Conclusion

As readers trained in ecosemiotics, we can take the literary entanglement of materiality and metaphor as a basic prerequisite for thinking about the ecotone as a structural model of the novel. In understanding the Arctic Ocean as an ecotone, the complex structure of voices and positions that the novel weaves together in a literary experimental field can be viewed as a whole. The irritating genre diversity of the book is a recurrent feature of many research contributions. With regard to the postcolonial theme, the novel has been accused of having a static architecture of opposing positions and indirectly perpetuating colonial dualism. Viewed through an ecosemiotic lens, however, the Arctic ecotone as a metaphor and aesthetic structural model offers a reading that reveals and explains the complexity and connectedness as well as the irresolvable nature of the various incompatible voices and positions that Smilla carries within herself. The ecotone thus serves as a model for rethinking identity, natural and cultural environments, space-time embodiment, and narrative.

> The places where cultures, natures, life worlds, experiences, and ideas collide and intermingle are ecotones; the places on the margins where mainstream culture is met and challenged by other forms are particularly fertile with possibilities for change. . . . [E]nvironmental studies is . . . about natural-cultural ecotones: the understanding of human-nonhuman mutuality, conflict, change, influence. These relationships comprise much more than the biophysical impact of one on the other; they concern the ways in which human lives are inextricably and variously intermingled with the lives of the animate creatures and inanimate objects with which we share time and space. Because . . . these relationships are social and cultural as well as biological and physical, good environmental studies needs sociology, history, and literary criticism as much as it does zoology, botany, and geology.
>
> (Mortimer-Sandilands 2004, 46f.)

In Høeg's thriller, ironically, it is the villain who expresses this importance of literary criticism and especially of literature for "possibilities for change" (which in our context means a consciousness of oceanic and environmental connectedness) when he says: "The writers see where we're headed before the scientists do" (1993, 392).

Notes

1 The book was first published in its original form in 1992 (Høeg 1992). Two English translations are available: a British (Høeg 1993) and a US-American version (Hoeg [sic] 1993). I quote from the British version.

2 DeLoughrey emphasises that "allegory and allegoresis are central to theorizing ontologies of the whole. Spatial totalities such as planet and ocean . . . are often only captured

by the allegorical telescoping between the part and the whole, allowing a disjunctive narrative of a partially glimpsed or illuminated figure of alterity" (2019, 135).

3 For a discussion of Tørk Hviid's ambivalent character (as both godlike and a representative of hell), see, for example, Seiler (2023, 90), and of his symbolic name causing associations like whiteness and drought/dryness, see, for example, Rühling (2002, 287).

Works cited

Brudin Borg, Camilla. 2024. "Ecocritical Spatial Analysis Methods." In *Contemporary Ecocritical Methods*, edited by Camilla Brudin Borg, Rikard Windgård, and Jørgen Bruhn, 243–64. Lexington Books.

Chen, Cecilia. 2013. "Mapping Waters: Thinking with Watery Places." In *Thinking with Water*, edited by Cecilia Chen, Janine MacLeod, and Astrida Neimanis, 274–98. McGill-Queen's University Press.

DeLoughrey, Elizabeth M. 2019. "Oceanic Futures: Interspecies Worldings." In *Allegories of the Anthropocene*, edited by Elizabeth M. DeLoughrey, 133–64. Duke University Press.

Hamm, Christine. 2025. "Dangerous Skepticism and the Challenge of Acknowledgment in Peter Høeg's *Smilla's Sense of Snow*." *CLUES. A Journal of Detection* 43 (1) (Spring): 37–47.

Heede, Dag. 2020. "Høeg, Peter: Frøken Smillas fornemmelse for sne." In *Kindlers Literatur Lexikon* (*KLL*), edited by Heinz Ludwig Arnold. J.B. Metzler. https://doi.org/10.1007/978-3-476-05728-0_4576-1.

Hoeg [sic], Peter. 1993. *Smilla's Sense of Snow*. Translated by Tiina Nunnally. Farrar Straus & Giroux.

Høeg, Peter. 1992. *Frøken Smillas fornemmelse for sne*. Rosinante.

Høeg, Peter. 1993. *Miss Smilla's Feeling for Snow*. Translated by F. David. Harvill Press.

Lefebvre, Henri. (1974) 1991. *The Production of Space*. Translated by Donald Nicholson-Smith. Basil Blackwell.

Maran, Timo. 2010. "An Ecosemiotic Approach to Nature Writing." *PAN: Philosophy, Activism, Nature* (7): 79–87.

Mortimer-Sandilands, Catriona. 2004. "The Marginal World." In *Every Grain of Sand: Canadian Perspectives on Ecology and Environment*, edited by J. Andrew Wainwright, 45–54. Wilfrid Laurier University Press.

Neimanis, Astrida. 2012. "Hydrofeminism: Or, On Becoming a Body of Water." In *Undutiful Daughters: New Directions in Feminist Thought and Practice*, edited by Henriette Gunkel, Chrysanthi Nigianni, and Fanny Söderbäck, 85–99. Palgrave Macmillan.

Ortlieb, Cornelia. 2010. "Eismeere: Zur Geschichte eines modernen Phantasmas." In *Das Meer, der Tausch und die Grenzen der Repräsentation*, edited by Hannah Baader and Gerhard Wolf, 123–45. Diaphanes.

Rosenbæk Reetz, Karl Emil. 2026. *Impasse Poetics at the Nordic Oil Coast. Contemporary Petrofiction from Denmark and Norway*. West Virginia University Press.

Rühling, Lutz. 2002. *Opfergänge der Vernunft: Zur Konstruktion von metaphysischem Sinn in Texten der skandinavischen Literaturen vom Barock bis zur Postmoderne*. Vandenhoeck und Ruprecht.

Seiler, Thomas. 2023. *Snøens formler: Et naturfenomen i litteraturen*. Cappelen Damm Akademisk.

Stougaard-Nielsen, Jakob. 2017. *Scandinavian Crime Fiction*. Bloomsbury Academic.

Thisted, Kirsten. 2002. "The Power to Represent: Intertextuality and Discourse in Smilla's Sense of Snow." In *Narrating the Arctic: A Cultural History of Nordic Scientific Practices*, edited by Michael Bravo and Sverker Sörlin, 311–42. Science History Publications.

Whitridge, Peter. 2004. "Landscapes, Houses, Bodies, Things: 'Place' and the Archaeology of Inuit Imaginaries." *Journal of Archaeological Method and Theory* 11 (2): *Recent Advances in the Archaeology of Place*, Part 2: 213–50.

PART IV

The Atlantic Ocean

13

THE "TROLL WHALE" DEMYSTIFIED IN THE NORTH ATLANTIC IN THE EARLY MODERN PERIOD

From Olaus Magnus' map to the witchcraft trials

Ronny Spaans and Liv Helene Willumsen

"Within its swirling waters the vast Ocean presents a wonderful spectacle to every nation", Olaus Magnus (1490–1557) writes in the opening of the book "On Sea-monsters" (book 21), in his famous *Historia de gentibus septentrionalibus* (*A Description of the Northern Peoples*, Olaus Magnus 1998, III, 1081). This work, which was printed in Rome in 1555, was based on his *Carta marina* (1539)—the first modern map of the Nordic countries. Both the quote and the name of the map reveal the incorporation of the sea monster in the North Atlantic. Olaus Magnus points here to the northern part of the Atlantic, which he associates with stories of wonderful whales. The comprehensive map and the detailed book by the Swedish bishop were created to disseminate knowledge about the Nordic region in Europe, but, ironically, these works had in part the opposite effect, at least with respect to Northern Norway. They contributed like no other work to the association of the North Atlantic with dangerous sea monsters (Daston and Park 1998; Van Duzer 2013; Spaans 2024).

We argue that some of these whales go back to an older notion of the "troll whale" in the vernacular. This chapter will show how the North Atlantic underwent a process of mystification through Olaus Magnus's portrayal of this "troll whale" as a marvellous, dangerous being. We then show how the North Atlantic is demonised when the whale is assigned a demonic character in the seventeenth century, before there is a demystification of the "troll whale" and the North Atlantic in the transition to the eighteenth century. We interpret the "troll whale" as an expression of oceanic agency, and in addressing the disrupted agency of this creature, an allegorical understanding of the sea is necessary. In doing this, we lean on critical ocean studies as advanced by scholars such as Elizabeth DeLoughrey, who notes that "[b]ecause the immensity of the ocean cannot be fathomed, allegorical figures of the micro- and macrocosmos become integral to representation" (2019, 139).

DOI: 10.4324/9781003647638-18

In our study, the troll whale becomes an agent of the unknown at the outer limits of the Atlantic Ocean. In other words, the sea monsters at the edge of the maps represent the barbaric and uncivilised, but at the same time, they attracted scientific fascination and curiosity. At the end of the early modern period, the demystification of the "troll whale" and the North Atlantic in general, can in turn be can be linked to the Protestant Christianity of Denmark-Norway, a point we will return to in the conclusion.

We scrutinise the threefold change in the semantic meaning of the term "troll whale" through widely different genres. Firstly, through a close reading of Olaus Magnus's works, and through a study of the text that was used as a basis for his works, namely Erik Valkendorf's description of Finnmark. Secondly, we examine the occurrence of "troll whale" in other sources in early modern Nordic literature, in both maps and illustrations, and in topographical literature, such as the works of Erik Valkendorf, Erik Hansen Schønnebøl, and Peder Claussøn. Occasionally, we include references to authors outside the Nordic countries, such as Sebastian Münster and John Milton. In addition, we read seventeenth-century court records of witchcraft trials.

Our examination is motivated by the semantics of the ambiguous term *troll*. It refers to the giant of folklore: at the same time, however, we recognise the word in *trolldom*, that is, "witchcraft", meaning evil, the performance of harmful deeds. "Troll whale" is in the early modern period, therefore, translated with terms such as *Teuffelwal* (Devil Whale) in German, and *Witch-Whale* in English.[1] This makes us curious about references to whales in court records of witchcraft trials. Thirdly, at the end of the chapter, we show how the concept "troll whale" loses its meaning denoting a dangerous creature and becomes associated with a specific type of whale, the sperm whale. We link this to the changing understanding of the sea, from a space of unknown wonders, to a space of malevolent agencies, to, finally, a space of scientific exploration and new knowledge. We discuss allusions and references to the whale in the poetry of Petter Dass as well as in topographical literature and natural history of the eighteenth century.

Wonderful and unknown seas: the first mention of the "troll whale"

The sources of Olaus Magnus's book on sea monsters are extensive. The Swedish Catholic bishop bases his stories of marvels on the church fathers, and classical sources such as Pliny the Elder.[2] However, as a bishop, he had travelled in the Nordic region, although not to Northern Norway, and relied on first-hand accounts and travelogues from his contemporaries. He himself mentions Norwegian fishermen as informants in the preface of his book (1998, III, 1082). With the publication of the 1539 map, a booklet was published, explaining the details of the map. On this map, off the coast of Iceland, two large whales are seen chasing a ship. Nearby, a similar whale is shown attached to a ship through the ship's anchor, and we also see people standing on the whale's back believing it to be an island (Van Duzer 2013, 48). The German version of Olaus Magnus's booklet, *Ain kvrze avslegvng vnd verklerung*, where these scenes are numbered with the letters K and L, describes and gives the name of this type of whale: "troll whale, that is Devil Whale in German" (Olaus Magnus 1539, fol. Aiiv).

Olaus Magnus's source to these scenes is provided in chapter V of book 21, "On the fearful monsters off the Norwegian coasts", in his *Historia de gentibus septentrionalibus*. Olaus Magnus tells of a letter written by Erik Valkendorf, archbishop of Trondheim in Norway, and sent to Pope Leo X in Rome about the year 1520: *Breuis et summaria descriptio . . . Findmarkia . . .* (*Short and concise description of . . . Finnmark . . .*). In paragraph 7 in Valkendorf's letter, we find the first textual mention of a "troll whale":

> In the sea which surrounds Finnmark live innumerable animals, and of marvellous size; among them are great whales, which in our language are called *Trollhval*, some 200, some 300 cubits long, immensely thick, and of unusual shape. A great multitude of these fish had once on this my journey nearly stopped and killed me.
>
> (Walkendorf 1901, 8)

FIGURE 13.1 An engraving of Olaus Magnus's *Carta marina* (1572) © National Library of oslo.

Valkendorf also writes that the "troll whale's" head is encircled by pointed horns. Contrary to his booklet, Olaus Magnus does not include the name "troll whale" in *Historia de gentibus septentrionalibus*. However, each chapter in "De piscibvs monstrosis" (book 21) contains a vignette illustrating the phenomena he describes. In some of these vignettes, we recognise the whale from Olaus Magnus's explanations of the letters K and L. In his explanation of letter L in the booklet, Olaus Magnus refers to the myth of the whale as an island, enticing the sailors to anchor on its back and thus exposing themselves to mortal danger. In *Historia de gentibus septentrionalibus*, the Swedish bishop elaborates on this myth in chapter 26 of book 21 (1998, III, 1104). We can also recognise the "troll whale" in the illustration accompanying chapter 32 (1998, III, 1115–16). In this text, Olaus Magnus dwells on the dangers of sea travel and gives advice on how to chase away dangerous whales by throwing barrels into the sea, and blowing a trumpet, in accordance with his explanations in the booklet. We also notice the pointed horns around the head of many monsters in the illustrations, see especially the text and illustration of chapter 5 (1998, III, 1086–87). It is therefore likely that the "troll whale" forms a foundation for some of Olaus Magnus's sea monsters.

The idea of the North Atlantic whale as a horrifying creature is spread to other books, and this time, the name *trolual* is included. We see this, for example, in Sebastian Münster's plate "Monstra Marina & Terrestria" in his *Cosmographia* (1572, 1004–5). Here, the German translation of the word is repeated: *Teufelual*. Münster's book went through many editions and exerted great influence.

The mystification of the sea has its origins in the scientific theories of this period. As Lorraine Daston and Katharine Park (1998) emphasise, *curiosa* and *mirabilia* in the renaissance science include emotions such as astonishment and even horror. There is no shortage of that in Olaus Magnus's works. The North Atlantic monsters "cause spectators to tremble and anyone who gazes longer at them grows terrified and numb" (1998, III, 1086) (Daston and Park 1998, 14–16, 365–66). During the same period, indigenous peoples in the High North are portrayed as beings more akin to animals than humans. The Sámi are depicted as people with a bear-like nature (Simek 2016). Like the terrestrial space, the North Atlantic is filled with mystery and danger. The North Atlantic thus became both a place of repulsion and attraction. Sailors steered clear of it for fear of sea monsters, while scholars were attracted by the mysteries of the sea. For these scholars, who were mostly based in Southern Europe, the mystical North Atlantic was therefore the perfect frontier geographically as well as epistemologically.

Troll—giant or devil?

There are two more important sources for the term "troll whale" originating in Norway, firstly a topographical description of Lofoten and Vesterålen by the bailiff Erik Hansen Schønnebøl (1535–1595), and secondly in the writings of the

FIGURE 13.2 Sebastian Münster's plate "Monstra Marina & Terrestria". The troll whale is depicted at the top left and at the bottom. From the German translation of *Cosmographia,* © State Library of New South wales.

clergyman Peder Claussøn (1545–1614). Schønnebøl's *Enn saare kortt doch sandru beskriuelse om Lofoten, Vestraallen . . .* (A Short but true Description about Lofoten, Vesterålen . . .) (1591) was not printed until modern times, but copies circulated among scholars. Schønnebøl lists different kinds of whales that gather at the coast, among them the troll whale. Some of them were dangerous, especially those overturning boats. Peder Claussøn is the first significant topographical author of Norwegian origin. He wrote *Om Diur, Fiske, Fugle oc Trær udi Norrig* (About Animals, Fish, Birds and Trees in Norway) (1599). In this manuscript, he mentions among the "troll whales": they are "angry, ugly and harmful fishes", which "never get full or satiated by manslaughter" (1881, 64). At the end of this chapter, Peder Claussøn writes that all these whales are described in the Old Norse work *Konungs skuggsjá* (*King's Mirror*). Whereas Olaus Magnus situated the whale in humanist culture, with its roots in antiquity and Catholic legends, the Norwegian priest gives the whale a Norse origin. But if we look in *Konungs skuggsjá*, we do not find the term "troll whale". It is an early modern invention. During the same period, the "troll whale" was transferred to other works through Peder Claussøn's *Norriges Bescriffuelse* (*A Description of Norway*), which became a model for similar historical-topographical descriptions and was translated into both Latin and German.

Through Schønnebøl and Peder Claussøn, the term spreads in Danish-Norwegian culture. There are also other occurrences of the term in written culture that seem to come directly from the vernacular. An example is the discussion of the "troll whale" in Lucas Debes's *Færoe et Færoa reserata* (1673, 184). In these books, the whale is mainly presented in the same ideological historical context as in Olaus Magnus: the stories display a fascination for *mirabilia* and *curiosa* on the outskirts of Europe. However, the English translation of Debes *Faroa et Færoa Reserata* (1676) is of particular interest in this context, as the translator introduces yet another sphere of ideas in his translation of the word: He translates "troll whale" as "the Witch-Whale" (1676, 184).

In the early modern period, "troll" was not a much-used word in English and other languages. We see the word used in the folklore of Orkney and Shetland in the form of *trows* (Marwick 2005, 32–39, 162–63). It is not until the nineteenth century that the term is in frequent use to mean "giant", as we know it from fairy tales and fantasy literature. However, as mentioned, "troll" also evokes associations with *trolldom*, "witchcraft", *trollquinde*, "female witch", and *trollkarl*, "male witch". We saw this in the translations "Devil whale" and "Witch-whale". What is really the meaning behind *troll* in "troll whale"? That is a question addressed in another source from the early modern period. In 1593, the Icelandic scholar Arngrímur Jónsson (1568–1648) published *Brevis commentarius de Islandia*, in which he criticises the works of numerous authors who have written what he believes to be false about his home country. He attacks Münster's understanding of the "troll whale". We quote the English translation from 1599: "Neither doeth *Troll* signifie the deuill, as you interprete it, but certain Giants that liue in mountains" (Arngrímur Jónsson 1599, 568), thus the trolls in the folklore.[3] In the letter of Valkendorf, and in other works of natural history and topography that we have discussed so far, the "troll whale" must primarily be understood in accordance with this definition. In "Monstra Marina & Terrestria", by far the largest whale is a "troll whale". It is portrayed as frightening and dangerous, but not demonic.

However, there are some sources that contradict Arngrímur Jónsson. That brings us closer to witchcraft. We refer here to another Icelandic book, *Píslarsaga* (*Passion Story*) (1658–1659) by the Icelandic Reverend Jón Magnússon. In this defence document, Jón Magnússon deals with the trials that befell him in 1655 when he fell ill due to witchcraft, as he claimed. The author mentions two witches who were burnt to death. He praises God that these *tröllhveli* were caught. "Troll whale" is here used as a metaphor for a witch (Jón Magnússon 1967, 119).[4]

Oceanic malevolence: the whale in witchcraft trials

This introduces us to the witchcraft trials in early modern Europe. In two trials in landscapes bordering the North Atlantic, references to the concept of a dangerous whale are found, but without including the first part of the word, "troll".

The meaning of the "troll whale" undergoes a change in the seventeenth century. The wonderful, but dangerous, whales of the North Atlantic had new qualities attributed to them. Due to the learned European doctrine of demonology—the intellectual study of demons and their powers—the whale is turned into an evil creature, a tool used by the Devil's ally—the witch. The whale is no longer an agent acting on its own but is used by a witch to perform *trolldom*, "witchcraft". According to demonology, a person could get access to witchcraft power by entering into a pact with the Devil. The demonological concept also included the collective performance of witchcraft and witches' gatherings. And here the whale, and by extension the North Atlantic, comes into the picture, due to another demonological element, shape-shifting. A woman interrogated during a witchcraft trial might tell that she used shape-shifting in order to reach a gathering place or to perform collective witchcraft. In accused persons' confessions, we hear about shape-shifting into animals, birds, and whales, the latter to overturn boats and cause shipwrecks (Willumsen 2013, 123, 210, 212, 254).

A number of demonological works were published in Europe from 1400s till 1600s.[5] The demonological doctrine was formed as a narrative. The ideas spread quickly from learned treaties to the mentality of peasants because of narrative structures, easy to learn and retell by people who knew how to tell a story.

In early modern Europe, there was an understanding that witchcraft and the North, where the Sámi people were living, were related. In Olaus Magnus 1555 volume, witchcraft and Sámi magic are connected. In the late 1500s, additional texts known within learned Europe mention witchcraft related to Lapland. William Shakespeare used the expression "Lapland sorcerers" in his play *The Comedy of Errors* (1594) (Page 1965, 215–32). The Scottish King James I (later James I of England) published in 1597 in London a demonological treatise, *Dæmonologie in Forme of a Dialogue*. The dialogue is a conversation between Philomathes and Epistemon. Epistemon is asked what the cause is "that this kind of abuse is thought to be most common in such wild partes of the worlde, as Lap-land, and Fin-land, or in our North Iles of Orknay and Schet-land". The answer by Epistemon: "Because where the Deuill finds greatest ignorance and barbaritie, there assayles he grosseliest" (James I 1597, 69).

Finnmark, the northernmost district of Norway, suffered under intense witchcraft trials during the 1600s, and information about these trials was widely spread. With regard to ethnicity, a few Sámi people were accused of witchcraft during the first two decades of the 1600s. However, during the rest of the Finnmark witchcraft trials, the bulk of persons accused and executed were Norwegian women (Willumsen 2024, 238–47). Ludvig Holberg asked in his comedy *Uden hoved og hale* (1725, act 1, scene 5), why the Devil would choose Vardøhus, the King's Fortress bordering the North Atlantic, as his residence. Thus the author underlined the link between witches, troll whales, and the geographical space of the North Atlantic.

We will now turn to two witchcraft trials wherein the whale is mentioned, not as a self-independent and wonderful whale in the tradition of Olaus Magnus, but as a creature steered by an alleged witch. One trial took place in Shetland and one in Finnmark. The Shetland case was the trial of Marion Pardoun in 1644. As part of the indictment, it is recorded that she

> be your said wicked, detestable, abhominable and divilish airt of witchcraft, being transformed in the lyknes of ane *pellock quhaill* . . . ye did cum under the said boat and overturne hir with thame, and drowned and devoirit thame in the sey.
>
> (Goodare & Willumsen 2025, 292)

The whale could overturn a boat. The closeness to the sea gives potential solutions when it comes to choice of shape of a creature as a tool for evildoing. However, the image of Olaus Magnus's magnificent whale has faded. The sublime sea creature is reduced both in size and in its operation of independent will and power. His shape is useful, not to invoke wonder, but as an instrument of demonic agency. However, still, he is feared.

The Finnmark case is the trial of Karen Jonsdatter in 1654.[6] She came from the village of Andenes in Nordland and confessed that she and her stepmother cast spells on two boats near her home village. At the time, she was shaped as a seal, and her stepmother was in the likeness of a whale. In her confession, the whale is not specifically named as a "troll whale" but has some qualities in common with the phenomenon. In the court records, the whale is still portrayed as dangerous. However, because a suspected witch has taken control, the self-confident whale is overpowered by malevolent forces. The fear of the witch's whale is the fear of the Devil's power. One of the sea's wonders has been drawn into the ongoing fight between God and Devil, and on the wrong side.

The two witchcraft trials show a whale overtaken by ideas of allegedly evil people. The demonological idea that a human being can obtain witchcraft from the Devil and use it to shape-shift has tipped the balance between nature's creation and human individuals; demonic power channelled through humans has trumped the unknowable wonders of nature, in the shape of a huge whale. Olaus Magnus says, with reference to the power of minute creatures, that "the presumptuous pride of humanity is mocked" (1998, III, 1117). In the witchcraft trials, the whale has lost its authority and become a handy vehicle for a witch. The "wondrous God", who should be "glorified through His creatures" (Olaus Magnus 1998, III, 1117), has competition from the Evil One. With the new understanding of sea creatures during the witchcraft trials, the North Atlantic becomes an ocean that the Devil can control.

The knowable ocean: the "troll whale" disappears

Fascination for the "troll whale" reaches a peak in poetry in the seventeenth century; by the end of the century, references to the sea creature have started to dry up.

We find an interesting rewriting of the term in the poetry of the Norwegian poet Petter Dass (1647–1707). But before we get that far, two authors of epic poetry will be mentioned who devote great space to the sea monster. The "troll whale" appears several times as a reference in *Hexaëmeron* (written between 1631 and 1637 but published in 1661) by Anders Arrebo (1587–1637), the father of Renaissance poetry in Danish-Norwegian literature. *Hexaëmeron* is inspired by *La première Semaine* (*The First Week*) by Guillaume de Salluste Du Bartas. Arrebo was one of Erik Valkendorf's Protestant successors to the bishopric of Trondheim, from 1618 to 1622. He included observations of the far north in his book, but although he occasionally relies on first-hand accounts, much of his depictions of Northern Norway are taken from Schønnebøl's description of Lofoten.

In his chapter about God's creation of fish and birds, Arrebo revels in Norwegian fauna. The wicked species of fish in the Atlantic exist to evoke reverence for the Creator. The "troll whale" is therefore presented in the same passage as the torpedo fish and Greenland shark: "You cruel troll-whale, you great Fish-King, / On the Atlantic vastness, you close the eyes of many / . . . So big and long you are, so cruel and evil in spirit" (Arrebo 1965, I, 192).

The best-known example of a link between sea monsters and the Norwegian coast is probably found in *Paradise Lost* (1667) by John Milton, the poem that explores the biblical story of the Fall of Man. In the scene where Lucifer is cast down from heaven by God, the fallen angel compares his situation, where he is lying in a sea of flames, with other monsters in the sea, as the Bible's Leviathan. Although he does not mention the "troll whale" by name, we recognise Milton's Leviathan from Münster's illustration—in the combination of Valkendorf's "troll whale", and with the myth of the whale as an island. In Milton's poem, a culmination of different definitions of the "troll whale" is formed: Olaus Magnus's Atlantic as the home of the marvellous is combined with the seventeenth-century demonisation of sea monsters.

Between 1670 and 1700, Petter Dass wrote his versified topographical description of Nordland, *Nordlands Trompet* (*The Trumpet of Nordland*), which was first published in 1739. In some parts of the poem, he uses Arrebo as a model. When Dass imitates Arrebo's apostrophe to the whale that I quoted earlier, he makes a small but significant change: "You grim Sea-King! you Troll in your Water!" (1763, 55). Dass rewrites the passage wherein "troll whale" occurs, so that "troll" simply becomes a metaphor for a "big whale".

Why this change? It is worth taking a closer look at Petter Dass. Dass has much in common with Arngrímur Jónsson: He was a priest of the parish Alstahaug, in Northern Norway, living close to the whales that the Renaissance writers, from their homes on the continent, wrote their imaginative stories about. As in *Brevis commentarius de Islandia*, *Nordlands Trompet* can be understood as a defence of the north, although it is not as polemical in tone as the Icelandic book. A key to understanding the intention of this book is expressed in the preface. Here, Dass reveals a critical source awareness which has been characterised as unique for its time. When it comes to phenomena in his home region, Helgeland, he only describes what he has seen with his own eyes, while observations from further north are from reliable informants (Spaans 2024).

Even so, the "troll whale" does not completely disappear from the North Atlantic. There are still references to this whale in the eighteenth century, but the term is filled with a new, secular content. A book that marks a departure from older natural history representations of Norway and the surrounding seas is *Det Kongerige Norge* (*The Kingdom of Norway*) (1763). It was published under the name of Erik Johan Jessen, but the book's real author is Hans Steenbuch. The first part of the book contains a critical review of both older mapping and textual depictions of Norway (Jessen-Schardebøll 1763, 1–136). With Hans Steenbuch comes the final reckoning of the sea monsters on Olaus Magnus's map:

> The most amusing thing, however, is that in the margin of the map he depicts a great heap of terrible sea serpents, whales and other strange animals and creatures to be found in the North Sea. . . . Such more strange things were those for which the Nordic kingdoms and especially Norway were famous at that time.
> (Jessen-Schardebøll 1763, 20–21)

A similar reckoning with the topographies of earlier times can be found in Knud Leem's *Beskrivelse over Finmarkens Lapper* (*Description of the Finnmark Sámi*, 1767). Leem also discusses the witchcraft trials, where he among other things mentions witches who claimed they shape-shifted into whales, but dismisses it all as "phantasy, delusion and dreaming" (1767, 463). In a chapter on fishing, where he discusses the types of whales in the North Atlantic, he includes the "troll whale", stating that "he is a whale fish overgrown with scales, and, according to reports, can be heard under the water with a terrible roar" (1767, 299). Leem's text is equipped with scientific notes by the naturalist Johan Ernst Gunnerus, the bishop of Trondheim, thus the successor of Valkendorf and Arrebo, and who was a renowned scientist. Gunnerus gives a more critical definition of the sea creature: "troll whales are otherwise understood to mean all large powerful whales that look ugly; just as all ugly fish and birds are called troll-fish and troll-birds" (Leem 1767, 299). We note that Gunnerus resorts to a third interpretation of *troll*, now meaning "ugly", probably based on the scales on the whale's skin. During this period, references to "troll whale" are linked to the sperm whale, *Physeter macrocephalus* in the taxonomy of Carl Linnaeus.[7] The same bishop delivers a reckoning with the belief in sea monsters in his article "Critiske Tanker om Kraken, Søeormen og nogle flere Vidunder i Havet" (Critical Thoughts on the Kraken, the Sea Serpent and Some Other Monsters in the Sea, printed posthumously in 1774). Nevertheless, sea monsters do not totally disappear. Even though the troll whale is mentioned less often, interest in the sea serpent and the giant squid kraken persists. This interest continues thanks to the well-known bishop Erik Pontoppidan's fascination with sea monsters. But they are disenchanted and seen more as large zoological creatures than mysterious beings.

Conclusion

The semantic and cultural development of the term "troll whale" can be interpreted in a cultural and ideological light. The project of Olaus Magnus, who launched his book and map of the Nordic countries with the aim of enlightening Europe, had a negative side effect: The North Atlantic was popularised as the home of fantastic but deadly sea creatures. His physical distance from the High North governs Olaus Magnus's understanding of it; he lived after the Reformation in Italy, belonging to a learned, bookish Renaissance culture, with no opportunity for direct observation of the coasts and seas of northernmost Europe. The North Atlantic was not only filled with stories of horror and terror that kept sailors and travellers away but also became a space that scholars in Southern Europe could fill with their scientific imaginations.

Peder Claussøn's view of the North Atlantic is also characterised by his distance from Northern Norway, as he lived in Southern Norway and differentiated between southern (harmless) and northern (dangerous) whales. You would expect Erik Schønnebøl and Lucas Debes, who lived along the coast of the North Atlantic, to give a more realistic view of the Norwegian Sea and its sea creatures. However, they were probably still too strongly influenced by a European culture that focused on wonder as a guideline for scientific endeavour.

In the seventeenth century, the view of the North Atlantic is affected by the spread of European witchcraft trials and the accompanying change in belief about magic. In European Christianity and jurisprudence, the demonological doctrine included the idea of shape-shifting: witches transforming into deadly whales. Now the North Atlantic becomes an ocean over which the Devil has power.

This shared scholarly and folkloric idea of the "troll whale" lost ground during the seventeenth century and towards the Enlightenment. But among fishermen and farmers, the magical understanding of nature and the creatures of the world is preserved. For example, "trolls" have remained in the folk tradition, but in the mountains, not in the sea. As is well known, the emphasis is on land-based peasant culture and its folk traditions when Norwegian culture is established during Romanticism in the nineteenth century. Coastal culture, with its "troll whales" and the "kraken", becomes marginalised. Bishop Gunnerus's Christian worldview, based on eighteenth-century ideals, involves a Christianisation of the North Atlantic. The Enlightenment's sense of scientific systematisation also led to the "troll whale" being redefined as a sperm whale. As a result, the North Atlantic loses its mystique and is nationalised for Danish-Norwegian maritime traffic.

Two names stand out in this chapter, the Icelander Arngrímur Jónsson and Petter Dass, who lived in Northern Norway. Their writings, which are polemical to varying degrees, criticise the Central European definition of the North Atlantic. As Elizabeth DeLoughrey points out (2022, 149), Critical Ocean Studies is about who knows and navigates the sea. Priests Arngrímur Jónsson and Petter Dass both reveal prejudices associated with sea creatures in the North Atlantic. *Nordlands Trompet* also has an ecocritical message, but about agriculture in the interior. The author nevertheless resorts to a maritime metaphor here. When the land is plagued by farmers, "they sink ships to the bottom, / Where they carry more than they can carry" (1763, 117). Petter Dass would likely have supported today's calls for a responsible harvesting of marine resources.

Notes

1 Cf. https://kalkarsordbog.dk/ordbog?query=troldehvalv. A search for *troll* in Scandinavian dictionaries yields several meanings, but our chronology of word meanings is based on historical sources referring to "troll whale".
2 See John Granlund's comments in Olaus Magnus 1998 (III, 1140–56). See also Nigg 2013. See also Szabo 2008 for the representation of the whale in pre-modern times in the North Atlantic world, but the "troll whale" is not mentioned in this work.
3 We quote the English translation: Arngrímur Jónsson 1599, 568. We thank Jon Gunnar Jørgensen for this reference.
4 Jon Magnússon uses a derivation of the word with the same meaning. Jc. also Svanberg 2005, 265–76.
5 One of the first being Johannes Nider's treatise *Formicarius*, written and circulated at the Council of Basel in 1437 and published in Cologne in 1475 (Rummel 2020, 65–85). The main demonological work is *Malleus Maleficarum* (*The Hammer of Witches*), published in Speyer in 1486, a manual in detection and interrogation of alleged witches (Kramer and Sprenger 2009).
6 Regional State Archives of Tromsø, The Archives of Finnmark District Magistrate, no. 8, Records of Court Proceedings 1648–1654, fos. 132v–137v.
7 See the definition of the word in Norwegian dictionaries.

Works cited

Arngrímur Jónsson. 1599. *A Briefe Commentary of the True State of Island*. London: George Bishop, Ralf Newberie and Robert Barker.
Arrebo, Anders. 1965–1984. *Samlede skrifter*. Vol. I–V. Edited by Lundgaard Simonsen. Copenhagen: Munksgaards forlag.
Dass, Petter. 1763. *Beskrivelse over Nordlands Amt*. Edited by Albert Dass. Copenhagen: Hartvig Godiche.
Daston, Lorraine, and Katharine Park. 1998. *Wonders and the Order of Nature, 1150–1750*. New York: Zone Books.
Debes, Lucas. 1673. *Færoe et Færoa reserata*. København: Matthias Jørgensen.
Debes, Lucas. 1676. *Færoæ & Færoa reserata*. Translated into English by J. S. London: William Iles.
DeLoughrey, Elizabeth. 2019. *Allegories of the Anthropocene*. Durham: Duke University Press.
DeLoughrey, Elizabeth. 2022. "Mining the Seas: Speculative Fictions and Futures." In *Laws of the Sea: Interdisciplinary Currents*, edited by Irus Braverman. New York: Routledge.
Goodare, Julian, and Liv Helene Willumsen. 2025. *Scottish Witchcraft Trials*. Suffolk: Scottish History Society.
Gunnerus, Johan Ernst. 1784. "Critiske Tanker om Kraken, Søeormen og nogle flere Vidunder i Havet." In *Nye Samling af det Kongelige Norske Videnskabers Selskabs Skrifter* 1. Copenhagen: Christian Gottlob Profts Forlag.
Holberg, Ludvig. 1725. "Uden hoved og hale." In *Hans Mikkelsens Comoedier Sammenskrevne for Den Danske Skue-Plads. Tredie Tome*. Copenhagen: J.J. Høpffner.

James, I. 1597. *Dæmonologie in Forme of a Dialogue*. Edinburgh: Robert Walder-graue Printer to the Kings Majestie.

Jessen-Schardebøll, Erik Johan. [Steenbuch, Hans]. 1763. *Det Kongerige Norge*. Copenhagen: Gottmann Friedrich Kisel.

Jón Magnússon. 1967. *Píslarsaga síra Jóns Magnússonar*. Reykjavik: Almenna Bókafélagið.

Kramer, Heinrich, and Jacob Sprenger. 2009. *The Hammer of Witches: A Complete Translation of the Malleus Maleficarum*. Translated by Christopher S. Mackay. Cambridge: Cambridge University Press.

Leem, Knud. 1767. *Beskrivelse over Finmarkens Lapper*. Copenhagen: G.G. Salikath.

Marwick, Ernest. 2005. *The Folklore of Orkney and Shetland*. Edinburgh: Birlinn Limited.

Milton, John. 1667. *Paradise Lost*. London: Peter Parker.

Münster, Sebastian. 1572. *Cosmographia*. Basel: Heinrich Petri.

Nigg, Joseph. 2013. *Sea* Monsters. Chicago: University of Chicago Press.

Olaus Magnus. 1539. *Ain kvrze Avslegvng vnd Verklerung der neuuen Mappen von den alten Gœttenreich vnd andern Nordlenden* [. . .]. Venice: Olaus Magnus Gotthus.

Olaus Magnus. 1998. *A Description of the Northern Peoples*. Vol. I–III. Edited by Peter G. Foote. London: The Hakluyt Society.

Page, Raymond Ian. 1965. “Lapland Sorcerers.” In *Saga-Book of the Viking Society 1962–65*, Vol. 16. London: University College London.

Peder Claussøn [Friis]. 1881. *Samlede Skrifter*. Edited by G. Storm. Kristiania: A.W. Brøgge.

Rummel, Walter. 2020. “ ‘I Confess That I Have Been Ignorant’: How the *Malleus Maleficarum* Changed the Universe of a Cleric at the End of the Fifteenth Century.” In *Demonology and Witch-Hunting in Early Modern Europe*, edited by Julian Goodare, Rita Voltmer, and Liv Helene Willumsen. Abingdon: Routledge.

Schønnebøl, Erik. 1895. “Enn saare kortt doch sandru beskriuelse om Lofoten, Vestraallen.” In *Historisk-topografiske Skrifter om Norge og norske Landsdele, forfattede i Norge i det 16de Aarhundrede*, edited by Gustav Storm. Christiania: A.W. Brøggers Bogtrykkeri.

Simek, Rudolf. 2016. “Monstra septentrionalia: Supernatural Monsters of the Far North in Medieval Lore.” In *Imagining the Supernatural North*, edited by Eleanor Rosamund Barraclough, Danielle Marie Cudmore, and Stefan Donecker. Edmonton: University of Alberta Press.

Spaans, Ronny. 2024. “ ‘My Eyes Have Never Yet Beheld Him’: Demythologising Arctic Sea Monsters in the Poetry of the Norwegian Priest and Fish Merchant Petter Dass.” In *Ichthyology in Context (1500–1880)*, edited by Paul J. Smith and Florike Egmond. Leiden: Brill.

Svanberg, Ingvar. 2005. “ ‘Du grumme Søe-konge, du trold o det Vand’: Troldhvaler i norrøn tradition.” In *Fólaleikur: Heidursrit til Jóan Paula Joensen*, edited by Andras Mortensen. Tórshavn: Føroya Fróðskaparfelag.

Szabo, Vicki E. 2008. *Monstrous Fishes and the Mead-Dark Sea*. Leiden: Brill.

Van Duzer, Chet. 2013. *Sea Monsters on Medieval and Renaissance Maps*. London: British Library.

Walkendorf, Erik. 1901. *Finmarkens Beskrivelse: brev til Pave Leo den 10de*. Edited by K. H. Karlsson and G. Storm. Kristiania: Det Norske Geografiske Selskab.

Willumsen, Liv Helene. 2013. *Witches of the North*: Scotland and Finnmark. Leiden: Brill.

Willumsen, Liv Helene. 2024. “Kjønn og etnisitet: Trolldomsprosessene i Finnmark, 1600–1692.” In *Steilneset minnested og kvinnedrap*, edited by Reidun Laura Andreassen and Liv Helene Willumsen. Stamsund: Orkana.

14

SEXUALITY AT SEA

The challenge of entanglement in Amalie Skram's *Betrayed* (1892)

Christine Hamm

Within Norwegian literature, Amalie Skram occupies a prominent position as one of the most significant writers of the nineteenth century. In her four novels focusing on marriage—*Constance Ring* (1885), *Lucie* (1888), *Fru Inés* (1891), and *Forrådt* (*Betrayed,* 1892/2018)—she offers a powerful critique of the oppression of women in bourgeois, patriarchal society. Her tetralogy *The People of Hellemyr* (1887–1898) explores the unrelenting misery of a family striving for social advancement. Returning from an involuntary psychiatric hold at a mental asylum, Skram published *Professor Hieronimus* (1895) and *At St. Jørgen's* (1895), which depict the treatment of a vulnerable female artist at the hands of a male psychiatrist. Through her detailed, often grotesque descriptions, Skram is regarded as a leading exponent of naturalism in Norwegian literature (Engelstad et al. 1988, 197; Andersen 2011, 269). Scholars have shown that many of her works may be seen as naturalist experiments; in accordance with Émile Zola's recommendations in *Le roman expérimental* (1881), Skram situates her characters within specific social and environmental contexts, observing their behaviour as shaped by biological and social inheritance (Beyer and Beyer 1970, 217; Engelstad 1992; Hegdal 1996; Hamm 2006). Most of her novels culminate in decline and death.

In *Betrayed*, Skram tells the story of the young and innocent Aurora Ingstad, who marries Captain Adolph Riber, a man fifteen years her senior and far more experienced. Aurora, affectionately known as Ory, has grown up in a sheltered bourgeois milieu, shaped by a typically Norwegian strict Protestant moral code. Riber, in contrast, belongs to Norway's socially significant class of tradesmen who, at the time, transported goods across seas in barks and ships.[1] After marrying Aurora, Riber begins to regard himself as a "sinner", hoping that he might be "purified" through his union with his virtuous wife. Aurora, however, is simultaneously repelled and fascinated upon learning of her husband's sexual past. During their first journey

DOI: 10.4324/9781003647638-19

together—a sea voyage from London to Florida Bay—Aurora becomes obsessively curious, subjecting Riber to a relentless series of probing questions.[2] The narrative ultimately leads to Riber's psychic collapse. As he comes to realise that Aurora's inquiries are increasingly destructive, he becomes tormented, eventually suffering hallucinations. In a final act of desperation, he ends his life near the American coast, throwing himself into the sea with weights tied around his neck.

The fact that the action is set aboard a ship—with the ocean as a constant, enveloping presence—is rarely mentioned in academic readings of Skram's novel.[3] However, as I shall demonstrate in the following analysis, a close examination of how the narrative unfolds across the shifting seascapes through which the ship passes reveals a profound connection between the struggle for knowledge of sexuality—initially conceived by Riber as "natural things" (2018, 55) —and the couple's relationship with the ocean: that vast, fluid, and perilous expanse that Riber and his crew must continually negotiate.

I argue that the development of the two central characters, Ory and Riber, is intimately tied to their differing relationships with the human body as part of human nature. At the same time, each character relates in fundamentally different ways to the sea that surrounds them. Drawing on Elizabeth DeLoughrey's (2022) explorations of how conceptions of knowledge intersect with ideas of materiality, I trace how Skram's novel constructs an analogy between knowledge of the sexed human body and knowledge of the ocean. Written over 130 years ago, *Betrayed* emerges as a fascinating text for scholars interested in the interrelations between imaginings of sexuality and depictions of the sea.

In my reading of *Betrayed*, I explore several key questions: What lies at the heart of Aurora's problem with sexuality, and how does the novel link her desire for sexual knowledge to contemporary efforts of exploring the ocean? What does the novel suggest about Riber's decline and his eventual loss of control over both his ship and his wife? Finally, how does Skram's text shape our imagination of different seascapes?

I will show that the novel problematises the concept of "nature" in multiple ways. Many contemporary scholars in the field of blue ecocriticism—including hydrofeminists inspired by New Materialism—are today influenced by Karen Barad's thoughts on knowledge production as they are elaborated in *Meeting the Universe Halfway: Quantum Physics and the Entanglement of Matter and Meaning* (2007). In this book, Barad studies how philosophers of science and others have imagined that human beings relate to the surrounding world. She proposes to replace "reflection" with "diffraction" when describing *entanglement*—the mutual engagement of subject and object (2007, 90). She wants us to fundamentally grasp that we are not interacting with nature but "intra-acting", that there are no clear-cut boundaries to be drawn by a subject that is not already part of the object to be studied and *vice versa*: "We are not outside observers of the world. Neither are we simply located at particular places *in* the world; rather, we are part *of* the world in its ongoing intra-activity" (Barad 2007, 184).

While Barad and those inspired by her thought in this way also question the principal difference between the human and the "more-than-human" or "posthuman", feminists working in this tradition also engage in eco-ethical argumentation, and claim that greater awareness of our entanglement is essential in order to foster a more sustainable relationship with nature, and with the oceans in particular (Alaimo 2016; Neimanis 2019). Skram's novel, however, illustrates that human entanglement may itself be the central difficulty with which humanity must contend. As I will show, entanglement in *Betrayed* is not proposed as a solution, but as the original challenge.

More specifically, *Betrayed* complicates the stereotypical ecofeminist idea of an innate, gendered relation to nature. Greta Gaard, for example, in her introduction to *Ecofeminism*, refers to Nancy Chodorow's and Carol Gilligan's studies, which suggest that "a sense of self as separate is more common in men, while an interconnected sense of self is more often common in women" (Gaard 1993, 2). More recently, hydrofeminists such as Astrida Neimanis have argued that the fluidity of the "blue" element challenges "phallogocentric" thinking (Neimanis 2017, 3). Yet in Skram's novel, it is precisely the male character, Captain Riber, who experiences a deep entanglement with the ocean, whereas his wife, Ory, actively resists any association with fluidity and water.

Skram's novel draws a striking parallel between knowledge production of the sexed body and of the ocean, while also demonstrating that sexuality is experienced differently by men and women. Simultaneously, *Betrayed* contributes to the formation of distinct imaginings of various seascapes. For instance, the novel portrays London harbour as dirty and unsightly, while the North Sea appears threatening and uncontrollable—mirroring Ory's perception of Riber's sexuality. In its final chapters, the Atlantic Ocean is depicted as a calm, blue surface, in contrast to the Gulf of Florida, which is presented as hot, infected, and death-bringing.

By intertwining descriptions of the ocean with the unfolding marriage plot, Skram's novel continues to shape readers' relationships with the natural world, resonating even today.

Ory's experience of London harbour: dirty sexuality and dirty water

Before proceeding to a close reading of the text, it is important to emphasise the novel's narrative technique: the text shifts between passages rendered from Ory's perspective and those from Riber's point of view. The third-person narrator alternately foregrounds Riber's attitude towards nature—understood in terms of sexuality and the surrounding sea—and Ory's contrasting perspective. While the chapters and scenes focusing on Riber's actions and reflections on his sexual life also reveal his view of the sea, those that render Ory's feelings about male sexuality simultaneously reinforce her perception of the surrounding waters. This shifting focalisation significantly shapes the reader's interpretation of the different seascapes described throughout the novel.

While the opening of *Betrayed* depicts Riber's irritation with certain guests at the wedding celebration, the first chapter closes with the scene in which Ory bids farewell to her younger siblings and her mother. She is apprehensive about having to share a bed with Riber in the future, a fear that surprises her mother. When Aurora reproaches her mother for not having enlightened her, her mother responds, "Why should I soil your imagination before it was necessary?" (2018, 12). This exchange leads Ory to associate sexual knowledge with impurity—a belief that proves crucial for the remainder of the novel.[4] In Riber's eyes, however, sexuality is linked instead to the threat of death. This connection becomes evident when he attempts to soothe Ory's anxiety by telling her ghost stories. After Aurora escapes Riber's sexual advances and the bedroom in her mother-in-law's house, Riber retrieves her and pacifies her with stories of strange apparitions on his ship. The first story concerns Riber's deceased father, who appears at the stern of the boat as Riber crosses the English Channel (2018, 21). According to Riber, his father's ghost glares at him threateningly before pointing towards the window. It is there, ultimately, that Riber will abandon both his ship and his marriage, and take his own life. Thus, from the outset, the threat of sexuality and the spectre of death by drowning are intimately connected in Riber's mind.

Whereas Riber's instinct compels him to fear the voyage he intends to undertake with Ory—both literally and metaphorically—Ory comes to see her ignorance of sexuality as intertwined with her ignorance of the sea. She believes she must gain mastery over both human nature in the form of sexuality, and the ocean and its creatures through the acquisition of knowledge. The analogy between the sea and sexuality is further underscored during the couple's stay in London, where Ory perceives both the harbour waters (2018, 71) and the sailors' sexuality as impure. Riber initially attempts to introduce his young wife to the diversions enjoyed by sailors awaiting their next voyage. Confident that Ory will learn to embrace her sexuality, he delights in the amusements offered in the streets near the harbour (2018, 39).

In Ory's eyes, however, human nature—namely sexuality—becomes increasingly intertwined with her notion of the ocean, the realm inhabited by the men she observes with growing disgust. It becomes impossible for her not to notice Riber's familiarity with the establishments they visit, prompting her realisation that he had an active sexual life prior to their marriage. Moreover, Ory prevents Riber from exerting cruel dominance over his crew and undermines his masculinity by challenging his authority as captain.

Finally, *Betrayed* depicts the departure from the harbour via the Thames. Readers are presented with this scene through Ory's perspective, as she observes the bark—the *Orion*—being manoeuvred from the wharf out onto the river: "then they were out in grey, muddy water with tiny dirty-grey waves and a strange swarm of ships, big and small, steamers and sailing ships, heavily loaded colossi, some with broken masts and smashed bulwarks" (2018, 76). The novel invites readers to envisage the murky waters of the Thames, surrounded by industrial buildings looming over the river, with the surface bustling with ships and smaller vessels. Skram thus conveys the image of London harbour as dirty and polluted through her narrative technique, specifically by channelling Ory's perspective.

Riber and the Bay of Biscay: struggling with entanglement

Whereas Ory lacks knowledge of both sexuality and the ocean, Riber possesses an intimate awareness of both and struggles to maintain control over them. He is plagued by anxiety attacks; even before meeting Ory, he has spoken of seeing ghosts—for instance, that of his father. Moreover, he recognises that his sexuality is at times beyond his control. This becomes apparent during an argument with Aurora, when Riber admits he was relieved when another captain left the room with a woman he might otherwise have pursued, at a time when Ory was waiting for him at a boarding house.

Riber is also acutely aware of the sea's perilous nature for men like him. During a storm off the Spanish coast, he remains outside on deck. Chapter IX opens with a vivid description that allows readers to conjure an image of the dangerous Spanish sea as Riber fights to steer the ship:

> Gale-force winds were howling from the southwest. Riber stood at the stern on the rolling deck holding on tight to the edge of the companion hatchway; from time to time he squinted up at the black sky filled with dirty yellow snowflakes. The dogwatch, already on deck, had secured the double reefed top sails and set the lower top sails. They had sailed against head winds in the Channel, clearing it that afternoon, and as they entered the Bay of Biscay, they were met by a choppy sea with large swells rolling from every direction. At the same time, the wind was increasing in strength.
>
> (2018, 89)

The storm is both a threat and a reprieve for Riber. It keeps him occupied on deck, leaving him almost no time to be with Ory, and it allows him to postpone the promised confession to her. Furthermore, the situation reinforces his authority among the crew. He instructs them: "Alert me if anything happens. I'm going below for a bit" (2018, 90). His position as captain is clearly underpinned by an image of masculinity; his command over the ship, the crew, and the ocean symbolises his power over his wife as well.[5]

At this point in the novel's plot, Riber acknowledges his own entanglement—he acknowledges human embodiment and understands that he needs to accept the human as being part of nature. He knows that it is impossible to have absolute control over nature, that he cannot keep it at a distance, as an object observed by a securely positioned subject. All his actions show his awareness of being intertwined with his body and the watery nature that surrounds him: As captain, he seeks to handle the stormy waters in times of danger, as Ory's husband, he tries to control his sexuality, while being aware of his bodily desire. Ory, however, perceives him not as entangled but as *unified* with the ocean. As she attempts to sleep below deck, Riber enters and tries to kiss her. The force of the storm seems to mirror his own strength; Ory senses this when she complains that he is so wet—he is soaked and tastes salty, like the ocean itself. Riber is, in Ory's eyes, not only part of nature but completely nature, and she feels compelled to restrain both his desire and the ocean's force, thereby escaping the challenge of entanglement. She even refuses to acknowledge the storm as dangerous and instead finds amusement in the chaotic scene around her: "'This is so much fun!', Ory called out to him. 'It's like the cabin has come to life and let loose, set on making a ruckus'" (2018, 90). Ory claims to enjoy herself, distancing herself as though watching a theatrical performance. This reveals that, in contrast to Riber, she avoids the awareness of her own entanglement that the close acquaintance with the sea forces her into. Through this, Skram draws an analogy between the female character's avoidance of insight into her own nature and her willed ignorance of the ocean's true character.

Over the Atlantic Ocean: Aurora's dream of transcending nature

Having survived the storm just off the coast of Spain, the *Orion* is now crossing the Atlantic, driven by the trade winds from astern. The sun is shining, and they are sailing at high speed. However, as the narrator's portrayal of Ory's thoughts reveals, Ory suffers deeply after hearing Riber's confession. In the fine weather, he has disclosed to her every detail of his past sexual experiences, and Ory can no longer tolerate his proximity; yet, paradoxically, she still desires to hear more. The tense atmosphere between the captain and his wife begins to affect the crew, who observe that Ory and Riber no longer share a bed. Consequently, Riber loses the respect of his men. Ory, on the other hand, loses all appetite, a change that does not go unnoticed.[6] Eventually, Riber turns to reading the Bible and starts drinking alcohol.

Riber's turn to the Bible is prompted by Ory's reaction to his confession: she had declared him to be a great sinner. Ory herself dreams of dying; she longs to rid herself of her corporeal existence and escape the challenge of nature entirely. She perceives the Atlantic merely as a surface and wishes to dissolve either into the waters or into the sky, to reach a place where everything is as she had always imagined it to be:

> She was gazing at the deep blue colour of the gentle swells sliding by and she imagined floating out there in the lovely water under the warming sun and

> endless sky, and being carried away on the rolling waves, farther and farther away, until she's be washed up on a foreign shore in a foreign land, where everything was as she had believed it to be.
>
> (2018, 96)

Ory seems to yearn for an escape from the ship; she simply wishes to forget her position as a person who cannot even swim well. Although still driven by a fascination with sexual knowledge, she cannot accept her entanglement, as, for instance, her own sexual needs. In her self-disgust, she desires to transcend her physical body and nature altogether (see also Hamm 2006, 150–55). She dreams of a country where everything is as it was before her marriage, in a state of sexual innocence akin to the Garden of Eden. This longing is accompanied by an idealised image of the Atlantic Ocean as blue, gentle, and soothing.

The marriage between Aurora and Riber has thus reached a state of crisis. In response to Ory's expression of disgust, Riber withdraws from his duties as captain. Through the narrator's depiction of Riber's thoughts, the reader understands that his attitude towards the sea has also shifted. He no longer attempts to acknowledge nature but instead wholly identifies with it. After being accused by Ory of being a great sinner in need of cleansing, Riber now considers himself to be nothing other than his sin—his sexual desire. While Ory denies her own sexual drives and dreams of transcending human nature, Riber fully accepts his, yet comes to regard his natural self with revulsion. Skram's novel presents both attitudes as ultimately destructive.

Florida Gulf: Riber's identification with his fallen nature

The final chapter of the novel opens with the bark lying motionless just off the coast of Florida, becalmed due to a complete absence of wind (2018, 115). The situation on board has by this point become almost intolerable. In a last, desperate attempt to stave off his total collapse, Riber becomes increasingly aggressive and tyrannical towards his crew: "He railed and barked orders from morning to night, nothing was to his liking" (2018, 115). On several occasions, he clashes violently with the first mate, who deliberately provokes him into losing his temper. Furthermore, the drinking water has become contaminated, leaving everyone afflicted with diarrhoea.

When the narrator focuses on Riber's inner state, it becomes clear that he perceives everything as being arrayed against him: "Mankind and animals and all the earth's bounty—nonsense—God and men and weather and wind" (2018, 117). He hears biblical phrases within himself and feels condemned, having betrayed Ory, even as he feels betrayed by her in return. In addition, he begins to hallucinate once again, imagining a creature "wet and with bowed head" (2018, 117). As several critics have observed, this creature bears an embarrassingly explicit resemblance to a phallic symbol (Engelstad 1992, 156), and Riber ultimately identifies with it: "We are one, you and I" (2018, 117).

Thus, Riber identifies not only with this figure from the sea but also with his own fallen human nature: he becomes his sexuality—the "wet penis"—and nothing more. Consequently, he longs to dissolve entirely into the ocean, which he perceives as a "shroud" (2018, 116). When Ory, oblivious to his despair, continues to question him, he winds the rope with its weights "twice around his neck" (2018, 120) before he "disappeared through the window" (2018, 121).

The novel's final sentences are clearly conveyed from Ory's perspective. She hears only "a splash and the sound of water swallowing and closing over his heavy body" (2018, 121). Ory is once again revealed as fundamentally detaching herself from all human entanglement. She does not want to see beneath the ocean's surface and perceives only the splash. Although she desires knowledge, she refuses to let herself be affected by it. She thus remains distanced—from sexuality, the sea, and her husband—whom she entirely fails to understand.

Conclusion: between epistemology and ethics

Working within the field of ecocriticism, scholars from diverse disciplines—including philosophy, anthropology, gender studies, and literature—have sought to articulate what, with reference to Karen Barad, is often termed an *epistemology of entanglement*. They call for a focus on the ways in which we know, grounded in the fact that, as Jeremy Davies has expressed it, we "live within the crisis" (Davies 2018, 194). Like Davies, Greta Gaard (1993), writing from a feminist perspective, has argued that we must reject any dualism that separates humans from "nature"; rather, we must recognise that we are part of it.

Hydrofeminists such as Astrida Neimanis have advanced similar arguments, notably in works such as *Bodies of Water: Posthuman Feminist Phenomenology* (2017). Drawing inspiration from feminist poststructuralist thinkers—including Donna Haraway, Elizabeth Grosz, and Stacy Alaimo—Neimanis emphasises the materiality of the body and the complexity of human embodiment, contending that an awareness of our ontological condition fundamentally affects modes of knowing:

> Watery embodiment thus presents a challenge to three related humanist understandings of corporeality: discrete individualism, anthropocentrism, and phallogocentrism. We also note that these three "isms" are all deeply entangled, mutually enforcing the claims of each other. The work of bodies of water is thus in part to remind us of this still-pervasive ontological Old Boys' Club. To imagine ourselves as bodies of water is to stage a clubhouse break-and-enter, a direct-action protest that floods up from the basement.
>
> (Neimanis 2017, 3)

By highlighting the fact that our bodies are composed primarily of water, just as much of the Earth's surface is covered by water, Neimanis urges us to consider that nature is not simply external to us but rather constitutes us—both as our other and in its otherness—and to imagine across boundaries. She proposes a posthuman feminist ethics, calling upon us to acknowledge our vulnerability as well as our responsibility towards the environment in which we exist.

While Davies seems to encourage us to just accept being part of a more-than-human historical development, the underlying premise in hydrofeminist texts is a turn to ethics and, in part, ecofeminist activism. Gaard, Barad, and Neimanis think that a new theoretical understanding of nature—in Neimanis' case, of the ocean as a fluid, dynamic element—might engender a new mode of thinking, which would in turn compel new forms of action. However, my problem with these different kinds of theoretical elaborations is the insight provided by a reading of novels as Skram's *Betrayed*: As I have shown, Skram's plot shows that the true challenge for humans in confronting their relation to nature lies not in the right or wrong way to conceptualise knowledge production, but in the difficulty of responding adequately to the knowledge we already possess, or to relate to the processes of knowledge production we are involved in.[7] A theory of entanglement—or, as Ory's desire for knowledge might aspire to; a new epistemology—will not necessarily help us to change our actions. What is needed would be a reorientation of how we deal with our knowledge.

To summarise, by bringing together two individuals with markedly different upbringings and experiences aboard a bark, Skram illustrates the profound difficulties they face in understanding one another. They embody two radically different attitudes towards nature, represented through sexuality and the ocean. Ory believes that she requires knowledge and seeks to control nature. This also entails her denial of her own nature; she refuses to accept her entanglement and the fact that her actions affect Riber. For Ory, the ocean appears merely as a surface phenomenon, not as something she depends upon or to which she must respond. Riber, by contrast, possesses an intuitive, embodied understanding of sexuality and the ocean. Yet, when pressed by Ory's questions, he begins to question his naturally entangled position. He experiences himself as one with his body, with his sexuality, and with the ocean, but in feeling disgust for himself, he ultimately takes his own life.

Both Aurora's and Riber's attitudes towards nature—both human and more-than-human—are presented by Skram as deeply problematic. Neither recognises the necessity of acknowledging their relation to nature: their own sexuality, their environment, and, crucially, each other. They fail to grasp that what is required is not the acquisition of more or better knowledge, but rather a willingness to respond to the knowledge they already possess. The novel thus demands a turn away from an exclusive focus on epistemology towards an ethical engagement. In the words of Stanley Cavell, formulated in another context, one might say that what is needed is understanding that "acknowledgement goes beyond knowledge": "Goes beyond not, so to speak, in the order of knowledge, but in its requirement that I do something or reveal something on the basis of that knowledge" (Cavell 1995, 257). Our knowledge is made visible only through our actions.

In my reading, Skram's novel becomes an exploration of human entanglement, demonstrating that we must both accept our vulnerability and natural condition and acknowledge nature's otherness. At the same time, we must act as humans and acknowledge each other as such; we are not only nature.

Notes

1 Amalie Skram was herself well acquainted with this group. Born in 1846 in Bergen—a lively seaport and Norway's second largest city—she was intimately familiar with the prominent social position held by a captain (*skipsfører*). Moreover, at the age of 17, Skram was compelled to marry an older man, Captain August Müller (see Tiberg 1910; Køltzow 1992).
2 Feminist scholars have interpreted Ory's desire for knowledge both as a consequence of her upbringing (Engelstad 1992) and as part of her psychological development (Krane 1961; Hamm 2006).
3 Research on Skram is no exception to the general neglect of the ocean in literary studies (see Mentz 2019; Frank 2022). Nevertheless, some scattered observations on Skram's works appear in Christopher Messelt's dissertation on novels depicting the Norwegian coast (2021). In the afterword to their translation, Katherine Hansson and Judith Messick (2018) remark upon the challenges of conveying every detail accurately. They emphasise that Skram possessed first-hand knowledge of sailing: in the late 1870s, she accompanied her husband, Captain August Müller, on voyages to Australia, New Zealand, Peru, and the Mediterranean. Several of her other novels are also set in exotic locations, notably *To venner* (1887) and *Fru Inés* (1891).
4 As feminist critics such as Krane (1961) and Engelstad (1992) have demonstrated, the novel thus opens with a critique of the double moral standard whereby women are expected to remain innocent—even ignorant—while men are permitted, indeed expected, to gain sexual experience from the age of 16. According to these scholars, this disparity in sexual knowledge ultimately leads to the collapse of the marriage at the novel's conclusion.
5 Since the 1840s, the Norwegian fleet had traded timber, fish, and other commodities to England and subsequently to the wider world. Norwegian trade flourished, and as a captain, a man held considerable power and responsibility, often conducting business on behalf of the shipowner (in many cases, the captain himself held a share in the ownership).
6 In my interpretation of the novel in *Medlidenhet og melodrama [Compassion and Melodrama]*, I have commented at length on Ory's lack of appetite (Hamm 2006).
7 This stresses that literature can contribute to what Timothy Morton (2009), for theoretical reasons, no longer wishes to describe as "nature". In a Scandinavian context, one might consider Søren Frank's *A Poetic History of the Oceans* (2022) or Katie Ritson's *The Shifting Sands of the North Sea Lowlands* (2019), to mention just two publications dedicated to the sea, which focus on specific literary motifs, plots, and styles. Both authors emphasise the necessity of examining not only texts from different countries but also those from various historical periods.

Works cited

Alaimo, Stacy. 2016. *Exposed: Environmental Politics and Pleasures in Posthuman Times*. University of Minnesota Press.
Andersen, Per Thomas. (2001) 2011. *Norsk litteraturhistorie*. Universitetsforlaget.
Barad, Karen. 2007. *Meeting the Universe Halfway: Quantum Physics and the Entanglement of Matter and Meaning*. Duke University Press.
Beyer, Edvard, and Harald Beyer. 1970. *Norsk litteraturhistorie*. Aschehoug.

Cavell, Stanley. 1995. "Knowing and Acknowledging." In *Must We Mean What We Say: A Book of Essays*. Cambridge University Press.

Davies, Jeremy. 2018. *Birth of the Anthropocene*. University of California Press.

DeLoughrey, Elizabeth M. 2022. "Mining the Seas: Speculative Fictions and Futures." In *Laws of the Sea: Interdisciplinary Currents*, edited by Irus Braverman, 145–63. New York: Routledge. https://dx.doi.org/10.4324/9781003205173-7.

Engelstad, Irene. (1984) 1992. *Sammenbrudd og gjennombrudd: Amalie Skrams romaner om ekteskap og sinnssykdom*. Pax forlag.

Engelstad, Irene, Jorunn Hareide, and Irene Iversen, eds. 1988. *Norsk kvinnelitteraturhistorie* 1. Pax forlag.

Frank, Søren. 2022. *A Poetic History of the Oceans: Literature and Maritime Modernity*, Brill.

Gaard, Greta. 1993. "Living Interconnections with Animals and Nature." In *Ecocriticism. Women, Animals, Nature*, edited by Greta Gaard. Temple University Press.

Hamm, Christine. 2006. *Medlidenhet og melodrama: Amalie Skrams romaner om ekteskap*. Unipub.

Hansson, Katherine and Judith Messick 2018. Afterword. In Amalie Skram: *Betrayed*. Norvik Press.

Hegdal, Åsfrid. 1996. "Mellom frihet og determinisme? Amalie Skrams Oline som naturalistisk karaktertype—en analyse av skikkelsens betydning i *Hellemyrsfolket*." *Edda* 3.

Krane, Borghild. 1961. *Amalie Skrams diktning: Tema og variasjoner*. Gyldendal.

Køltzow, Liv. 1992. *Den unge Amalie Skram: Et portrett fra det nittende århundre*. Gyldendal.

Mentz, Steve. 2019. *Break Up the Anthropocene*. University of Minnesota Press.

Messelt, Christopher. 2021. "Båten, fiske, fyret, naustet, stranden: Fem topos fra den norske kystromanen." Diss., NTNU Trondheim.

Morton, Timothy. 2009. *Ecology without Nature: Rethinking Environmental Aesthetics*. Harvard University Press.

Neimanis, Astrida. (2017) 2019. *Bodies of Water: Posthuman Feminist Phenomenology*. Bloomsbury Academic.

Ritson, Katie. 2019. *The Shifting Sands of the North Sea Lowlands: Literary and Historical Imaginaries*. Routledge.

Skram, Amalie. 2018. *Betrayed*. Translated by Katherine Hanson and Judith Messick. Norvik Press.

Tiberg, Antonie. 1910. *Amalie Skram som kunstner og menneske*. Gyldendal.

15

THE ATLANTIFICATION OF FAROESE LITERATURE

An archipelagic reading of Jørgen-Frantz Jacobsen's *Barbara*

Bergur Rønne Moberg

An archipelago in the North Atlantic

In the book that William Heinesen (1900–1991) produced with photographer Gérard Franceschi, *The Faroe Islands: The Magic Islands* (1971), he places the Faroe Islands in a larger geographical context, that is to say, the archipelagic edges of Europe:

> Together with Brittany, Cornwall, Ireland, the Hebrides and Iceland, the Faroe Islands form a part of the world's western extremity. It is the long, rocky foaming frontline, where Europe, this fingered offshoot of the immense Asian mainland, meets the equally immense Atlantic Ocean.
>
> (Heinesen and Franceschi 1971, 13–14, translated by Heinesen and Bergur Rønne Moberg[1])

Together with Jørgen-Frantz Jacobsen (1900–1938), Heinesen is the most famous Faroese writer; both of them wrote in Danish. Heinesen highlights the Faroe Islands as an archipelago located north of the Scottish island system in the middle of the North Atlantic Ocean, a space of contrast to the world's biggest—Euroasian—landmass. Heinesen's and Jacobsen's writings offer many examples of literary settings that emphasise the Faroe Islands' location in the middle of the most vast, blue, and oceanic North Atlantic. Jacobsen and Heinesen treat the vast oceanic distance in this area as an archipelagic gap, something that both outsiders and Faroe Islanders are constantly struggling to overcome. In an archipelago, land mass is constantly both interrupted and connected by water mass, and this duality is deeply reflected in their writing. It is the basis for cultural notions of broken lines and lines of

DOI: 10.4324/9781003647638-20

attraction between the Faroe Islands and Denmark/Europe, and between the Faroe Islands and the rest of the world.

The chapter describes this archipelagic gap in Jørgen-Frantz Jacobsen's novel *Barbara* (1939), which constitutes the author's major breakthrough, and the breakthrough of Faroese literature in general. The novel immediately became an artistic and commercial success in the Nordic countries as well as in Europe and beyond. At the same time, it is the most striking example in Faroese literature of an archipelagic dynamic. *Barbara*'s purposeful use of archipelagic motifs contributes to the modernisation and oceanisation of Faroese literature which will be examined through inter-archipelagic, intra-archipelagic, and trans-archipelagic perspectives. The chapter argues that archipelagic aspects are critical to *Barbara*, and that the novel can be read as a response to the terracentrism of European and world literature in this period. The world literary discovery of the "new" Faroese archipelago occurs for the first time in this novel; the archipelagic serves to both disrupt and expand the imagined geography of the region. In the first part of the chapter, discontinuity will be explained as disruptions of centralising structures understood as the influence of a more or less monolithic, abstract European modernity and "imperial geography of reason" (Yountae 2024, 145). The second part of my approach describes how the novel renegotiates the Faroese cultural case as seen from a trans-archipelagic point of view. From this perspective, the Faroe Islands are expanded by the inclusion of the sea, creating a dynamic zone of encounter. Reading the novel with the sea is thus a productive way of uniting the different levels of the novel.

Three archipelagic dimensions

This chapter draws on Ottmar Ette's distinctions between different types of archipelagic connections and disconnections: firstly, *intra-archipelagic* relationships concerning relationships unfolding between different islands of the same archipelago, and secondly, *inter-archipelagic* relationships unfolding between different archipelagos. In these first two archipelagic relations, disruption dominates, with no mutual exchange. I will explore these archipelagic connections with special attention to disconnections and missing links. Within these categories, Ette differentiates between horizontal and vertical disconnections. Thirdly, Ette operates with *trans-archipelagic* relationships between different archipelagos, which—in contrast to the inter-archipelagic—represent reciprocal connections between the islands and the world in the form of fruitful and transformative cultural encounters (2011, 28). According to Ette, the metaphorical archipelagisation of the world is a modern development arising from the relationship between the local and the global, and has made both disconnections and transformative reconnections more apparent (2016, 283). Worldwide complexity, Ette argues, should be understood trans-archipelagically (2016, 284), because the simultaneity of distinct localisations and distinct globalisations corresponds to islands as contradictory geographical units. The trans-archipelagic space represents the middle world, which can only be interpreted by integrating different segments of cultural and linguistic knowledge (Ette 2016, 326). Reciprocity is achieved by encounters between gazes from within and gazes from without. Altogether, Ette's idea of the archipelagic is of cultural encounters that contain both disruptive structural inequalities and transformative connectivities; the former do not block, but in many respects dynamise the latter.

The national versus the oceanic in Faroese literary history

Danish history and influence are present in most aspects of Faroese culture. This is seen in the bilingualism of Faroese literature, which is divided into a literature written originally in Faroese and a literature written in Danish. Jacobsen and Heinesen spearheaded Faroese literature in Danish; their work constitutes a breakthrough in the context of the monotonous, nationally minded Faroese literature in the early twentieth century. Their work is informed by a heightened awareness of the interconnected archipelagic world of the eighteen islands that make up the Faroes. Although Faroese geography is archipelagic and strongly influenced by its oceanic location, Faroese literature, surprisingly, was not born oceanic. On the contrary, it was created as a national romantic phenomenon in the years surrounding 1900, certainly with a notion of the sea as a significant other, but mostly without deeper implications. This national framing was in the main still central to Faroese literature for several decades into the twentieth century. As with nation-building in general, Faroese literature was initially conceived terracentrically, from the land, with the family and the farmer as symbols of the nation's healthy core. In contrast to the dynamics of archipelagic self-understanding, national romanticism is usually based on a static, conventional understanding of space.

Jacobsen's and Heinesen's renewal of Faroese literature draws on their archipelagic understanding, and their use of Faroese marine geography becomes defining for Faroese literature, creating a

> transnational space . . . moving beyond the nation-state container. . . . For much of human history, the great seas did not function as isolating bodies of water, but rather allowed for contacts and interconnections across political and cultural boundaries . . . oceanic spaces . . . mediating between national and global processes.
>
> (Conrad 2016, 118–19)

The atlantification of Faroese literature by Heinesen and Jacobsen moves Faroese literature from terracentric to oceanic thinking. As Conrad notes, "The focus on maritime encounters posed an important challenge to conventional landlocked notions of national space" (2016, 135). The national space becomes atlantified in Jacobsen's and Heinesen's oceanic thinking, creating dynamics between a gaze from within and a gaze from outside. They were close friends and poets who both grew up in a Faroese-Danish environment in Tórshavn, and thanks to them, this setting has an exceptionally robust literary legacy. The Danish language gave them access to a larger audience, and they both declared that they wrote about the Faroe Islands for a Nordic audience. At the same time, this gives their narratives the character of travelogues from an exotic geography. Their distinct—geographical—view of the Faroe Islands in relation to the Kingdom of Denmark, Europe, and the world in general is thus closely linked to the fact that they wrote in Danish for a non-Faroese audience. For the same reason, Faroese geography is more prominent in their work than in that of other Faroese writers, and this is particularly evident in the way they include the sea. They combine the view from the land with an oceanic worldview from the outside, which together can be characterised as an archipelagic in-between world: islands–sea, islands–continent, and island–island. Their archipelagic imaginations expand Faroese geography, and thus they turn Faroese literature into a world phenomenon and world literature.

Archipelagic studies and postcolonialism

Discussing literary and cultural issues only on the basis of centre and margin is insufficient in an archipelagic context. The binarism of centre-periphery overshadows other "positive" archipelagic aspects such as the importance of relative geographical size, distance, and the unifying character of the place. Thus, as in postcolonialism, the archipelagic view resists centring structures—both focus instead on how "linearity gets lost" (Dash 1989, xxxiii) when confronted with different kinds of writing back. However, as Rod Edmond and Vanessa Smith remark, the archipelagic view is more concerned than postcolonialism with concrete, relational matters of geography, as "islands had tended to slip the net of postcolonial theorizing" (Edmond and Smith, quoted in McMahon and André 2018, 297).

It is important to point out the differences between archipelagic studies and postcolonialism in this context. The overall difference between postcolonialism and archipelagic studies and other spatial studies is the fracture-oriented understanding of spatiality within postcolonialism and the more open and nuanced approach to spatiality within spatial studies. In this chapter, the archipelagic—explained here using different spatial studies (Ette 2005, 265) and the adjacent field of island studies—therefore distances itself from postcolonialism's rather abstract and un-dialectical focus on centre and periphery as metaphors for power and the absence of power, respectively. The archipelagic incorporates not simply a decentralised gaze, but the gaze from an intermediate world between sea and land, continents and islands. The archipelagic is a modern spatial version of classical complementary oppositions between isolation and connection and a more place-sensitive version of Edward Said's concept of "imagined geography" (1978).

It is important to emphasise this difference here, because archipelagic studies is radical in its realisation of the middle world between land and sea as an independent, paradoxical universe that, in its islands, recalls the polylogic of literature.

Islands as paradoxical universes of contradictions are congenial with the spirit's contradictory middle world of inexhaustible tensions and contradictions that deepen each other. The archipelagic, in alignment with this, recognises oceanic agency and the "dual capacity" (Patke 2018, 10) of the paradoxicality of islands floating between contradictory meanings and complex emotions such as isolation and openness, prison and paradise, robustness and vulnerability, inferiority and superiority, and dependence and independence.

Barbara is pervaded by all these contradictions; the universe in the novel is depicted as an enclosed "underworld" (Jacobsen 2004, 30), while at the same time extremely open to foreigners; as both a prison and paradise; and as a robust, independent testing zone for the men from the outside, while at the same time as a vulnerable, dependent, and inferior place in need of assistance from the outside. This way, *Barbara* reproduces the kindred spirit between literature and island imaginaries (McMahon and André 2018). As noted by Ottmar Ette in his reflections upon archipelagic studies, literature is experiential knowledge that "teaches us to think polylogically" (2016, 5). This endows literature with a special power and force, which Ette calls "the power of literature—or better: its living, rebellious counter-power . . . its capacity to invent worlds" (Ette 2011, 50).

Archipelagic discontinuity in *Barbara*

Barbara is a love story about the Faroese woman Barbara Christina Salling and the Danish priest Pastor Paul. Barbara has twice been a priest's widow before marrying another priest, Paul Aggersøe. The plot turns primarily on Pastor Paul, who arrives in the Faroe Islands at the beginning of the novel. He has been summoned to rectify the poor moral conditions in the Faroe Islands and has been promised a subsequent position in Denmark. He falls in love with and marries Barbara, but immediately after their wedding night, they begin to drift apart. Pastor Paul is forced to stay on the island of Mykines for eleven days due to bad weather: marital drama erupts when Barbara has an affair with the Faroese student Andreas Heyde in Pastor Paul's absence. Her third husband ends up being "condemned by the court for misconduct and desperate behaviour" (Jacobsen 2004, 220). Barbara is also betrayed in the end when Andreas returns to Copenhagen without her. She disrupts Pastor Paul's moral-Christian mission in the Faroe Islands and Andreas' enlightenment project, which is directed by his uncle, Judge Johan Hendrik Heyde, who has set out to bring the Faroe Islands into the modern age. These men are the modernisers, whose projects are disrupted by Barbara's actions.

Barbara is the prime example in Faroese literature of the archipelago as a site of disruption. This theme is touched upon already in the opening paragraph where the lamps in the Royal Stores Warehouses are "nearly blowned out by draughts that came with each blast of the gale" (Jacobsen 2004, 11). The introduction highlights the novel's grasp of two worlds: a dark Faroese world and light from the European Enlightenment era, which is the period in which the novel takes place. Furthermore, the novel is one of the best examples in Faroese literature of expansion and new beginnings for the Faroe Islands, which is also a theme throughout the novel (Moberg 2025). *Barbara* can be read as a story about the vexed arrival of modernity in the Faroe Islands, including social, cultural, and artistic modernity. The Faroe Islands is depicted as a country on the cusp of being seriously included in European culture, and this is most evident in the novel's openness to all sorts of ideas, periods, schools of thought, and cultural contexts, including: Christianity, natural science, the Middle Ages, Rococo, Enlightenment, (natural) philosophy, Faroese folk beliefs, the market and materialism, common sense, and radicalism. The novel expresses Jacobsen's view of the reception and observation of European culture. This drama unfolds like a baroque drama, where all the ideas are like bubbles that are inflated, then burst, only to reappear over and over again. They represent the novel's openness towards history, but also its hesitation towards a radical historical project, the decisive limitation of which is that it denies myth and nature.

The Faroese archipelago is depicted as an alternative conception of both time and space in relation to continental influence. The archipelagic disruptions, in historical terms, appear as a resistance to totalising Eurocentric pretensions of progress and what has been referred to by Glissant as "the march of history" (Dash 1989, XXX); their disruption of Eurocentric space and homogenising sameness constitutes a deterritorialisation of Danish-European influence, corresponding to a dissolution of a stable geographical-territorial order and thus of conventional geographies. The temporal and spatial disruptions, which occur both in the interactions of the characters but also on the structural level of the novel, reveal the archipelagic otherness of the Faroe Islands, insisting on a world that cannot be equalised by monolithic systems of either Christian or secular nature.

Foremost disruptors are the main character Barbara and the sea. *Barbara* tells how monolithic European systems, ideas, and influences are doomed to failure in a "new" archipelago. Regarding Pastor Paul, she explains her own "life's dodges and short-cuts" as "coloured stones" (Jacobsen 2004, 116), an archipelagic expression of the rocky outcrops in the sea. Not only do the stones shine in all sorts of colours, they also flash fleetingly and remind us of Barbara's dynamic, shifting in-between world. Pastor Paul's geography of reason falls short in his attempt to assimilate the Faroes into the geography of Europe. But characteristic for the novel's articulation of archipelagic middle world—which is reflected in the relationship between Barbara and Pastor Paul—is that it is precisely Paul who sees Barbara's mind as coloured stones in a fairy-tale sack. He has a sense of the power and magic she represents, and even of her suffering, but is unable to seriously grasp her otherness. He represents a fixed, system-driven world that neither can nor will adapt to Barbara's volatility. That is why he fails both as a human being and as a priest. But it is also precisely for this reason that he manages to rise above the many superficial gazes at Barbara.

Inter-archipelagic disruptions: the Faroe Islands and Denmark

Pastor Paul's encounter with the Faroe Islands offers a parade of disruptions which intensify in force. Upon his arrival, he is confronted with a dark, undeveloped, and gloomy society that forms the strongest possible contrast to the world he comes from in Copenhagen. He arrives with tremendous cultural capital in the form of his theological education and a promising future. Pastor Paul has been sent on a singular mission, as he has committed "himself as Christ's soldier to remote Faroe" (Jacobsen 2004, 48). This soldierly mission refers not only to his individual mission but also to homogenising church power and to his role as an educated Western subject striving for abstraction "that often alienates the thinking subject from place" (Yountae 2024, 144). Upon arrival, his head is filled with French vocabulary and memories of Copenhagen as a powerful cultural and religious centre. The new priest therefore becomes a luminous event in stark contrast to the dark, poor Faroese "underworld" (Jacobsen 2004, 30). His confidence is literally described as a shining capital against the backdrop of "the uttermost darkness" (Jacobsen 2004, 53) he has arrived in. Out in this darkness sits Barbara, representing the Faroe Islands as radical otherness. She does so in a literal sense when the newly arrived Andreas shows off new things from Copenhagen in a living room, where Barbara sits outside the circle of light.

Pastor Paul's encounter with the Faroe Islands is more specifically an encounter with an oceanic uncertainty. Barbara's volatile mind is described in oceanic archipelagic terms, effectively blocking Pastor Paul as a representative of a centralised bureaucracy and a systemised form of dogma. Barbara's impulsive nature is constantly shifting like "flood-tides and ebbs" (Jacobsen 2004, 116) and like the sand which "widened and then narrowed . . . remained firm and ripped like the waves" (Jacobsen 2004, 116–17). Pastor Paul becomes "dizzy" and "horrified" (Jacobsen 2004, 118) in the face of Barbara's fluid geography and personality, and he vainly seeks solace in his devotional books, because "his heart was empty and responded only to the shell-curved waves" (Jacobsen 2004, 116). Pastor Paul's capital, which has been drawn from monolithic church power supremacy and is based on a fixed terracentric, theological standpoint, is tossed about in the turmoil of Barbara's water world. He is lost in Barbara's archipelago, and his project of controlling this "new" geography dissolves. His broken love brings almost everything else down with him: Christianity as a church institution, Enlightenment, marriage, philosophy, and local common sense. However, these are not just bubbles that burst and are gone; they constantly take on new life, renewing themselves and taking part in a plurality of possibilities in which new questions and new solutions appear at the same time. Representing neither a monolithic European system nor an isolated local sign system, the archipelagic contains all the many questions of the novel while at the same time comprising pragmatic adjustments and negotiations.

Barbara, on the other hand, is a clearer example of the archipelago writing back against its reduction to a mere disruptor of European hegemony. Her name, which means stranger, is an indication of her disruption of the familiar. She continuously allies herself with the foreign by marrying one Danish priest after another, resulting each time in divorce and chaos. She is a disruption through and through; in her alignment with the sea, she challenges the terracentric understanding of a fixed, uniform reality. Barbara questions the men's projects when asking Andreas: "Is it right?" (Jacobsen 2004, 289). Her question specifically concerns whether it is right that she can join Andreas when he is leaving for Copenhagen, because she senses that she is being deceived by the enlighteners, who don't want her to lead Andreas further astray. But the question also refers to a deeper layer in the novel concerning the theme of goodness. Barbara is associated with "the good", referring to love and not rational enlightenment. Therefore, the translation could just as well have stayed closer to the original and rendered "Er det godt?" as "Is it good?" instead of "Is it right?"

The challenge posed by Barbara to the men's singular projects raises significant questions. As Pastor Paul observes the sleeping Barbara, he longs to understand her mystery and suffering, and one of them is: "Who was Barbara?" (Jacobsen 2004, 181). While the novel can render Barbara's archipelagic capacities by rising above the often objectifying male gaze, and seeing her as both nature and human, Pastor Paul is not always capable of this. The influence of the church, his career, and his marriage suddenly lose meaning for him, and the day after the wedding, he feels "as though a great soap bubble had burst" (Jacobsen 2004, 183). Instead, he finds himself standing "under full sail out to the sea" (Jacobsen 2004, 187) and then sinking in the middle of Barbara's restless, dangerous, unpredictable, and violent currents. At the same time, the reader learns that the novel's intellectual character, the Judge, has chosen to protect himself from life and his preoccupation with Barbara, because "a philosopher never suffers absolute shipwreck" (Jacobsen 2004, 187). Shipwrecked figures are familiar motifs in postcolonial cultural encounters between the West and the "rest", including between the West and the Faroe Islands as a rest in the West (Moberg 2014). They are a means of describing the dramatic and traumatic European encounter with the foreign, which on the one hand represents pleasure and untamed desire and on the other hand destruction, madness, death, and darkness (Selnes 2010, 76). The European system's encounter with the foreign thus ends in the ultimate disruption of the geography of reason. This story of shipwrecked love unfolds with the sea and archipelago at its centre. Barbara merges allegorically with the Faroese archipelago as an eros character manifesting the dynamics of the fragmented archipelagic geography.

Intra-archipelagic disruptions: lost in the Faroese archipelago

Let us look more closely at some examples of disruptions in the novel's intra-archipelagic infrastructure. The Faroe Islands' internal archipelago is described as almost even more impassable than the route to the Faroe Islands from the outside. The novel's most significant intra-archipelagic disruption is when

Pastor Paul sets out “to the lonely island Mykines” (Jacobsen 2004, 127) and ends up marooned by the weather, and during this stay loses Barbara to Andreas. This island trip is an example of what Ottmar Ette calls an intra-archipelagic discontinuity (2011, 28), which manifests itself as a rift in Pastor Paul’s life, illustrated geographically as an interruption of communication within the Faroese archipelago. On Mykines, he loses his connection for weeks to the rest of the Faroe Islands, which becomes fatal for his marriage and is the beginning of the end for Pastor Paul.

This disconnection unfolds horizontally. Another aspect of the intra-archipelagic disconnection is vertical: early on in the marriage, Pastor Paul begins to lose control of himself because he cannot reconcile God and Barbara. With his dualistic Christianity, he has to choose between them, which leaves him stuck somewhere between heaven and earth:

> Pastor Paul was like a bird-catcher who hangs on his line and dares look neither up or down. Because if he looks up towards heaven his line seems lost in the distance, and it is as though he is hanging by a severed thread. But if he looks down, he becomes aware of the horror of the abyss.
>
> (Jacobsen 2004, 120)

Pastor Paul is sent on a horizontal mission that is superficially power-promoting and order-creating, without any sense of the depth of the place and the situational context. However, as this quotation shows, the novel insists on verticality, which “even in the era of disenchantment and flatness continues to hold metaphoric resonance in narrative poetics” (Schilling 2014, 226). Barbara represents the magical “reality” outside the systems that prevents the world from becoming flat. While before there was a tantalising tension between the dimensions of God and Barbara, Pastor Paul has now lost both of them in his inability to negotiate between the dimensions of the spirit (“high”), the flesh (“low”), and the demands of geography.

The judge and the trans-archipelagic

While the inter- and intra-archipelagic describe the encounters between situational and local knowledge and the world outside it, the trans-archipelagic constitutes a new transformative connection to the world through the local, national, and regional specificity of Faroese culture. The novel first disrupts the conventional geography of reason and then, as we will see, establishes trans-archipelagic connections to the world outside the Faroe Islands. The trans-archipelagic transformation places the Faroe Islands in a larger European context that is also sensitive to the country’s place in the world locally, nationally, regionally, and globally. The novel is filled with references to (1) the local, in the form of the special Faroese-Danish culture in Tórshavn; (2) the national, as an awareness of Faroese folk traditions, folk characters, and (the development of) Faroese culture in relation to Danish culture, and also with code-switching in terms of Faroese-influenced turn of phrases in the novel; (3) the regional, with the ubiquitous importance of the sea; and, finally, (4) Faroese inflections of globalised culture, such as Christianity and the Enlightenment.

The Judge's position mediates between outside influences and the local culture. Faroese but from a Danish family, he represents a contact zone between Faroese culture and the Danish bourgeoisie in the Faroe Islands. He belongs to the Danish elite in Tórshavn, but the difference between him and the Danish upper class is that he has a different relationship with the Faroese than they do. Unlike "the better folk . . . and the *fine*" (Jacobsen 2004, 62), the Judge was "in tune with the people" (Jacobsen 2004, 64). It is against the backdrop of the tension between the fine ones and the "many" (Jacobsen 2004, 63) that the novel portrays the Judge as a representative of mixed heritage. This contrasts with the Danish bourgeoisie, which perpetuates an empire-like culture with a sharp division between the country's two national cultures. With his openness, the Judge has the privilege of looking from both the outside and the inside. There is "something that kept him apart" from people of his own background as they "conversed in Danish", but he has "nothing against them" (Jacobsen 2004, 64).

The Judge's special position between Faroese and Danish culture, the people and the elite, influences his ideas as an enlightened man and his plans for the Faroe Islands, which are rather nuanced. His relationship with the Danish bourgeoisie involves both the disruption of the monolithic Danish-European influence and the productive union of Danish elitism (comprising influence, learning, and opportunities for taking action) with local culture and situated knowledge. In launching the Faroe Islands as a project, he purposefully draws on both assets, engaging Andreas as the protagonist of the enlightenment project, while at the same time being a good listener, a mature observer, and a deep thinker. He can situate the Faroe Islands as a space in-between, neither in the stagnation and isolation of myth nor the Enlightenment's radicalised conception of history, where "the secret of history was revealed" (Schanz 2020, 73, my translation). The trans-archipelagic is also found in the figure of Barbara, who like the sea is herself an archipelagic in-between world connecting myth and modernity and representing an alternative, adapted enlightenment demanded by local knowledge.

Following *Barbara*, there are also other Faroese prose works, especially novels, that have continued the atlantification of Faroese literature and lifted it further into the maritime. The novels reflect the North Atlantic archipelago as the bluest and wettest part of the Nordic space, in stark contrast to the metropolitan southern parts of the region. William Heinesen's *Det gode Håb* (1964; *The Good Hope* 2011), Jens Pauli Heinesen's *Rekamaðurin* (1977; The Beachcomber), Gunnar Hoydal's *Undir Suðurstjørnum* (1992; *Under Southern Stars* 2004), Carl Jóhan Jensen's *Eg síggi teg betur í myrkri*: *Forspæl til ein gleðileik* (2014; I See You Better in the Dark: Prelude to a Comedy) and *Terningar: Søgur av tilvild* (2019; Cubes: Stories of Coincidence) all constitute significant new interpretations of the sea in Faroese literature. Heinesen and Jensen integrate the specifically Faroese inter- and trans-archipelagic aspects into a blue Nordic context. Their novels draw on travel writing and take place wholly or partly on board ships and out at sea. The sea is the North Atlantic and the North Sea and the voyages are from Norway to the Faroe Islands, between Iceland and the Faroe Islands, etc. The sea is in most cases in the foreground as a significant trans- and non-national arena. One of the most striking features of the novels is their representation of the wild, borderless, risky, and mostly forgotten or repressed space as a new cultural arena in the light of the Faroe Islands as a modern nation.

The novels convey an understanding of the sea as a larger entity than land, a maximised arena for connections. In this expanded perspective, the sea becomes a starting point for a confrontation with the terracentric peasant culture as an outdated reality. Carl Jóhan Jensen belongs to a younger generation than Jacobsen and Heinesen, and the shipwrecks we see in his novels point to phenomena such as Nazism, thus differing from Jacobsen and Heinesen's fundamental cultural optimism and stories of progress. The form also changes to polyphonic postmodernism. But fundamentally, these works can be read as examples of a world literary use of the novel to shed light on the Faroe Islands in a Nordic and global maritime context governed by changing perspectives from within and without.

Barbara is thus the first example of a distinct trans-archipelagic literature from the Faroe Islands. The titular Barbara stands for the fluid emotional world of the archipelago, where no meaning can be locked down and where the most powerful ideas and institutions—nationalism, Christianity, enlightenment, marriage, philosophy, market and money, local common sense, etc.—are confronted with each other's inadequacies and then sent out into the waves. This makes the novel a drama of ideas (Moberg 2014, 2025) and an obvious choice for a "blue" reading.

Conclusion

The novel *Barbara* marks the beginning of a turn towards the blue; my chapter argues that the novel has contributed decisively to atlantifying and modernising Faroese literature. The arguments are primarily drawn from archipelagic studies, which is used to explain the novel's archipelagic features as a special form of seaborne "imagined geography" which contests the binarism of centre-periphery and the logic of the self-contained island. I draw on Ette with the concepts of the inter-archipelagic, the intra-archipelagic, and the trans-archipelagic: the first two aspects are described as representing disruptions of or resistance to the brutal transfer of influence from continent to archipelago. This is shown by Barbara's disruptions of the outsider-men's singular missions to rationalise and enlighten. Barbara epitomises the Faroe Islands as a (con)testing zone. The sea in this archipelagic context challenges the terracentric understanding of a fixed reality by deterritorialising and making it unstable through the uncontrollable element of water.

In the trans-archipelagic, on the other hand, the sea provides a space for transformative cultural encounters, containing both disruptions of the monolithic influence from the outside and magnifying tensions within the archipelagic in-betweenness of land and sea, islands and continents. The trans-archipelagic is demonstrated as a whole "new" entangled geography hidden from the monolithic geography of reason—a cultural dynamic between the local, national, regional, and global. The trans-archipelagic draws its dynamism from the classical paradoxical meaning of islands, the opposing dynamics of modernity's imagined geographies, and ultimately from the ambiguity and polylogic of literature itself. It opens up the unifying meaning of the sea as a watery counterpart to the continental, terracentric gaze on islands as antipode to continents.

The whole novel is embedded in a geographical imagination that highlights the existence of the Faroese "underworld" in a European (world literary) context. The archipelagic in-betweenness forms a geographical model for the novel's hybrid understanding of culture and identity. More specifically, it brings together the trans-archipelagic threads in the novel's communicative universe of contradictions. This centripetalisation is represented by Barbara and The Judge, who each in their own way expresses the novel's striving for a diverse unity in the form of a trans-archipelagic integration of myth and modernity, disconnection and relations, hard and soft enlightenment, and experience and knowledge. Through them, development and reason are adapted to the special circumstances in the Faroe Islands as a "new" cultural geographical case.

Note

1 The quotation is my own rendering of the Danish and English versions of the text from this trilingual publication, which are not identical.

Works cited

Conrad, Sebastian. 2016. *What Is Global History*? Princeton University Press.

Dash, Michael. 1989. "Introduction." In Edouard Glissant: *Caribbean Discourse: Selected Essays*, xi–xxiii. University Press of Virginia.

Ette, Ottmar. 2005. *ZwischenWeltenSchreiben: Literaturen ohne feste Wohnsicht*. Kulturverlag Kadmos.

Ette, Ottmar. 2011. "Insulare ZwischenWelten der Literatur: Inseln, Archipele und Atolle aus transarealer Perspektive." In *Inseln und Archipele: Kulturelle Figuren des Insularen zwischen Isolation und Entgrentzung*, edited by Anna E. Wilkens, Patrick Ramponi, and Helge Wendt, 13–57. Transcript.

Ette, Ottmar. 2016. *TransArea: A Literary History of Globalization*. De Gruyter.

Heinesen, Jens Pauli. 1977. *Rekamaðurin*. Tórshavn.

Heinesen, William. 1964. *Det gode Håb*. Gyldendal.

Heinesen, William, and Gerard Franceschi. 1971. *De magiske øer: Gandaoyggjarnar; The Magical Islands*. Rhodos.

Hoydal, Gunnar. 1992. *Undir Suðurstjørnum*. Árting.

Jacobsen, Jørgen-Frantz. 1939. *Barbara*. Gyldendal.

Jacobsen, Jørgen-Frantz. 2004. *Barbara*. Translated by George Johnston. Norvik Press.

Jensen, Carl Jóhan Jensen. 2014. *Terningar: Søgur av tilvild*. Sprotin.

Jensen, Carl Jóhan Jensen. 2019. *Eg síggi teg betur í myrkri: Forspæl til ein gleðileik*. Sprotin.

McMahon, Elizabeth, and Bénédicte André. 2018. "Literature and the Literary Gaze." In *The Routledge International Handbook of Island Studies: A World of Islands*, edited by Godfrey Baldacchino, 296–311. Routledge.

Moberg, Bergur. 2014. *Resten i Vesten: Verdenslitteratur i modernismens margin*. Spring.

Moberg, Bergur. 2025. "Barokke boblestudier i et æstetisk omtumlet værk: William Heinesens redaktion af manuskriptet til *Barbara* med særligt henblik på Goethe-prægede harmoniseringer samt på *Barbara* som idédrama og Færø-allegori." In *Barbara i høj sø: En antologi om Jørgen-Frantz Jacobsens Barbara*, edited by Bergur Rønne Moberg and Erik Skyum-Nielsen, 116–57. Multivers.

Patke, Rajeev S. 2018. *Poetry and Islands: Materiality and the Creative Imagination*. Rowman & Littlefield.

Said, Edward. 1978. *Orientalism*. Pantheon.

Schanz, Hans-Jørgen. 2020. *Menneskene og alt andet*. Klim.

Schilling, Derek. 2014. "On and Off the Map: Literary Narrative as Critique of Cartographic Reason." In *Literary Cartographies: Spatiality, Representation, and Narrative*, edited by Robert T. Tally Jr, 216–28. Palgrave Macmillan.

Selnes, Gisle. 2010. *Det fjerde kontinentet: Essays om America og andre fremmede fænomener*. Vigmostad Bjørke.

Yountae, An. 2024. *The Coloniality of the Secular*. Duke University Press.

16

ATLANTIC CROSSINGS, DEPARTURES, AND RETURNS IN MARJAM IDRISS' *HALVT*

Lill-Ann Körber and Hanna Rinderle

The Atlantic in Scandinavian history and literature

The seafaring history of Scandinavia is extensive and significant, characterised by a succession of notable voyages and vessels, beginning with the Vikings and continuing to the Vasa. However, it is only in recent years that other aspects of Scandinavian overseas activities have come to the fore, specifically the engagement of Scandinavians in both overseas colonialism and the transatlantic slave trade (Naum and Nordin 2013; Weiss 2016). From this perspective, the period of seafaring, which is celebrated in Denmark as "den florissante periode" (the Age of Flourishing) and in Norway as "seilskutetiden" (the Age of Sail), appears in a different light. The Atlantic Ocean, in particular, has emerged as a focal point in the resurgence of interest in colonial seafaring. The non-European shores of this vast body of water were initially accessed for trade purposes in the aftermath of the so-called voyages of discovery, which subsequently gave rise to territorial claims. The most significant consequences were probably incurred with the establishment of the transatlantic triangular trade system between Europe, West Africa, and the Americas and Caribbean, which included the trafficking of millions of people for the purpose of supplying labour to the newly established plantation colonies. It is evident that Scandinavian companies, merchants, and royal houses participated in the lucrative Atlantic trade, with trading bases located on the so-called Gold Coast, as well as in plantation colonies and free ports in the Caribbean. Concurrently, the Atlantic Ocean became a site of mass burial, not only for seafarers but especially for deported and enslaved Africans who perished during the transatlantic voyage, particularly on the so-called "Middle Passage". In the Scandinavian case, this term refers to the journey between the Scandinavian forts on the Swedish and Danish Gold Coast (1650–1663 and 1658–1850, respectively) and the colonies in the

DOI: 10.4324/9781003647638-21

Lesser Antilles, namely St. Barthélemy (under Swedish administration 1784–1878) and the Danish West Indies (St. Thomas, St. John, and St. Croix; under Danish administration from 1672, 1718, and 1733, respectively).

The Danish West Indies were ceded to the United States in 1917. The years surrounding the 2017 centenary of the transfer marked a peak in an ongoing "post-colonial turn" in public discourse, scholarship, and the arts in Scandinavia, with a particular focus on the colonial history of the Atlantic Ocean. Despite the union between Norway and Denmark lasting until 1814, the topic of overseas colonialism has historically received minimal attention within Norwegian national discourse. However, recent years have witnessed a shift towards a more comprehensive examination of the role of Norwegians in colonialism and slavery (Hove 2024). Marjam Idriss' novel *Halvt* (2024; Half) is representative of this tendency, and of a cross-Scandinavian current of colonial-historical fiction (cf. Körber 2025). Idriss' approach to the genre is innovative in that the novel posits a *longue durée* of raw material extraction on the shores of the Atlantic—a continued pattern that Dominic Boyer (2023, 12) has recently described as the "*fossilization* of sucropolitics", an "ecocidal trajectory of global capitalism", starting with sugarcane plantations and developing over carbon-driven industries to petroleum-based artefacts. In their novel *Halvt*, Idriss combines an exploration of Norway's historical involvement in colonial agricultural production, trade, and slavery with a reflection on its contemporary role as an oil-producing nation. Engebret (or Engebrecht) Hesselberg (1728–1788), a plantation owner, jurist, and colonial official, personifies colonial extractivism in the novel. During his tenure on St. Croix, Hesselberg brutally suppressed a planned uprising of enslaved plantation workers (Vedeler Nilsen 2020). In a significant development in the narrative, Hesselberg is revealed to be the ancestor of the protagonist, Inge, who is employed by the Norwegian oil corporation Equinor. Consequently, Idriss' novel lends itself to a combined postcolonial and petrocritical reading, and this chapter aims to demonstrate how it links the Atlantic shores of Europe and the Caribbean via a flow of produce (sugar and oil) and people.

A notable aspect of contemporary Scandinavian colonial-historical fiction set in the Atlantic sphere is the diminution of the ocean's significance, in terms of both its material presence and the experiences it encompasses, in comparison to accounts from the "Age of Sail" (cf. Frank 2022, 15). *Halvt* is a good example of this tendency. The protagonists arrive in St. Croix at the novel's offset, and the view of the Atlantic and Caribbean seas from above is only mentioned briefly. Subsequently, the narrative is set in the coastal town of Christiansted or the island's hinterland. Despite the proximity to the ocean, throughout the book, there is no mention of any of the characters looking at the sea, going swimming, or on a boat trip. The protagonist maintains communication with her relatives in Norway via video calls or live streams. Contemporary means of transport and communication technology have resulted in a significant reduction in the perceived distance across the Atlantic Ocean. The characters in question have no physical or identity-bearing relationship with the sea. The only individual in the novel who refers to the waters, and their subsequent pollution, is a local woman.

Despite its continued role as a node in flows of goods, people, and knowledge, the Antillean archipelago functions in Scandinavian fiction as an imaginary space, a remote location in a distant past that holds little direct relevance for individuals today. This tendency is analogous to the tendency observed in Scandinavian colonial-historical fiction to conclude its account with the official cessation of colonialism, thereby leaving the question of contemporary transregional links unaddressed (Körber 2025). In a similar vein, Scandinavian accounts of colonialism in the Caribbean could be described as "land-based imaginaries" (DeLoughrey 2019, 135). The ocean as experiential space is central to both local livelihoods and historical developments, but less so to current Scandinavian-based explorations of colonial legacies. Idriss' novel partly conforms with and partly deviates from this paradigm. The characters arriving in St. Croix from Europe do not have any personal connection with Caribbean waters. Their job is to protect the economic interests of the oil corporation Equinor, which are compromised by pollution and subsequent local protests. In *Halvt*, a business consultant's statement, "no one makes money on coral reefs" (Idriss 2024, 68; translations are our own), is symptomatic of the sublimation of the physical ocean whilst maintaining its flows of money and goods.

It is precisely this "turning away from" or "forgetting of" the sea that we see as the point of departure of the novel's exploration of a continuity of capitalist ventures linking the shores of the Atlantic Ocean. In the following three sections, we will explain how the novel examines Norway's maritime engagements in the Caribbean from the era of Dano-Norwegian colonialism to the present-day oil industry. We argue that the dual emphasis on Norwegian-Antillean connections provides novel insights into both periods. According to our reading, *Halvt* sheds light on how late modern petrocultures perpetuate colonial structures and highlights the need to rethink the Norwegian "Age of Sail". Idriss' novel posits the notion that to understand the contemporary moment, one must "remember" the Atlantic Ocean in its multifaceted aspect: as a delicate ecosystem, a realm for economic activity, a cultural and social space that has both unified and segregated individuals, and a metaphorical, imaginative sphere.

Atlantic (hi)stories in Idriss' *Halvt*

In *Halvt*, Idriss takes up the historical circumstances of Engebret Hesselberg's actions on St. Croix and their mnemonic afterlife on the island. Idriss weaves these threads into a transgenerational family narrative that connects past and present, Norway and St. Croix. The protagonist, Inge, is a young woman employed by the Norwegian energy company Equinor, where she has swiftly climbed the corporate ladder and landed a prestigious position due to her mother's roots in St. Croix. At the beginning of the novel, set in 2023, Inge travels to the island with her boss Geir, and Michael, a young business consultant, to testify at a hearing on behalf of the company, following a major fire in one of the company's oil facilities. The incident has had a significant impact on both the local population and the environment. It soon becomes clear that Inge has been brought along primarily as a token: She is meant to symbolise the Norwegian company's connection to St. Croix and its people, even though Inge has not been to the island in years and has had little contact with her maternal family there.

However, this changes as the plot unfolds and another detail of Inge's family history comes to light: on her father's side, she is a descendant of Engebret Hesselberg. The Norwegian side of Inge's family history alters the dynamics of the hearing as Michael, in particular, shifts the public's attention to these aspects of her background, diverting focus from Equinor's own scandal. Inge is now perceived not as a descendant of the island, but as a neocolonial exploiter. Ironically, the situation ultimately brings her closer to her estranged family on St. Croix: by the end of the novel, they reconnect in the island's natural surroundings. Idriss structures this family narrative, which spans centuries and continents, into three parts: the first and third parts focus on Inge's experiences on St. Croix in 2023, while the middle part recounts the events of 1759. Under the heading "Variation", Idriss combines historical sources, particularly Hesselberg's legal writings, with her own—fictionalised—account of the planned rebellion.

Tidalectic protagonists

In *Routes and Roots: Navigating Caribbean and Pacific Island Literatures* (2007), Elizabeth M. DeLoughrey offers a comparative literary analysis of the Caribbean and the Pacific Islands, introducing the concept of "tidalectics" as a methodological framework. Borrowing from Kamau Brathwaite, she emphasises the cyclical and dynamic relation between land and sea to illuminate the histories of migration and Indigeneity in the Caribbean and the Pacific. These literatures, she argues, deconstruct colonial myths of isolation and peripherality.

Such tidalectic dynamics also shape Idriss' *Halvt*, particularly through the figures of Inge's parents. They meet on St. Croix, where the African Caribbean mother, Liz, is from, and where the white Norwegian father arrives to research his ancestor Hesselberg. Their love story initially appears to overcome colonial binaries. As Liz recalls,

> It was beautiful that the two of them—descendants of colony and colonial power, enslaved and enslaver, two people who had had nothing to do with each other for nearly three hundred years—should meet again at an overpriced pizza restaurant and fall in love.
>
> (Idriss 2024, 123)

On the one hand, Idriss thereby undermines what DeLoughrey (2007, 20) calls "the myth of island isolation": St. Croix is not a peripheral backdrop but the narrative centre, while Norway becomes marginal. Moreover, agency lies not with the European father but with Liz, who migrates northwards. *Halvt* thus becomes a narrative "of transoceanic migration to undermine the myth of the confined islander" (DeLoughrey 2007, 20). On the other hand, the novel exposes the racism Liz encounters in Norway, which ultimately drives her to return to St. Croix and to leave both her marriage and her daughter behind. In doing so, Idriss reveals how even the seemingly peripheral former colonial power continues to exert authority. The migrations and relationships across the Atlantic in *Halvt* are therefore neither one-directional nor simple but fluid and dynamic.

This tidalectic, fluid, and dynamic mode of narration and characterisation culminates in the figure of Inge. The encounter between the descendants of coloniser and colonised, and their daughter Inge, forms the novel's experimental core. The fluidity embodied by this character is already anticipated in the novel's title: the paradigmatic term *Halvt* ("half") encapsulates both the potential and the limitation of fluidity. This dualism first becomes apparent to the reader through the protagonist's name. Her full name, Ingeborg—a female name—is shortened to the name Inge, which is typically a male name in Norway. This nominal shift signals a destabilised identity: Inge emerges as a queer character whose hybridity extends beyond race or ethnicity.

Closely connected to this fluid identity on the level of gender is the novel's insistence that both skin colour and identity are constructed rather than innate. While Inge's mother observes that her own presence "made Inge browner" (Idriss 2024, 75), her father notes that "no one in Norway believed him when he said she was his daughter. They only saw the brown skin and the curls and could not accept that she might also be his" (Idriss 2024, 98). The act of "not believing" her father marks a refusal of relational and fluid identities: whiteness remains the default of national belonging, while brownness disrupts kinship (cf. Joof 2024). Inge thus refigures the colonial "tragic mulatto" trope, through which Idriss links contemporary racism to the legacies of Norway's colonial entanglements. As with the character of Liz, Inge's portrayal makes visible both the possibilities and the limitations of an entangled transatlantic history. This queering of cultural and gender identity, moreover, invites an intersectional reading of hybridity.

In shaping her characters, Idriss thus resists the colonial "rooted discourse" (DeLoughrey 2007, 96) often associated with island literatures. Instead, she portrays both Inge and Liz as tidalectic figures who move across the Atlantic, rendering the novel a "vital counter-narrative of transoceanic routes and diaspora" (DeLoughrey 2007, 96). Especially in Inge's case, the movement across the ocean is mirrored in a fluid construction of identity that nonetheless avoids collapsing into simplistic narratives of hybridity or multicultural inclusion: By questioning or even negating Inge's fluid identity, the novel exposes the limits of tidalectic protagonists.

Tidalectic historiographies

As previously mentioned, the novel is divided into three parts. While the first and third sections—set in the narrated present—are simply titled Part 1 and Part 3, Idriss assigns a specific time and place to the second: "St. Croix, Dansk Vestindia Advent, 1759" (2024, 127). This middle section recounts, in the form of an analepsis, a planned revolutionary uprising on St. Croix and its brutal suppression by Engebret Hesselberg. Two formal aspects are striking here. Firstly, there is a shift in focalisation: whereas *Halvt* employs an extradiegetic, non-focalised narrator in Parts 1 and 3, the second part features multiple focal points, including several Crucians.

Secondly, this section differs structurally through its extensive use of epigraphs, most of them drawn from Hesselberg's legal and historical writings. Seven quotations originate from *Juridisk Collegium*, a long-standing reference work in Danish and Norwegian law, and another seven from *Species Facti*, Hesselberg's own report on the uprising and its suppression. In addition, Idriss includes a quotation from Genesis (9:20–25), and in Part 3 returns to this technique: under the heading "Appendix", she reproduces an excerpt from *Species Facti* in which Hesselberg not only documents the rebellion and its suppression but also lists the convicted Crucians by name, outlines the accusations, specifies the punishments, and notes the duration of each person's suffering until death, down to the minute.

The novel's historical dimension is thus framed by a dense network of paratexts that foreground the mechanisms of a colonial record-keeping. Through this technique of referencing, Idriss draws attention to the historical authentication and supposed factuality of the narrated events, and to the global relations that shape island histories. The paratexts constitute what may be called a colonial-Christian archive: a system of textual authority that legitimises imperial power through documentation. By carefully selecting these materials and situating them in precisely defined times and places, Idriss exposes how colonial archives transform administrative writing into historiographical authority, but also how our perception of sources depends on their framing—the novel thus underlines their relationality.

Furthermore, the colonial archive is not the novel's only mode of historiography. Between the appendix from *Species Facti* and the acknowledgements in the afterword (2024, 283), Idriss inserts a brief section titled "A Grave Robbery", whose narratological status remains ambiguous. Although a first-person narrator speaks here, it is formally unclear whether this "I" belongs to the author or remains part of the fiction—an indeterminacy that functions as a poetological statement. In this passage, the "I" reflects on the use of historical sources within the novel: "[Hesselberg's] report is the principal source of information about what occurred on St. Croix, Advent 1759. As for Cudjo, William Davis, Benjamin Bean, Mathias and Tornhull, I have access only to Hesselberg's descriptions, in which they play minor roles" (Idriss 2024, 279–80).

On the one hand, this self-reflexive narrator confirms the factual existence of Hesselberg and the Crucians allegedly involved in the rebellion; on the other, it questions the extent to which such sources can transmit historical truth. Instead, the passage insists that all historical records are themselves narrative constructs. Within this framework, *Halvt* situates itself as another form of historiography, one that diversifies rather than confirms dominant narratives:

> All the information I've had access to here was documented by the dominant party. I've taken liberties, gotten lost, fixated on some details, ignored others. I believe Hesselberg did the same. He had just as much reason to lie as I do.
>
> (Idriss 2024, 280)

Even though the sea no longer plays a literal role in this part of the novel, Idriss' counter-narrative to the fixed colonial archive functions as a tidalectic approach to history itself: She juxtaposes colonial historiography with what DeLoughrey terms an "alter/native historiography" (2007, 22), without privileging one over the other. The archival quotations and Idriss' creative variations stand side by side as parallel models of the past. Just as her characters are rendered tidalectic—moving fluidly across the Atlantic—so too does the island's historiography become fluid and dynamic.

Continuities

A significant feature of *Halvt* is the novel's emphasis on continuities between the heyday of Scandinavian colonialism in the Caribbean and economics of the late modern present. Inge's partner Karo summarises:

> The worst thing is that the link is surely there . . . the link they're talking about between you and Hesselberg, between Norway now and St. Croix then. It's there, of course, but not through you, it's not your responsibility, not personally.
>
> (Idriss 2024, 244)

Karo's comment can be understood to represent a key to the novel, whose main focus seems to be the "link" itself (in Norwegian "kobling"). Although the link is personified in the novel by Inge, the focus is less on the protagonist's development than on a discussion of what she embodies and what is projected onto her: the spatial, temporal, and personal connections rooted in colonial endeavours and their continued impact. While other colonial-historical novels close with the official end of the colonial era, Idriss' interest in the various legacies of Atlantic colonialism begins right there. In *Halvt*, the colonial heritage manifests itself in the form of family history and family relations; in today's racism with its roots in colonial-era ideologies and practices; and in the adoption of the plantation economy's ownership structures and land use by today's tourist and extractive industries in the

Caribbean. It thus also includes the continuity between different forms of energy extraction described by Boyer (2023, 11–44).

Both Hesselberg and Norwegian merchants wrung resources and wealth from the Caribbean colony: Hesselberg with the plantation he ran in addition to his service as colonial official, and the merchants and trading companies through their participation in the slavery-based sugar cane trade. Today, Norway is a global stakeholder in the fossil fuel industry, which creates a tension between the positive image of pristine nature and progressive social development on the one hand, and the dirty image of an exploitative "petroculture" on the other. In short, *Halvt* positions Equinor as heir to the Dano-Norwegian colonial traders. By juxtaposing the two phases of Norway's economic history, the novel raises the question of how the colonial plantation economy and today's fossil-based economy are linked via the exploitation of human and natural resources, and how both are based on social inequalities.

A first instance where the novel posits a manifest continuity between colonists and today's global corporations is the transformation of the Crucian landscape in the form of deforestation. In the beginning of the novel, when Inge's mother picks her and her colleagues up from the airport, she tells them that the "Danes and Norwegians cut down all the forest, . . . but then it grew again" (Idriss 2024, 18). To Inge's despair, caught between her desire to reunite with her mother and her loyalty to her employer, her mother continues in a similar vein the next morning at breakfast in the hotel where the Equinor team is staying:

> 'People are worried', says Inge's mother, her voice stern. 'The oil industry doesn't exactly attract tourists. It also cleared a lot of the forest, and then it went to hell anyway, the drinking water turned black, completely black, ash everywhere, after the fire.'
>
> (Idriss 2024, 67)

In reality, there is no forest in the middle of the island to pass through on the way from the airport, and no oil refinery operated by Equinor. Instead, you pass the industrial ruins of a gigantic American-Venezuelan oil refinery that prospered from the 1960s to the 1990s, so in the process of fictionalisation, the oil plant has become "norwegianised". The only forest area that has survived the actual clearing for the colonial plantations is located in a mountainous area in the north-west of the island, Maroon Ridge. It is named after people of African descent who tried to escape slavery by settling in the impenetrable forest but were constantly persecuted. The relocation of the forest to the centre of the island in the novel's fictional universe allows the central motif of the "link", or *kobling*, to be condensed into the short distance that the Equinor team travels in Inge's mother's car between the airport and the hotel King Christian in Christiansted (named after a Danish king and existing in real life), immediately after landing on St. Croix. In the novel, the forest and its repeated destruction represent the economic development of the island by overseas actors since colonisation, the ecological, social, and ethical costs of this development, as well as local resistance. The place where Inge seems to find her way back to her maternal family at the end of the novel might just be a place where the forest "grows again".

Another instance where a transatlantic link between Norway and St. Croix materialises in *Halvt* is the pollution of waters. This time, it is Inge's aunt, her mother's sister, Linda's turn to speak:

> Did Inge know that they spill oil in the Caribbean almost two hundred times a year? . . . Soon it will be all oil and no water here. Where will the tourists swim then? If the beaches are black and slippery, frying in the sun and breading them like fish in a frying pan. It's not going to work. We need the sea, the forest too. (Idriss 2024, 91)

Linda establishes a transatlantic connection between the Caribbean and Norway via oil, whereby streams of oil and money seem to flow in different directions: "No, we're not going to sell ourselves for oil money that will never come. That money will disappear somewhere else anyway. The money is going to Norway, they know that. The only thing that comes here is the dirt" (Idriss 2024, 90). The image of transoceanic trails could be extended to again include a temporal connection in addition to the geographical one: after all, both enslaved African people and crude oil have been described as "Black gold" by profiteers, each as a preferred commodity of the respective era. Both trade systems—the colonial transatlantic trade with its massive displacement of labour and the global oil industry—have been engines of both unfathomable development and destruction (see also Boyer 2023, 11–44). In addition to the pollution of vulnerable ecosystems in the Caribbean, the comments also reflect the current domestic debates in Norway on maritime issues: should new oil fields be developed and deep-sea mining be allowed? How threatened are Norway's fjords by agricultural discharges, salmon farming, and mining waste? The comment that there soon "will be all oil and no water here" can thus be understood, in a concrete and figurative sense, as a comment on Norway's dependence on oil extraction and the exploitation of maritime resources.

A third instance where the novel connects the past and the present as well as the two sides of the Atlantic occurs when Equinor's rhetorical strategy is uncomfortably reminiscent of the colonial ideology of a "civilising mission", the idea that a privileged European class would be responsible for the progress of less developed regions and peoples. Inge's boss Geir and the consultant Michael are the representatives of such a paternalistic and capitalist view in *Halvt* when they tie a promise of prosperity to the future of the oil industry on St. Croix:

> 'The economic impact can hardly be measured', says Michael. 'Coral reefs and wetlands are all well and good, but no one makes money on coral reefs. If

> St. Croix is going to catch up with the mainland, if it's going to have a reliable mobile network, if it's going to reopen its libraries, it has to say yes to this.'
>
> (Idriss 2024, 68)

In contrast to images of Norway as an agent of peace, sustainability, and equality, a Norwegian state-owned company is in *Halvt* a representative of a colonial ideology and a perpetrator of environmental destruction. Both white Norwegian men in the novel, the Equinor employee Geir and Inge's father, appear as liberal and initially friendly towards the Crucian women in Inge's family, yet they exploit an unequal distribution of power in terms of gender, wealth, and race. They display, biologically or figuratively, kinship with Engebret Hesselberg. By asserting the *longue durée* of capitalist, paternalistic, and ultimately violent ventures since the heyday of transatlantic colonialism, the novel takes a critical look at exceptionalist Norwegian self-images of innocence, benevolence, and gender equality.

Conclusion

Marjam Idriss' novel is groundbreaking in its combination of postcolonial and ecocritical perspectives on the Norwegian past and present, linking (or "kobling") Norway and St. Croix via a common transatlantic heritage of regimes of resource extraction. The connection between colonial history and contemporary petroculture proves efficient to examine the legacies of colonialism pertaining to both the economic and social history of Norway, Europe, and the former colonies. With its insistence on temporal and geographical continuities, the novel contributes to a variety of topical issues in literature and public discourse.

Firstly, our "tidalectic" reading of the novel emphasises how the island of St. Croix and the Fennoscandinavian peninsula are connected via global flows of humans, goods, and water. *Halvt* discusses the extent to which fluid identities emerge from colonial transatlantic entanglements. It thus echoes approaches from critical theory originating in the Caribbean in which identity is understood as composite, thus rejecting ideas of purity, authenticity, or "rootedness" in relation to national belonging, and assuming hybrid or creolised collectives as the premise of Caribbean—and by extension Atlantic—realities (Gilroy 1993; Glissant 1989; Hall 1989). The very title of the novel, and then the narrated experiences of the main character and her mother, invite the audience to adopt such an understanding of identity for Norway and Nordic literature as well.

Secondly, with its "tidalectic" historiography, the novel not only frees sources (a word that already implies fluidity) from the colonial archive that reveals the extractivist legacies of contemporary petroculture but also reflects on the necessity to integrate the archives and sources that have been separated over the course of the last few centuries of Atlantic history.

Thirdly, *Halvt* reminds us of the continuities in regimes of energy and wealth extraction. The novel thus contributes substantially to the necessary consideration of colonial legacies within topical debates in Norway about so-called clean energy and the ethical and social dimensions of oil wealth. *Halvt* can be read as an example of literary petrocriticism in that the novel focuses on economic inequality and ecological decline in the wake of the extraction of fossil fuels. The novel's special quality is that the main character embodies both sides of an "oil encounter" (Ghosh 1992) and must deal with the resulting internal and external conflicts: on the one hand, the privileged oil producers benefitting from a welfare state financed by morally dubious investments; and, on the other hand, the local population on St. Croix who must live with the social and ecological consequences of a failed oil industry project.

Halvt contributes to exploring Norway's position in transatlantic colonialism and colonial heritage. The novel asks to what extent an unwillingness to confront racism is linked to an unwillingness to consider Norway's contribution to the slavery-based colonial economic system of the past that has evolved into today's exploitative and polluting industries. Young Inge never learns the truth about Engebret Hesselberg despite her father's research into their ancestor. Instead, the affectionately named "Enge" figures in pleasant bedtime stories. Inge wonders how the stories (including the similarity of the name) have contributed to her identification with the paternal, Norwegian side of the family (Idriss 2024, 100–3). The goodnight stories end with "Enge's" sojourn in Copenhagen before his travels to the Caribbean: "The father didn't get any further than that before Inge grew too old for bedtime stories" (Idriss 2024, 103). This element of the story can be read allegorically as Norway's "sanctioned ignorance" (Habel 2012, 99) about colonialism and slavery, and the replacement of stories of violence by positive national narratives and self-images.

To summarise, we read the novel as a commentary on late modern globalisation and its connection to colonialism as the first phase of global trade and extractivism. In the age of the "global village" with its means of communication and transportation, the Atlantic Ocean seems to have been conquered, and the distance collapsed, materially and metaphorically. In *Halvt*, the whole world including the protagonist's partner in Norway can live-stream Equinor's hearing on St. Croix. Compared to the era of tall ships, mutinies, and transports of enslaved humans, Atlantic crossings today are less associated with exotic adventures, dangers, fears, and horrors. However, the novel demonstrates that the distance between the north-east Atlantic and the tropical Atlantic still exists in the form of unequal power relations and the exploitation of people and resources. Whereas in the "Age of Sail", the Atlantic was a symbol of the encounter between human enterprise and the forces of nature, today it is an example of a natural space impacted by global capitalism and extractivism. Idriss' novel illustrates how the supposed gap between the colonial and postcolonial period does not exist, pointing to a *longue durée* of transatlantic trade systems and their impact on humans and waters.

Works cited

Boyer, Dominic. 2023. *No More Fossils*. University of Minnesota Press.

DeLoughrey, Elizabeth. 2007. *Routes and Roots: Navigating Caribbean and Pacific Island Literatures*. University of Hawai'i Press.

DeLoughrey, Elizabeth. 2019. *Allegories of the Anthropocene*. Duke University Press.

Frank, Søren. 2022. *A Poetic History of the Oceans: Literature and Maritime Modernity*. Brill. https://doi.org/10.1163/9789004426702_001

Ghosh, Amitav. 1992. "Petrofiction: The Oil Encounter and the Novel." *The New Republic* (March 2): 29–33.

Gilroy, Paul. 1993. *The Black Atlantic: Modernity and Double Consciousness*. Verso.

Glissant, Édouard. 1989. *Caribbean Discourse: Selected Essays*. Translated by Michael Dash. University Press of Virginia. Originally published as *Le Discours antillais*, 1981.

Habel, Ylva. 2012. "Challenging Swedish Exceptionalism? Teaching While Black." In *Education in the Black Diaspora: Perspectives, Challenges, and Prospects*, edited by Kassie Freeman and Ethan Johnson, 99–122. Routledge.

Hall, Stuart. 1989. "Cultural Identity and Diaspora." *Framework: The Journal of Cinema and Media* 36: 222–37.

Hove, Jon Olav. 2024. "Trøndelag, Atlanterhavet og transatlantisk slavehandel: Histori og kulturminne." In *Midtnorske historier*, Vol. 2, edited by Emil Mård Vaadal Eliasson, Magne Brekke Rabben, and Knut Ola Naastad Strøm, 39–69. Novus forlag.

Idriss, Marjam. 2024. *Halvt*. Gyldendal.

Joof, Camara Lundestad. 2024. *I Talk about It All the Time*. Translated by Olivia Noble Gunn. University of Wisconsin Press. Originally published as *Eg snakkar om det heile tida*, 2018.

Körber, Lill-Ann. 2025. "Contemporary Scandinavian Colonial-historical Fiction: Slavery and Exceptional Whiteness." In *A Comparative Literary History of Modern Slavery: The Atlantic World and beyond*, Volume II: *Slavery, Memory and Literature*, edited by Karen-Margrethe Simonsen, Madeleine Dobie, and Mads Anders Baggesgaard, 188–213. John Benjamins. https://benjamins.com/catalog/chlel.37.11kor.

Naum, Magdalena, and Jonas Nordin, eds. 2013. *Scandinavian Colonialism and the Rise of Modernity: Small Time Agents in a Global Arena*. Springer.

Vedeler Nilsen, Live. 2020. "Dommeren og slaven." *Nasjonalbiblioteket*. https://www.nb.no/historier-fra-samlingen/dommeren-og-slaven/.

Weiss, Holger, ed. 2016. *Ports of Globalisation, Places of Creolisation: Nordic Possessions in the Atlantic World during the Era of the Slave Trade*. Brill.

PART V

The Global Sea and the Undersea

17

"ALL THE WORLD WAS BLUE"

The sea in Isak Dinesen's stories

Annegret Heitmann

When the entire world around you is perceived as blue, you are likely surrounded by water somewhere on the high seas. Due to the reflection of the sky, the transparent surface of the water appears as a blue sphere—just as the Earth, which is over 70 per cent covered by water, is perceived as the "blue planet" from the perspective of outer space. Since antiquity, these maritime spaces and aquatic dimensions of life have been a significant subject in literature. In recent times, an ecologically oriented field of literary studies has turned its attention to this blue world, placing the interaction between humans and the sea at the centre of its interest under the concept of "blue humanities": "Covering more than two-thirds of the earth, the inhospitable sea represents the most powerful nonhuman actor in world history", Steve Mentz writes in *An Introduction to the Blue Humanities* (2024, 2). But the wet element is always in flux and its colourlessness can appear in different guises, leading Margaret Cohen to differentiate between blue, brown, and white water in her discussion of the affinity of maritime spaces with the Bakhtinian concept of chronotopes. She understands the chronotope of "blue water" as referring to the oceans with their fundamental lawlessness and their defiance of human expectations (2006, 650). Due to its changeability and its significance for humanity, this blue world became an inexhaustible source of literary images and descriptions.

As such, it also appears in different guises in the narratives of the Danish author Karen Blixen, who first achieved fame under the pseudonym Isak Dinesen with stories written in English, which she later translated into her first language and published as Karen Blixen. Sometimes, the visual impression of the sea is in the foreground: "the surface of it glimmered like silver" (Dinesen 1942, 196),[1] sometimes the auditory effect of "the incessant, songful murmur of the sea" (Dinesen 1942, 196) is called upon; the proximity of the maritime sphere can even appeal to the olfactory sense, when "the rank briny smell was intoxicating" (Dinesen 1934, 2).

DOI: 10.4324/9781003647638-23

There can also be talk of the power of the "untamable waves" (Dinesen 1934, 1), or a realm of the dead is evoked: "this flatness and whiteness of the sea was very strange, like the breath of death over the world" (Dinesen 1934, 232). At other times, the vastness and endlessness of the seascape remind humans of their finiteness: "It was the end, and the beginning, of the world" (Dinesen 1942, 200); the sea evokes a different timeline from human linear time.

The sea is perceived with all the senses and given multiple meanings; it can be friend or foe, uncanny or transparent, reflective surface or unfathomable depth. All these poetic formulations are ascriptions and symbolic charges by the human narrator. However, their originality and conciseness make it seem worthwhile to take a closer look at the functionalisation of the maritime, but above all at the position of the human in relation to the sea and the possible interaction between humans and the sea. What is the nature of the relationship between the aquatic habitat and the characters in Dinesen's texts? Do her texts reveal alternatives to the dominant anthropocentrism? Does the encounter with the ocean lead to revisions of the human position or even a different humanity? These questions will be pursued in four stories with iconic human–sea relationships, such as the shipwreck.

"Tempests"—shipwreck and the theory of elements

Isak Dinesen is not primarily known as a maritime author, even though she wrote most of her texts at her house in Rungstedlund with its view of the waters of the Øresund. She published a relatively small oeuvre between 1934 and 1960, and given her status in between the national literatures of Denmark, England, and the United States, as well as her rich international intertextuality and her unconventional, somewhat stilted language, she is difficult to place in terms of literary history and genre traditions. Her "short stories" are long and full of anecdotes, digressions, and different plotlines, and she does not follow genre rules or current literary trends. In spite of her archaic language, her fantastic or "gothic" elements, her historical and seemingly old-fashioned settings, and her often aristocratic and strange or "queer" characters (Heede 2001), she is a thoroughly modern author who works with provocations, ambivalences, unconventional perspectives, and calculated strangeness.

It would be wrong to assume that Blixen/Dinesen[2] had a decidedly ecocritical intention[3]; unlike today's authors, she does not report on the threats to the world's oceans from global warming, pollution, and microplastics, nor on the hazards to archipelagos from rising sea levels. It is obvious from her texts and various statements, though, that she perceived her natural surroundings very consciously and with respect. What is most remarkable about her understanding of nature, however, is not an environmental but a quasi-philosophical impetus. A decisive confession can be found in a letter to Aage Henriksen: "I cannot at all see that there is a dividing line between nature and humankind" (Blixen 1966, 213, letter dated 13

May 1954, my translation). Dualistic thinking is alien to her; for her, humans are part of nature. In several places, her texts reveal that she draws on the traditional concept of the theory of elements in her reflections on the relationship between culture and nature. First systematically developed by Empedocles around 2,300 years ago, and effective for a long time until it was marginalised by the modern natural sciences around 1800, the concept represents a holistic understanding of the world order with all its manifestations. The four elements of fire, water, earth, and air—arranged as a tetrad—their understanding as forces, the techniques associated with them, and their transfer to the human body were depicted and consolidated for centuries in visual art, mythology, and literature. "No other scientific theory has had such a long period of validity", write Gernot and Hartmut Böhme, explaining the influence of the theory by the fact that the understanding of nature as an element reveals a reciprocal power structure between humans and nature: "In the elements, the power of nature was experienced most emphatically, and in the elements, the power of man over nature was established most clearly" (2004, 19–21).

Dinesen repeatedly refers to this doctrine in her work, for example, by assigning natural phenomena or even her fictional characters and their actions to one of the elements. The idea is developed most extensively in the story "Tempests" (in *Anecdotes of Destiny*), a tale about a shipwreck, a natural disaster "in the worst weather which . . . has ravaged our coast" (Dinesen 1958, 89). The shipwreck is a classic topos of the confrontation between humans and the sea; people's efforts to make use of the sea as a means of transportation fail in the face of the forces of nature: "Shipwreck stories represent the human experience of natural hostility, narrating humankind's failed attempts to navigate an uncertain world" (Mentz 2015, xxv). The title of the last story the author completed before her death contains a double meaning: The plural form simultaneously refers to a weather phenomenon and to Shakespeare's drama *The Tempest*.[4] In the course of the plot, a planned production of this play is thwarted by a natural disaster in which the protagonist Malli heroically steers a ship threatened by storm and shipwreck through the dangers into a safe harbour. At the highest point of danger, when "the roaring sea [was] breaking over them house-high", the ship runs aground on a rock, "immediately shipped a quantity of water" (Dinesen 1958, 90), and threatens to sink, but 19-year-old Malli is able to save it through her courage and her commands. She achieves this through the self-suggestion that she is the powerful air spirit Ariel, whom she was supposed to embody in *The Tempest*. The fiction produces an empowerment that resists or fails to perceive the sublime and powerful nature of the raging sea, providing a compelling example of the performative power of art. However, the feeling of victory over nature is short-lived; the tragic death of Ferdinand, the young sailor fighting alongside Malli, exposes her alleged victory as a catastrophe and ultimately as an arrogant fantasy of omnipotence, which in return casts doubt on the power of art. In this sense, Dinesen's last story—just like the Shakespearean intertext—has mostly been read as a self-reflexive, sceptical review of her own work (Selboe 1996, 128–31; Holmberg 1995, 193–97). The story reflects on fiction as presumption and shows how a contingent event has been misunderstood as providence through the assumption of human power over a natural catastrophe.

However, one can also focus more strongly on the other meaning of the title word, on the storm as a weather phenomenon, and recognise in the text's reference to climate a conception of the world in which humans and nature are united. The text is set in a maritime environment, in a harbour town where the Hosewinckel family of shipowners has become rich from overseas trade. Malli is connected to the oceans by her absent father who lives his life by the words "stop wave-beat, stop heart-beat" (Dinesen 1958, 79), and she is characterised by her "close contact to the elements" (Holmberg 1995, 187). It is the interaction of the elements of water and air that stirs up the sea, threatens the ship, and causes the disaster. In her role as the air spirit Ariel, Malli believes she can take on the waves while the others on board "came within an ace of giving up the struggle with the elements" (Dinesen 1958, 91). The text thus explicitly mentions the elemental status that seems to characterise nature and humans alike.

In the natural sciences, the theory of the elements has of course long become obsolete, and yet talk of the elements has persisted in colloquial language and in art; indeed, "only a few symbolic forms [show] a similar persistence in our culture" (Böhme and Böhme 2004, 19), because the elements function as "media of perception" (Böhme 2010, 463). When Dinesen now insists on the elementality of nature, and represents water and air as elements, she achieves two things. Firstly, nature is presented as an actor in its own right; the sea and the weather are—as in ancient times—attributed agency, a power that modern science, technology, and politics generally deny them, and a restriction they then attempt to break in Dinesen's story. Secondly, this mythical language has the effect of integrating humans into nature rather than assigning them a position superior to nature, which would in turn present nature as an Other. In the theory of elements, humans and nature are part of a whole. In this respect, the narrative contrasts the will to dominate that characterises modern anthropocentrism with the older worldview of the elements as forces of nature. Malli's heroism is an attempt to conquer the sea and subordinate it to the needs of humans. Her renunciation of this quest at the end of "Tempests" is, then, an admission of an inappropriate anthropocentrism based on the desired dominance of culture over nature.

"The Sailor Boy's Tale"—unity of human, animal, and sea

Confrontations with the sea in Dinesen's work do not always thrust humans into such precarious situations. On the contrary, the many travellers, sailors, captains, skippers, and pirates who appear in her texts make use of the wet element to travel or trade, showing that "worldwide blue-water trade routes became essential to European economies" (Mentz 2015, 2). It was the world's oceans that connected previously remote places, creating links between countries and continents and thus facilitating the global movement of goods, which in turn promoted the rise of capitalism and colonialism (Sloterdijk 2006, 19). There are traces of this economic and political significance of the sea in many of Dinesen's stories; goods such as ivory, rhinoceros horn, silk, tea, vanilla, cinnamon, pepper, and even tropical birds like parrots, cockatoos, and monkeys are mentioned, ferried in from overseas and sold at a high profit.[5] Probably the most prominent example of global trade is the famous menu in "Babette's Feast" (in *Anecdotes of Destiny*), which is composed with ingredients from colonial trade and the exploitation of the colonies. This unifying dimension is underlined by the fact that all of the world's oceans play a role in Dinesen's textual universe: the Atlantic, the Indian and Pacific Oceans, as well as the North Sea, the Caribbean, the South China Sea, the Baltic, and the Mediterranean: "It was the same water", the narrator says of "The Deluge at Norderney", "which rolled, out to the west, a hundred fathoms deep, and washed the white feet of the cliffs of Dover" (1934, 4). The liquid medium creates connections and contact zones, for better and for worse.

The little ship's boy Simon in "The Sailor Boy's Tale" (in *Winter's Tales*) is also part of this maritime world of trade and global connections. However, he is introduced in a situation where he finds himself all alone "on the wet, swinging deck" (Dinesen 1942, 3) of a ship on its way from Marseille to Athens. He is a lonely individual on unsteady ground, belonging only to the vastness of the sea. Surrounded by high, grey waves, the human being stands out as tiny and insignificant. His origins are never specified, making him appear homeless and without roots. But then he discovers a peregrine falcon in the rigging of the ship, a creature that lives on all continents of the world and travels enormous distances. The boy experiences "a fellow-feeling" with this animal: "That bird is like me" (Dinesen 1942, 3), and he decides to free the falcon trapped in the rigging of the mast. Human, animal, and sea form a unity here (Mortensen 2026, 31–35).

After this opening scene, a fairy-tale-like adventure narrative begins, which has often been described as a coming-of-age story, but its goal is not the expected form of individuation. The adventures take place two years later in the remote northern Norwegian harbour town of Bodø, which is described as an "international meeting place of herring fishermen and merchant ships from all corners of the world; . . . a forest of masts, and on shore a turbulent, irregular display of life with many languages spoken" (Dinesen 1942, 4). The sailor boy's homelessness is thus set in a busy world networked by the sea. When he goes ashore on an evening trip, he encounters the hustle and bustle of the merchants and the colourful life of a port town. Moreover, he is confronted with sexuality, love, violence, and death; he meets a girl and snatches a kiss, but the girl Nora remains separated from him by a fence during their two encounters. Much closer to him is Ivan, the Russian sailor, who also kisses him and violently harasses him, causing Simon to draw his knife and stab him. In doing so, he kills an opponent, a challenge the hero must often master in mythopoetic initiation stories. It is important that his escape is successful so that the adventures can be recounted afterwards. Simon thus becomes a narrator;

the text concludes with a homage to the fairy-tale-like formula: "and lived to tell the story" (Dinesen 1942, 15).

His rescue is made possible by an old Sámi woman, who turns out to be the falcon in a different guise; such shape-shifting is also a motif taken from fairy tales, but it likewise figures in Scandinavian folk ballads and the animist traditions, sometimes ascribed to the Sámi.[6] The boy experiences a deep commonality with this bird-woman. Sunniva helps him and saves him; like him, she is a wanderer and a stranger: "That falcon was me. We Lapps often fly in such a manner, to see the world. When I first met you I was on my way to Africa" (Dinesen 1942, 14). The community into which Simon's initiation leads is not determined by the expected awakening of an identity determined by love, sexuality, and sedentariness, but his life is governed by a unity with nature—the sea and a falcon—and a commitment to permanent mobility, which connects him with the peregrine falcon, the nomad woman, and the sea. His loneliness turns into independence, which predestines the character to be a narrator. After all, tales of the sea, the dangers of travel, and maritime adventures have always been at the heart of the narrative repertoire (Frank 2022).

This includes the fact that, as a narrator and as a person belonging to the sea, Simon transcends the usual boundaries of autonomous human identity. Peter Mortensen understands the text as an expression of a "human-animal entanglement . . . where Simon intuits a strange kinship with his surroundings" (2026, 34) or, as the narrative puts it, "as if the sea and the sky, the ship, the bird and himself were all one" (Dinesen 1942, 4). The transgression of a human being that is identical with itself is also underlined by a sexuality that conflicts with heteronormativity (Heede 2001, 127). In the male-dominated world of the sailors, Simon resists homosexuality while his heterosexual desire fails, and he remains alone. Here too, the person is not anchored in the typical human relationship system but becomes part of the weave of nature, as represented by the Sámi people and Simon's affiliation with the sea.

While Malli was antagonistic towards the sea in "Tempests", Simon fully belongs to the maritime world: as a sailor, as a narrator, and as a "little bird", as Sunniva calls him. Like the sea and a bird, he is part of nature, constantly on the move, and free from the constraints and norms of human society.

"The Blue Story" in "The Young Man with the Carnation"—the horizontal view

Lady Helena's relationship with the sea is even more intimate as we learn from an interspersed anecdote recounted by the writer Charlie Despard, protagonist of the story "The Young Man with the Carnation" (in *Winter's Tales*). The short, inserted narrative tells of an English lord whose purpose in life was to collect rare blue porcelain: "To that end he travelled to Persia, Japan and China, and he was everywhere accompanied by his daughter, the Lady Helena" (Dinesen 1942, 33). When a fire breaks out on board in the Chinese Sea, the ship is evacuated and the Lord is separated from his daughter in the disaster. She is rescued by a young English sailor with whom she drifts about in a small lifeboat for nine days before a Dutch freighter picks them up. Back in England, the father makes sure that his daughter recovers, and that the young sailor is sent "to go shipping in the other hemisphere and never come back" (Dinesen 1942, 34). The young woman then falls into depression and no longer feels comfortable on land, neither in her father's castle nor in the cities of the continent, but wants to go out to sea again: "So she began to sail from one country to the other" (Dinesen 1942, 34) to collect blue porcelain like her father. She is looking for a very special blue colour that is supposed to contain something "from the time when all the world was blue" (Dinesen 1942, 35). Finally, shortly before her death, she finds this particular blue in an old jug plundered from the Chinese emperor's summer palace.

The anecdote thus includes maritime trade, a shipwreck, days of drifting in a lifeboat, and a feeling of unity with the sea. It has mostly been read as an example of Charlie Despard's new aesthetics, which is no longer dominated by social criticism, as his successful first book was, but calls upon symbolist traditions; the search for the blue colour is reminiscent of the Romantics' longing for the blue flower and Mallarmé's poem "L'Azur" (Langbaum 1964, 160). What has been less discussed is that the realism Charlie has favoured so far is by no means completely absent from the anecdote. The rich Englishman buying up precious antiques in Asia is a very real figure of global trade that profits from the inequality between the West and the rest of the world. He can do favourable business in the near and far East, and his daughter is even able to acquire antiques from the plundered palace of the Chinese emperor. Transcontinental trade is usually "directed" and exploits asymmetries; the plot of the story is therefore based on a business model. Furthermore, Helena's longing is conditioned by the repressive Victorian double standards and class system; the father removes the young sailor from his daughter's proximity because, firstly, he is not befitting of the Lord's station and, secondly, he might reveal juicy details from the nine days at sea. Helena is a victim of the patriarchal order; she perceived the time at sea as total freedom.

While the anecdote as a whole can thus be read as a clash of poetological positions, its most astonishing potential lies in Helena's highly original conception of the world. Firstly, she rejects the fact that the oceans have a bottom and imagines the earth as a bubble floating in the ether: "The water, which is the noblest element, does, of course, go all through the earth, so that our planet really floats in the ether, like a soap-bubble" (Dinesen 1942, 35). Secondly, as she imagines that the planet is made of water at its core, she believes that her eternal journey, the search for the blue colour and lost time with the sailor, is driven by a second ship:

> And there, on the other hemisphere, a ship sails, with which I have got to keep pace. We two are like the reflection of one another, in the deep sea, and the ship

> of which I speak is always exactly beneath my own ship, upon the opposite side of the globe.
>
> (Dinesen 1942, 35)

While the father banishes the sailor to the "other" hemisphere from what he considers to be his central point of view, for Helena, the world is a unity that is expressed in the interplay between the two ships, joined in a reciprocal cosmic relationship of attraction, "as the moon draws the tides" (Dinesen 1942, 35). The idea is somewhat analogous to Jules Verne's fantasy of a "journey to the centre of the earth", but here the protagonist, as a geologist, was primarily interested in the solid core of the globe. By contrast, Helena's idea of the world is based on water as the unifying (and, as she says, noblest) element; she sees her movements on the water as being controlled by cosmic forces. Although her theory appears naïve and unscientific, it does correspond to the fact that 71 per cent of the planet called Earth is covered by water, which naturally has an impact on the position of the antipodes: Only 4 per cent of the Earth's surface has both antipodes on land, and 46 per cent of the Earth's surface has both antipodes in water (Erthel and Stockhammer 2020, 166). In this respect, Helena's dream of the parallel ship with the sailor on board, whom she hopes to meet again after sinking "in the midst of the world" (Dinesen 1942, 35), rests on the recognition of the globe as an "ocean ball".

This idea of the world is characterised by coherence, lightness, and transparency, and therefore alludes to freedom. During the nine days in the lifeboat with the sailor, Helena experienced the water as her only point of reference: She experienced the world as the sea, the sea as a whole, and herself as part of it. From the small lifeboat, she had a horizontal view of the ocean, so she saw the world neither in its manifold details nor from an elevated position: "The horizontal view of the sea stands for the attempt *not to* view a whole in a totalizing way" (Erthel and Stockhammer 2020, 169). It is reserved for the sailor (or the swimmer) to look out across the plane seascape and into the distance and discover the uniformity of the world and the insignificance of humans. Helena's view—in contrast to her father's worldview—is holistic, horizontal, and egalitarian. The shipwreck and her time with the sailor in the small boat led her to this view; being at sea opened up an alternative view of what is really important: love, equality, and the trivial position of human beings in the world.

There is no heaven, no afterlife in her blue world. The hope of a reunion is not bound to another world or a striving for transcendence, which is illustrated by the fact that the blue colour of longing finally takes on a material form in an old Chinese jug. As a hollow vessel, the jug resembles the small boat that carried the castaways for nine days. The blue colour remains tied to the experience at sea and leads to a commitment to materiality, which culminates in Helena's wish that her heart be kept in the hollow of the blue jug after her death. Her heart is thus transferred to a world of blue that is equally material, and the reunion she desires is achieved not in heaven but in the blue here and now. In this story, the blue colour is not a sign of transcendence but a sign of the maritime experience, an immersion in an alternative natural world without hierarchies and power.

"The Diver"—unsettling the perspective

While a horizontal view of the sea is possible from a small boat, experiencing the water's depths is reserved for the diver: "For many things happen to those who dive to the bottom of the sea" (Dinesen 1958, 12), as Saufe, the protagonist of the story "The Diver" (in *Anecdotes of Destiny*), explains. He leads a contented life; being a diver in the Chinese Sea has even made him wealthy. Because he is able "[to] dive down into greater depths and stay down there longer than any other fisherman" (Dinesen 1958, 13), he can bring up the most beautiful pearls. Thus, he exploits the sea, but his luck has made him fat, and his serenity implies that no story can be told about him: His life contains no crises, changes of location, or turning points. Four times, the narrative emphasises that Saufe's harmonious life as a diver "makes no story", "no story at all" (Dinesen 1958, 16). Therefore, it is logical for the complexly structured narrative to break off at this point, and the untold story is replaced by the speech of a boxfish (Holmberg 1995, 163–71; Brantly 2002, 182–85).

The "old cowfish with horn-rimmed spectacles" (Dinesen 1958, 16) is a veritable philosopher and the diver's mentor: It was not the material wealth but her teachings that made him realise his good fortune. When she claims that the fish "among all creatures is the one most carefully and accurately made in the image of the Lord" (Dinesen 1958, 17), she invokes the biblical account of creation, but in the same breath she banishes man from the central place assigned to him therein as the crown of creation. While humans are bound to the earth and move laboriously on its surface, the fish leans "confidently and harmoniously upon our element" (Dinesen 1958, 18), they move "in all dimensions" and are carried by water from all sides.

The inversion becomes particularly clear in the motif of the Flood, which from the point of view of the fish meant abundance "and that blessed plenty appeared with fullhorn" (Dinesen 1958, 19), but not the transgression, danger, punishment, and a new beginning that it encompasses for humans. When God had created the heavens and the earth, the boxfish summarises the biblical account of creation, the Fall of man immediately followed: "Man, capable of falling, fell almost immediately . . . But the fish did not fall, and never will fall, for how or whereto would we fall?" (Dinesen 1958, 18). While the fish enjoy the Flood caused by sinful humans as a time of "abundance", humans experience the implications of the Fall as the establishment of a difference, the beginning of desire and shame as well as the freedom to act. It is only the fall that enables human self-creation—and establishes the narrative of self-empowerment.

Through this "unsettling" of perspective, the human position is robbed of its conventional exceptionalism; like the fish, humans are part of nature, which is affected by the Flood as a whole. Their specificity is not given per se (or by God) but results from transgression, which enables them to act and make a new start. The fish, on the other hand, stand for an existence of harmony and sinless living, illustrated by its free-floating existence in the ocean. When the sea is not being antagonistic towards people—as in a shipwreck or a storm surge—but being viewed from a non-human perspective, it presents itself as an image of balance and defiance of gravity. Of course, this is not a scientific examination of the marine ecosystem or the life of fish but a thought experiment that unhinges the usual anthropocentric view of humans and adopts an alienated perspective.

To challenge the accustomed superior position of humans, Dinesen employs her thinking in terms of the elements, which encompass both nature and humanity. When humans attempt to transcend the boundaries of the elements—when Malli believes she is an air spirit and can defy the sea, or when the diver in his former life as a student of theology tries to build wings to fly—it appears as hubris. The carelessness of the coastal inhabitants in "The Deluge at Norderney" (in *Seven Gothic Tales*) also represents an inappropriate transgression of the human attitude towards the powerful sea. Imitating the contemporary trend of seaside tourism, even the coastal residents forget their respect for the forces of the ocean: "The peasants and fishermen of Norderney themselves learned to look upon the terrible and faithless gray monster westward of them as some kind of *maître de plaisir*" (Dinesen 1934, 1).

In contrast, the elements are indeed attributed with transgressions; nature is often depicted in extreme and dynamic situations: The "faithless gray monster" ignores its boundaries and floods the land. Dinesen's texts contain various situations in which the four elements—earth, air, water, and fire—unleash their powers; there are descriptions of devastating storms, floods and ship disasters, earthquakes, and conflagrations.[7] Such natural events uproot the world of humans and upend their expectations: The seemingly immovable earth moves, a land area becomes a body of water, the previously calm and invisible air transforms into a storm, and fire destroys human dwellings. It is the moments of unsettling human security that Dinesen's stories show. "A world turned upside down" is the title which Susan Hardy Aiken chooses for her chapter on "The Deluge at Norderney" (Aiken 1990, 84), and Charlotte Engberg talks about "Verden på vrangen" (the world upside down) (Engberg 2019, 229). The catastrophic flood described in the story transforms a large coastal area into a mass of water, and humans literally lose their footing. The flooding brings about a crisis situation for the local population and an existential challenge for the fictional characters involved.

In general, the unforeseen and disruptive nature of these natural disasters aligns with Dinesen's poetic principle of unpredictability, of surprising twists and provocations, which also characterise her plot design, deviation from genre norms, character conception, and even her archaic language. These are situations in which the unforeseen and the utterly unknown destabilise established perspectives, uproot the usual course of life, and displace the seemingly secure position of humanity. The unpredictability recalls the fragility of their anthropocentric perspective, which the characters in Dinesen's stories experience in their confrontation with the maritime sphere, as shown in the four examples above. With their technique of reversing the gaze of characters and readers alike and their surprising twists and turns, the texts help to cast doubt on the position of human exceptionalism. Dinesen creates unsettling counter-stories and imagines situations in which it is not the human being but the sea that takes centre stage. It is not given a voice but the capacity to defy humans: "What the sea was doing you could not tell" (Dinesen 1934, 3).

Notes

1 I am quoting the American first editions of Dinesen's works, as listed in the bibliography. Since her own translations of her English texts in Danish are creative rewritings, her oeuvre must be considered bilingual. In the present English-language context, I am nevertheless refraining from rendering the philologically correct quotations in both languages.
2 In English contexts, it is usual to speak of the author as Isak Dinesen, as I am doing in this chapter. As this is a pseudonym, it is, however, not quite correct to use this name where her non-literary statements are concerned; these she made in her own name.
3 See the differentiated discussion of Dinesen's attitude to nature and ecological issues in Mortensen (2026, 1–8 and 25–31).
4 A brief, but well-informed and balanced account of all Dinesen's texts is to be found in Brantly (2002). About "Tempests", see esp. Selboe (1996, 119–31).
5 See, for example, "Syndfloden over Norderney" (1934, 38), "Drømmerne" (1934, 271), "The Supper at Elsinore" (1934, 217, 225), "The Monkey" (1934, 109), and "Converse at Night in Copenhagen" (1957, 316).
6 About "Sámi magic, shamanism, and thieranthropy (human-animal transformation)", see Mortensen (2026, 34–35).
7 See among others: "The Dreamers" (1934, 313), "The Deluge at Norderney" (1934, 1–4), *Out of Africa* (1937, 294–96), "The Cardinal's First Tale" (1957, 14–15), "Echoes" (1957, 161–62), and "Tempests" (1958, 89–92, 97–100).

Works cited

Aiken, Susan Hardy. 1990. *Isak Dinesen and the Engendering of Narrative*. Chicago University Press.
Blixen, Karen. 1966. *Karen Blixen i Danmark: Breve 1931–62*. 2 vols. Edited by Frans Lasson and Tom Engelbrecht. Gyldendal.
Böhme, Gernot, and Hartmut Böhme. 2004. *Feuer, Wasser, Erde, Luft: Eine Kulturgeschichte der Elemente*. Beck.
Böhme, Hartmut. 2010. "Natürlich/Natur." In *Ästhetische Grundbegriffe*, edited by Karlheinz Barck, Martin Fontius, Dieter Schlenstedt, Burkhart Steinwachs, and Friedrich Wolfzettel, 7 vols, Vol. 4, 432–97. Metzler.
Brantly, Susan C. 2002. *Understanding Isak Dinesen*. University of South Carolina Press.
Cohen, Margaret. 2006. "The Chronotopes of the Sea." In *The Novel, Volume Two: Forms and Themes*, edited by Franco Moretti, 647–66. Princeton University Press.
Dinesen, Isak. 1934. *Seven Gothic Tales*. Harrison Smith and Robert Haas.
Dinesen, Isak. 1938 [1937]. *Out of Africa*. Random House.
Dinesen, Isak. 1942. *Winter's Tales*. Random House.
Dinesen, Isak. 1957. *Last Tales*. Putnam.
Dinesen, Isak. 1958. *Anecdotes of Destiny*. Michael Joseph.

Engberg, Charlotte. 2019. *Latter og lettere beruset: Om at læse Karen Blixen*. Syddansk universitetsforlag.

Erthel, Thomas, and Robert Stockhammer. 2020. "Weltmeer." In *Welt-Komposita: Ein Lexikon*, edited by Robert Stockhammer and Thomas Erthel, 166–70. Fink.

Frank, Søren. 2022. *A Poetic History of the Oceans: Literature and Maritime Modernity*. Brill.

Heede, Dag. 2001. *Det umenneskelige: Analyser af seksualitet, køn og identitet hos Karen Blixen*. Odense Universitetsforlag.

Holmberg, Hans. 1995. *Ingen skygge uden lys: Om livets veje og kunstens i nogle fortællinger af Karen Blixen*. Reitzel.

Langbaum, Robert. 1964. *The Gayety of Vision: A Study of Isak Dinesen's Art*. Chatto & Windus.

Mentz, Steve. 2015. *Shipwreck Modernity: Ecologies of Globalization, 1550/1719*. University of Minnesota Press.

Mentz, Steve. 2024. *An Introduction to the Blue Humanities*. Routledge.

Mortensen, Peter. 2026. *Isak Dinesen's Ecological Power*. Brill.

Selboe, Tone. 1996. *Kunst og erfaring: En studie i Karen Blixens forfatterskap*. Odense Universitetsforlag.

Sloterdijk, Peter. 2006. *Im Weltinnenraum des Kapitals*. Suhrkamp.

18

HOW TO READ A WAVE

August Strindberg in the surf

Sylvain Briens

> DEAREST DOCTOR:—"I will, I will be mad!"
> . . . "More rightly wilt thou live, Licinius, if thou wilt not always steer
> forth upon the high seas, nor, cowering fearfully before the storm,
> cling too closely to the treacherous coast."
> In the mean time, all hail to madness!
>
> YOUR STRINDBERG.
> (Scheffauer 1913, 205)

In this letter to Nietzsche, written on 31 December 1888, August Strindberg associates the sea with his correspondent's inner storm, as he does on several occasions in his drama, notably in *Dödsdansen* (1900; *The Dance of Death* 1912). However, what is particularly interesting here is how life on land is related to maritime forces.[1] These forces of attraction and aspiration that the open sea exerts upon terrestrial life, which can be seen as a form of hydro-motive energy caused by the waves, will be the central focus of this chapter.

Strindberg's relationship to the sea is too rich to be summarised in a few pages. The author's biographical context and his intimate relationship with the Stockholm archipelago are well documented and analysed in Björn Meidal's *Strindberg och havet: öarna och horisonten* (2025; Strindberg and the sea: the islands and the horizon). Meidal's main contribution is to show that the sea is not a mere setting but a central element in Strindberg's life, imagination, and work as well as to reveal how the sea reflected his moods, nurtured his creativity, and shaped modern Swedish culture.

DOI: 10.4324/9781003647638-24

The richness of maritime themes in Strindberg's works—including the novels *Hemsöborna* (1887; *The People of Hemsö* 2012) and *I Havsbandet* (1890; *On the Seaboard* 1913), the short stories "Skärkarlsliv" (1984 [1888]; Life in the Skerries) and "Fagervik och skamsund" (1994 [1902]; Fairhaven and Foulstrand), the plays *Dödsdansen* and *Ett Drömspel* (1902; *A Dream Play* 1912), and his poetry—create a literary seascape, mainly centred around the Baltic Sea and Stockholm's archipelago. These literary seascapes invite a reading attuned to what blue humanities calls an oceanic perspective: One that, like surfing, moves with the sea's unpredictable rhythms and acknowledges the entanglement of human and elemental forces.

From this standpoint, the oceans are not merely geographical or physical spaces but also political, economic, infrastructural, and more-than-human social domains that have shaped and been shaped by human history. They also belong to, and generate, cultural ecosystems, as dynamic environments in which meaning, practice, and imagination circulate.

The essential dimension of blue humanities is the decentering of our cultural viewpoint by adopting an ocean-centric perspective. This shift allows previously underprioritised aspects of our epistemological and aesthetic relationship with oceanic environments to emerge. The fluid nature of such environments challenges our usual conceptual and perceptual reference points. In *Oceano, una navigazione filosofica* (2022; Ocean, a philosophical navigation), Roberto Casati advocates reducing the cognitive distance that separates us from the sea. He calls for an ethical imperative to *rethink* the ocean by renegotiating concepts related to it. Our language itself lacks the vocabulary required to describe the experience of immersion and the loss of semiotic reference points that any deep-sea experience evokes. Casati provocatively explains that we are illiterate when it comes to thinking the ocean.

In *Ocean* (2020), Steve Mentz suggests replacing traditional academic terms with words drawn from the oceanic environment: "current" instead of "field", "water" instead of "land", "flow" instead of "progress", "ship" instead of "state", "seascape" instead of "landscape", and "distortion" instead of "clarity" (2020, xv–xvi). Building on Mentz's impulse, I propose a reconfiguration of Strindberg's poetics in which reading and writing become gestures of surfing, defined as acts of movement, attunement, and response to the sea's shifting forces. In this sense, I argue in favour of a theory of reading and writing as surfing, one that understands the creative act as a continuous negotiation of balance and rhythm, an art of engaging with the unpredictable rather than mastering it.

By integrating movement, perception, and material contact, this conception aligns with the aims of blue humanities to decenter terrestrial and disembodied modes of thought, foregrounding instead the lived, multisensory, and ecological entanglements between human and sea. Within this framework, waves are not merely metaphors but material and affective phenomena that shape identity, memory, and knowledge. Blue humanities thus offers a methodology for understanding the ocean not as an abstract or symbolic expanse but as something felt and experienced through the body, a dynamic medium that values sensory engagement with current and motion, and that emphasises the fluid, relational, and affective dimensions of being in and with water.

I am not the first to compare Strindberg to a surfer. Literary scholar Arnold Weinstein intuited this comparison in an analysis of *I Havsbandet*: "Borg's tone is that of a surfer-engineer" (2019, 37). But Weinstein does not develop this idea further. In "North swell: Tracing the surf connection in the expressionist paintings of nineteenth-century polymath August Strindberg", California-based artist Axel Weinstein highlights the affinities between Strindberg's paintings and surfing:

> Strindberg seems to tap into something that's fundamental with surfers: the deep, longing, exploratory stare. Surfers spend an extraordinary amount of their lives simply staring at the ocean. The eyes caress it and the mind races over wave faces and through collapsing sections that we observe with a wanton but clinical focus. Our observation from shore is akin to touch, and we project ourselves constantly to sea. Once there, the immersion is complete and the relationship is almost carnal, driven more by instinct and reaction than anything resembling forethought. Strindberg adopted a directness of painting that is strikingly similar in process.
>
> (2022, n.p.)

I argue that the deep connections between Strindberg's creativity and surfing concern not only paintings but also various literary genres and lead to a poetics of surfing. By "surfing", I refer to the human activity of riding the wave, attuning the body to the sea's shifting rhythm, and moving in harmony with its unpredictability. Surfing is an art of responsiveness: It requires reading the ocean's movements, engaging with its dynamic materiality, and navigating its continuous transformations. In this sense, to surf is to occupy a liminal space between stability and flux, between the solidity of land and the fluidity of the sea. This perspective places the human within the oceanic medium itself, answering Mentz's and Casati's calls to adopt an oceanic rather than terrestrial point of view.

I will focus on Strindberg's gesture of writing as an embodied and expressive act rooted in the Greek *poiein* (to make) and echoing Aristotle's view of poetics as the *mimesis* of action. Inspired by Axel Weinstein's analysis, I will study Strindberg's poetics as a creative, world-shaping gesture based on the three phases that construct the practice of surfing: the initial phase consists of reading the waves; the second phase begins when one engages physically and spiritually with the waves; finally, once back on the beach, the surfer can reflect on his or her experience of the sea. These three moments will structure my reading.

Firstly, inspired by Italo Calvino's "Lettura di un'onda" ("Reading a Wave") from *Palomar* (1983; *Palomar* 1986), I analyse how Strindberg reads a wave to show that his engagement with the sea functions as a form of knowledge-making that fuses observation and imagination. Secondly, by studying Strindberg in the surf through the lens of Jack London's "Learning Hawaiian Surfing: A Royal Sport at Waikiki Beach, Honolulu" (1907), I argue that bodily immersion and confrontation with the ocean's force transform his naturalist gaze into an experiential, performative mode of writing. Finally, extending this exploration from bodily immersion to stylistic transformation, and reading Strindberg alongside Julien Gracq's reflections on André Breton, I show that these moments coalesce into a distinct "poetics of surfing", in which stylistic experimentation arises from the act of reading and translating the sea's movements into writing. Strindberg's wave-inspired syntax generates frequency, intensity, and interval, setting the narrative in motion and revealing an aesthetics grounded in flux, risk, and the fluid reciprocity between human and sea.

Reading a wave

In "Lettura di un'onda", Calvino initially focuses on how the protagonist Palomar observes a wave:

> Mr. Palomar is standing on the shore, looking at a wave. . . . Since what Mr. Palomar means to do at this moment is simply *see* a wave—that is, to perceive all its simultaneous components without overlooking any of them—his gaze will dwell on the movement of the wave that strikes the shore until it can record aspects not previously perceived.
>
> (1986, 4)

The observation mobilises scientific concepts and semiotic tools. Taking "the pattern of the waves as a model" (1986, 5), Calvino expresses the intention to reflect on how to "read something that is not written" (Lucente and Calvino 1985, 248). In the search for nature's hidden signature and for "the infinite coherence in the apparently great disorder" (Strindberg 2003, 268; my translation), Strindberg develops a narrative technique that can be seen as an act of *reading a wave* in search of the unwritten text.

In *A Book of Waves*, anthropologist Stefan Helmreich situates oceanographic knowledge at the intersection of scientific, ecological, geopolitical, and cultural discourses. He argues that "the wave should be approached not only as natural phenomena but also as culturally significant entities" (2023, xviii). Referring to Calvino's text, Helmreich explains that it invites a dual reading of waves: one of meditative poetic reverie and another of hermeneutic analysis supported by theoretical tools (2023, 21). Strindberg's engagement with the sea operates in a similar dual register, at once poetic and analytical, where the act of observing waves becomes a mode of aesthetic contemplation and a means of deciphering the structures of nature and thought.

This approach also resonates with a longer historical and scientific understanding of the sea as the cradle of planetary life. Since the early nineteenth century, evolutionary thought—from Lamarck's theory of transformism to Darwin's *On the Origin of Species* (1859)—has traced the adaptive transformations of life through natural processes, describing the gradual movement of species from aquatic to terrestrial forms over geological time. Strindberg's attention to the sea can thus be read as an attempt to reimagine the connection between land and sea as a continuum rather than a division: by reading the waves, he seeks to recover an oceanic perspective that re-situates human perception within the fluid continuum of life from which it once emerged.

In *I Havsbandet*, Strindberg reads a wave like Palomar; that is, he converts his concrete observations and analysis of the wave as a natural phenomenon into a way of writing. As Brandon Kaaz observes (2020), Strindberg's novel not only represents the sea but allows it to act as a textual force that unsettles the boundary between nature and culture, granting the ocean its own literary agency. Strindberg's interest in the oceanic environment reflects his endeavour to produce knowledge at the cutting edge of modern science. His vision of the sea, enriched by oceanographic observations and experiments, shapes an ecological awareness that resonates in his work on botany, chemistry, mineralogy, and meteorology. In the novel's opening, Axel Borg takes command of the boat as it crosses the archipelago on its way to Österskär and uses both intuition and scientific calculations to assess the risk of an uncontrolled monster wave threatening the ship. He first estimates by calculation the phenomena: "The commissioner had taken his bearings on the two lighthouses and used the lorgnette quadrangle glass as a distance measure" (1913, 10). After observing the wave formation, he uses his intuition to anticipate how it would break: "he, in a fraction of a second, could make a sure estimate of how great the danger was" (1913, 11). Finally, Borg was able to "count out the only way of escape, whereupon he cried out, 'Hold all!'" (1913, 11), foreseeing the danger and saving the boat and its crew.

For Strindberg, the sea represents a system of forces, movements, and currents, with the wave (here a rogue wave) as one of the central and paradigmatic driving forces together with the wind. A rogue wave is an anomaly in the wave sequence, coming unexpectedly and often dangerously. Metaphorically, a rogue wave also represents a sudden, overwhelming force. In *I Havsbandet*, announcing a coming storm, it represents the character's emotional state and his subconscious.

Strindberg also intuits that the rogue wave is symptomatic of a broader crisis in his relationship with nature. The sea ceases to function as a stable or knowable entity; instead, it becomes an unpredictable and autonomous force. To be "in the surf" presupposes a transient, exposed, and contingent relation to the ocean. This unstable encounter between human and sea anticipates an ecological awareness of the industrial era: A recognition that the natural world, increasingly subject to technological intervention, remains resistant to human control. In this sense, the rogue wave functions not only as a literal oceanographic event but also as a figure for epistemological and affective uncertainty. It embodies a crisis of modernity, in which nature no longer offers assurance of continuity or reciprocity but instead reveals its own agency and unpredictability.

This interpretation resonates with Helmreich's formulation of rogue waves as "newly cast characters in an Anthropocene drama" (2023, 20). Alongside storms and tsunamis, they appear as "avatars of an ocean encroaching in an age of climate change", manifestations of a planetary hydrosphere newly animated by human-induced instability. Strindberg's vision of the sea, therefore, may be read as an early expression of the estrangement that underlies contemporary ecological thought, where the ocean emerges as both medium and metaphor for a crisis of relation between humanity and the more-than-human world.

More precisely, Strindberg's relation with the sea can be connected with a reflection on the impact of human action on the environment, as he describes the sea as a reservoir for industrialisation's wastes in several texts. Here, Strindberg anticipates one of the major environmental problems of our contemporary societies, as we inhabit what Marco Armiero terms the *Wasteocene* (2021), an era shaped by "wasting relationships" that produce not only material waste but also marginalised people, places, and ecologies. In the prose poem "Solrök" ("Haze"), the poet discovers on the beach the remains of artefacts produced by society: "Here the sea has gathered everything that it could not swallow, sorted it, washed it, and polished it: broken oars, legions of corks, bark, reed-pipes, stave-hoops, barrel-hoops. There he sits and regards a loosely detached ship's plank" (1883, 91; my translation). This image of the sea as both archive and agent of transformation resonates with Curtis Ebbesmeyer and Eric Scigliano's *Flotsametrics and the Floating World* (2009), in which the drifting of lost consumer goods across the oceans reveals the sea as a dynamic system that registers, circulates, and reconfigures the material traces of human activity.

In a similar way, in *Hemsöborna*, the sea functions as city sewers and landfill:

> He gazed for a while into the water, which the paddle wheels had churned into a muddy froth; on its surface floated clumps of soot and patches of oil, shimmering in rainbow colours like an old windowpane. In just a moment the monster had managed to spew out every kind of filth and befoul the clear green water: beer corks, eggshells, lemon peels, cigar stubs, burnt matchsticks, scraps of paper that the bleak and small fry played with. It was as if the whole city's gutter had come out to disgorge its insults and refuse all at once.
>
> (2012, 17; my translation)

However, in the poem "Sömngångarnätter på vakna dagar" (Sleepwalking nights in broad daylight), the contrast between a polluted culture and a clean nature conveys a striking ecological message, as the sea and its waves appear to possess the power to purify the waste of civilisation:

> And now one hears the roar where the waves of Mälaren
> flow out to the sea to be washed clean
> of cultured filth and civilised dust.
> (1883, 159; my translation)

This image of the sea washing itself free from "civilised dust" suggests a dynamic of absorption and regeneration, a natural mechanism through which the ocean takes in, transforms, and ultimately neutralises the residues of human activity. Strindberg's vision thus corresponds to an early ecological conception of the sea as a self-regulating system, one that absorbs impurities not in order to conceal them, but to reintegrate them into broader natural cycles of renewal and decay. This intuition of pollution as a temporary disturbance within a self-restoring balance, however, was later challenged by twentieth-century ecological theory, which reconceptualised ecosystems as dynamic and adaptive rather than as entities tending towards a fixed equilibrium.

This poetic insight is aligned with the scientific understanding emerging in the work of Ernst Haeckel, who coined the term *Ökologie* to describe the interrelations between organisms and their environment. Like Haeckel's marine biological studies, particularly his research on radiolarians (*The Radiolaria*, 1862) and his later visual synthesis in *Art Forms in Nature* (1899–1904), Strindberg's sea functions both as a medium through which life's cyclical transformations become visible and as a metaphor for the continuity between organic and elemental forms. His description of the ocean's capacity to cleanse and reabsorb human waste can thus be read as an expression of the principle of natural balance, central to nineteenth-century evolutionary and ecological thought.

For Strindberg, the collective power of the waves is both regenerative and destructive, giving nature the role of monitoring the vital cycle. The last sentence of *I Havsbandet* places the sea as both the origin and the final force in life: "out over the sea, the mother of all, from the womb of whom life's first spark was kindled, the inexhaustible spring of fecundity and love, life's origin and life's foe" (1913, 300). On his way to the open sea, Axel Borg finally confirms his fusional relationship with the mother sea and his hope to find a free space there. His final dissolution into the maternal sea resonates with Ibsen's conceptualisation of the sea in his preparatory notes for *The Lady from the Sea* (1888), where the ocean's "magnetic power" and its evolutionary kinship with humankind position it as both a material and metaphysical origin to which certain individuals remain psychically bound (see Frederike Felcht's chapter on Ibsen in this volume). The waves finally appear in Strindberg's novel as an epiphanic event. Inspired by the experience of being in the surf, Strindberg shapes a wave epistemology at the intersection of science, ecology, and poetics.

Strindberg in the surf

In "Solrök", Strindberg describes a man who finds himself in the middle of the waves:

> and so he is up again in the midst of the roaring and struggling waves; and he wrestles with them until he gets tired and lies down on their backs; and they throw him up, as if they wanted to throw him right into the sky, and they drag him down into dark valleys as if they wanted to take him into the abyss; he ceases to want, he ceases to desire, he makes no resistance.
>
> (1883, 90; my translation)

This scene reads as a form of *bodysurfing*, which is to be distinguished from navigation or simple swimming. The human body is no longer an agent of control or direction but becomes integrated into the living motion of the sea. The man in the waves learns to yield rather than to dominate, transforming his struggle into a form of surrender, an early ecological intuition of reciprocity with the natural world. The act of "making no resistance" evokes a mode of environmental becoming grounded in the capacity to move *with* rather than *against* natural forces.

Such an art of non-resistance recalls Jack London's description of surfing in "Learning Hawaiian Surfing" (1907), where he presents wave-riding as an embodied philosophy of coexistence: "The whole method of surf-riding and surf-fighting, learned, is one of non-resistance . . . . Yield yourself to that undertow. Swim with it, not against it, and you will find the pressure removed" (1911, 93). For London, the practice of surfing is not merely a sport but a lesson in ecological attunement, a means of inhabiting dynamic relations with the ocean through bodily intuition, patience, and rhythm. In the following lines of "Solrök", Strindberg observes the remaining artefacts that the sea has brought to the shore. He finds among these artefacts a board. It would be an exaggeration to interpret this scene as evidence that Strindberg recognised the possibility of using this board for surfing, but the parallel with London's existential meditation on surfing as "the art of non-resistance" is striking.

In the surf, the senses are disturbed, an effect Strindberg (1894–95) calls "deranged impressions", and that is directly inspired by the roar of the sea. One of the last scenes in *Ett Drömspel* is set in Fingal's Cave, a sea cave in the Inner Hebrides of Scotland known for its unique acoustics, creating the impression of being inside a seashell. Agnes and the poet listen to the sounds of the wind and the waves and experience the *transience* of the ephemeral. The floating element that appears in the cave after the song of the waves is a lifebuoy, presented as the guardian of the sea, surrounded by a breaking or uncontrolled wave that threatens the cave: "Look, how the sea is rising—how high the waves are—soon we shall be unable to get out of the cave!" (1912, 89). Here, the spatial and temporal scale is defined by the dynamics of the waves, unstable and unpredictable. It is an echo of the general setting of the play, a dreamplay where "time and space do not exist" (1912, 24). In fact, it is another time and space, determined by the sea and the waves. Time and space do not exist for us earthlings, because we are illiterate when it comes to the sea. But the poet can decipher the dialectic of the waves in this "wet ontology" (Steinberg and Peters 2015).

Strindberg also associates the oceanic environment with emotional fluctuations, transforming the force of the waves into vital power. In *I Havsbandet*, Axel Borg explains to Maria the mechanism of this transfer: "Don't you feel what great power is within you, as you ride upon a wave, when it with a thousand pounds of weight would press you down into its depth?" (1913, 130). To experience a wave is also to feel something greater: the impression or desire to feel at one with the universe, what Romain Rolland in a letter to Sigmund Freud, 5 December 1927, calls "the oceanic feeling" (Vermorel and Vermorel 1993, 304). This oceanic feeling is an impression of eternity. Rolland saw it as the most important experience in Asian religions; so did Strindberg, who placed Agnes in a filial relationship with Indra, the king of the gods in Vedic religion. In Strindberg, the sea becomes a dynamic medium of emotional and spiritual transformation, its vital power aligning with Rolland's "oceanic feeling" and London's "art of non-resistance"; yet, unlike Ibsen's magnetic and binding sea, Strindberg's ocean invites dissolution into a cosmic flux that situates his modernism within a wider transoceanic poetics of surrender and renewal.

Poetics of surfing

In the painting series *Vågen* (1892–1902; The Wave), Strindberg depicts the motif of wild waves in seascapes where sea and sky dominate, and few elements of land appear. No people are present, contributing to a sense of loneliness and rawness, echoing inner struggle and spiritual chaos. The white foaming water produced by waves is represented by a line formed along the horizon. Strindberg sketches the oceanic horizon with the following words: "The northern horizon of the sea is only an illusion" (2003, 98; my translation). This illusion is what Marcel Proust calls the origin of metaphor. In *À l'ombre des jeunes filles en fleur* (1919; *In the Shadow of Young Girls in Bloom* 2015), Marcel finds himself in the studio of the painter Elstir. He looks at a seascape and reflects on how the boundary between land and sea disappears and everything in the landscape becomes sea. He notes that the representation is subjective: that is, it blurs the boundary of the horizon between land and sea, creating what he calls a "metaphor" (2015, 451). It is by observing the movement of the waves in one of the seascapes painted by Elstir that Marcel learns the art of writing. Similarly, Strindberg's wave paintings translate the instability of perception into an aesthetic of motion and attunement inspired by his reflections on the illusory horizon. Each painting implies that the very possibility of living and creating arises from sensitivity to environmental rhythms, a principle that Strindberg transforms into dramatic form.

In fact, in *Ett Drömspel*, Agnes interprets the song that the waves make, *the Music of the Waves*, as a poem.

> Hush! Now the waves are singing.
> *Recites to subdued music*:
> We, we waves,
> That are rocking the winds to rest—
> Green cradles, we waves!
> Wet are we, and salty;
> Leap like flames of fire—
> Wet flames are we:
> Burning, extinguishing;
> Cleansing, replenishing;
> Bearing, engendering.
> We, we waves,
> That are rocking the winds to rest! (1912, 85)

The structure of the sentences corresponds to Julien Gracq's idea of "the breaking sentence" (*la phrase-déferlante*), an expression coined in an analysis of André Breton's style: "In contrast to the 'conclusive' sentence, I would readily call it the *breaking* sentence. Its use is similar to that of 'surfer-riders' who maintain their balance on a board at the crest of a wave until it finally crashes" (Gracq 1948, 157).

This specific syntax called "the breaking sentence" is visible in the style of *Ett Drömspel* and can be considered as one of the most important experiments in Strindberg's poetic gesture. In that specific scene in Fingal's Cave, the instability and movement of the sentence open up a space of suspension that refuses to conclude or stabilise. This suspension is already created in the opening gesture revealing the song of the waves, which propels the reader forward, listening not to an ending, but to a swelling. Surfing the wave in repetition ("We, we waves"), the sentence embodies the role of the surfer finding harmony in rhythmic instability, as Gracq describes it. "Rocking the winds to calm them" paradoxically makes the waves stable, but this stability comes from perpetual motion. An epiphora ends the stanza and forms a crest that declares without concluding.

In the central part of the song, the stanza depicts the dizzying balance: there is no stable semantic basis, only a forward propulsion punctuated by contradictory poetic images (wet, salty water moves like fire; fire is wet; the flames burn while extinguishing themselves); each line undoes the logic of the previous one. Like a surfboard at the crest of a wave, this section remains suspended in a permanent movement that defies gravity. Even the punctuation (semicolons, dashes, and colons) fragments the sentence into whirlpools and eddies. It is syntactically open, leaning but without falling: "Burning, extinguishing; / Cleansing, replenishing; / Bearing, engendering". This is the climax of the poem's surge. The three pairs of verbs without anchors offer a tension between destruction and creation. There is no subject, no resolution, only excessive energy. Through a final repetition ("We, we, the waves"), the syntax loops back rather than moving forward. "To rest" might appear to be a final destination, but it is only a fleeting calm—like a wave that collapses only to begin again, in a perpetual cycle. This is the distinctive mark of the breaking sentence: not to explain, but to break.

I argue that this stylistic experimentation is directly inspired by the reading of waves and the transformation of this reading into a writing gesture. As a surfer, Strindberg uses the waves as dynamic forces that sculpt a new syntax and set the narrative in motion, creating frequency, intensity, and interval.

Another example of this poetics of surf can be found in a description of the seascape in *I Havsbandet*:

> Hear now the roaring of the waves, as the poets summarily call this symphony of sound. . . . You at first hear a buzzing which resembles the noise heard in a machine shop or a big city. It is the masses of water dashing against each other; next you hear a hissing; it is the lighter, smaller water particles which are lashed to foam. And now a grating as of a knife against a grindstone; it is the wave tearing against the sand. And now a rattling like the dumping of a load of gravel; it is the sea heaving up small stones. Then a muffled thud as when you clap the hollow of your hand to the ear, it is the wave which presses the air before it into a cavern; and lastly this murmuring as from distant thunder, it is big boulders, rolling on the stony bottom.
>
> (1913, 151)

This passage is another rich example of *breaking sentence*, but here in prose. The syntax and structure mimic the very waves it describes, not just through content but through a surging, rhythmical buildup that refuses conventional resolution. Here, the gesture of writing consists of trusting

> blindly in the momentum of the rising wave that carries the sentence away, to stay 'in the flow' at all costs, to cling to the mane of foam with a miraculous feeling of freedom, to follow it wherever a last burst of life takes it.
>
> (Gracq 1948, 157)

Both in Gracq's and in Strindberg's texts, the idea of listening better with eyes closed invites immersion, a gesture akin to the surfboarder leaning into the wave, surrendering linear sight for bodily intuition. The syntactic repetition in Strindberg's writing fosters a *rising rhythmic structure*. Here begins the surf-rider motion, when each sentence *lifts* off the previous: "next . . .", "and now:", "and now:", "and then . . .", "and lastly . . .". There's no full stop of thought, only pauses (semicolons and colons), letting the sentence gather and ride. This long syntactic wave is carefully crafted: a rhythm of images and sounds, of comparisons and explanations. The listener never lands as he is propelled from one sensory crest to the next.

The wave is here both physical and syntactic. As opposed to the "conclusive sentence" (Gracq 1948, 157), Strindberg's wave sentence is leading towards collapse but never ends. With the sentence "it is big boulders, rolling on the stony bottom", the description reaches a breaking point, at the edge of a collapse. "As long as we silence the mind-numbing mechanisms within ourselves", as Gracq states, "a lifeline also runs through the sentence, leading it not to its end but to something more exhilarating, which is a catastrophe, whipping the words into a dance and a light of foam" (1948, 157–58). The final crash of the wave isn't an end but a transformation leading to the feeling of freedom (1948, 158).

This description of the symphony of waves by Strindberg evokes waves not only as an image or subject but also as a model of the movement of language that can be compared to the style developed by Proust in *Jean Santeuil* (1952; *Jean Santeuil* 1955) when Proust explicitly associates style with the movement of the waves:

> And with each blow there came to him a momentary sense of relief, as, when we stand on the sea's edge our expectancy, strained by the slow building up, the gathering accumulation of strength, the topping curve of each incoming roller, dissolves with a feeling of relief, when the wave breaks, only to form again when the process is repeated.
>
> (1955, 211)

In a similar manner, Strindberg uses the wave structure not only to describe but also to shape language itself, demonstrating that meaning is not fixed but breaks, reforms, and recedes as the sea reshapes itself and the next wave arrives. A long, fluid sentence with clauses that undulate creates a linguistic swell. The cycle, the sequence, the intervals—combined, they create a sensory vertigo for the reader, what Gracq calls a "vertiginous balance" (1948, 157).

Conclusion

Taken together, these readings show that Strindberg's sea writing not only reconfigures the boundaries between knowledge and sensation but also anticipates an oceanic modernity attuned to instability, motion, and the porous entanglement of human and elemental forces. Strindberg writes like a surfer: with balance and responsiveness, with aesthetic patience and a keen sense of timing, guided not by a fixed destination but by perpetual formation. In his writing, waves are at once physical phenomena, lived experiences, and metaphors for creation itself. The sentence is written by a pen that, like a surfboard, is fragile and yet a vehicle for balance, not controlling the wave, but surfing it until it breaks.

Note

1 For a study of the resonance between Nietzsche and Strindberg in their relationship to the oceanic environment, see Stern (2008).

Works cited

Armiero, Marco. 2021. *Wasteocene: Stories from the Global Dump*. Cambridge University Press.

Calvino, Italo. 1986. *Palomar*. Translated by William Weaver. Harvest Brace Jovanovich.

Casati, Roberto. 2022. *Oceano, una navigazione filosofica*. Einaudi.

Ebbesmeyer, Curtis, and Eric Scigliano. (2009) 2010. *Flotsametrics and the Floating World*. Harper Paperback.

Gracq, Julien. 1948. *André Breton, quelques aspects de l'écrivain*. Corti.

Helmreich, Stefan. 2023. *A Book of Waves*. Duke University Press.

Kaaz, Brandon. 2020. "Das Meer als Meer lesen: Zu August Strindbergs *I havsbandet* (1890)." *NORDEUROPAforum: Zeitschrift für Kulturstudien* 29 (1): 119–36. https://doi.org/10.18452/30838.

London, Jack. 1907. "Learning Hawaiian Surfing: A Royal Sport at Waikiki Beach, Honolulu." In *Woman's Home Companion* 34 (7) (July): 8–9, 38–40.

London, Jack. 1911. "A Royal Sport: Surfing in Waikiki." In *The Cruise of the Snark*. Macmillan.

Lucente, Gregory L., and Italo Calvino. 1985. "An Interview with Italo Calvino." In *Contemporary Literature* 26 (3) (Autumn). University of Wisconsin Press, 245–53. https://doi.org/10.2307/1208025.

Meidal, Björn. 2025. *Strindberg och havet: öarna och horisonten*. Ocean Books.

Mentz, Steve. 2020. *Ocean*. Bloomsbury Academic.

Proust, Marcel. 1955. *Jean Santeuil*. Translated by G. Hopkins. Weidenfeld and Nicolson.

Proust, Marcel. 2015. *In the Shadow of Young Girls in Flower*. Edited by William C. Carter. Yale University Press.

Scheffauer, Herman. 1913. "A Correspondence between Nietzsche and Strindberg." *The North American Review* 198 (693) (August): 197–205.

Steinberg, Philip E., and Kimberley Peters. 2015. "Wet Ontologies, Fluid Spaces: Giving Depth to Volume through Oceanic Thinking." *Environment and Planning D: Society and Space* 33 (2): 247–64. https://doi.org/10.1068/d14148p.

Stern, Michael J. 2008. *Nietzsche's Ocean, Strindberg's Open Sea*. Nordeuropa-Institut.

Strindberg, August. 1883. *Dikter på vers och prosa*. Bonniers.

Strindberg, August. 1892–1902. *Vågen. Oil on canvas*. Musée d'Orsay (Paris), Nordiska Museet (Stockholm), and private collections.

Strindberg, August. 1894–95. "Sensations détraquées: 1894–1895." In *Le Figaro Littéraire*. 17 November 1894, 26 January 1895, and 9 February 1895. No pagination.

Strindberg, August. 1912. *Plays by August Strindberg: The Dream Play, The Link, The Dance of Death, Part I, The Dance of Death, Part II*. Translated by E. Björkman. C. Scribner's Sons.

Strindberg, August. 1913. *On the Seaboard: A Novel of the Baltic Islands*. Translated by E. C. Westergren. Grosset and Dunlap.

Strindberg, August. 1984. *Skärkarlsliv. Samlade Verk 26* [nationalupplagan]. Almqvist & Wiksell.

Strindberg, August. 1994. *Klostret: Fagervik och Skamsund. Samlade Verk 50* [nationalupplagan]. Norstedts.

Strindberg, August. 2003. *Naturvetenskapliga skrifter II. Samlade verk 36* [nationalupplagan]. Norstedts.

Strindberg, August. 2012. *The People of Hemsö*. Translated by Peter Graves. Norvik Press.

Vermorel, Henri, and Madeleine Vermorel. 1993. *Sigmund Freud et Romain Rolland*. PUF.

Weinstein, Arnold. 2019. "'May the Force Be with You': Strindberg's Paintings." In *August Strindberg and Visual Culture: The Emergence of Optical Modernity in Image, Text and Theatre*, edited by Jonathan Schroeder, Anna Westerstahl Stenport, and Eszter Szalczer, 35–51. Bloomsbury.

Weinstein, Axel. 2022. "North Swell: Tracing the Surf Connection in the Expressionist Paintings of Nineteenth-Century Polymath August Strindberg." *The Surfer's Journal*, No. 30.5. No pagination. https://www.surfersjournal.com/editorial/august-strindberg-north-swell/.

19

"A SEA WITHOUT DIRECTION"

The deep in contemporary Icelandic literature

Auður Aðalsteinsdóttir

On the world map, Iceland is positioned in the North Atlantic Ocean. On zooming in, however, maps start showing a more complex—and diverse—picture, with Iceland located at the meeting point of several seas: the Arctic Ocean, Greenland Sea, Norwegian Sea, Denmark Strait, and the Iceland Sea (the latter term existing only on some maps and in some human minds). The difference between such maps indicates how the categorisations and borders of the world's seas not only represent specific material ecosystems but are, at the same time, abstract, relative, and ever-changing cultural products. The geographical mapping reflects the horizontal thinking of land-centred humans: flattening the depths beneath the surface into a land-level view and dividing them into sections with political borders. However, marine sciences' mapping of ocean layers typically draws up a vertical cross section of the sea, similar to geographical block diagrams. They present an area that is normally out of human sight, but which is no less important for marine creatures. And, as we are becoming increasingly aware of in times of climate change, this is a space that also affects the human species. It is, effectively, an increasingly politicised space where humans have a presence, through pollution and deep-sea mining, for example, but also through protective measures.

In the poetry collection *Vistarverur* by Haukur Ingvarsson, the title having been translated into English as *Ecostentialism*,[1] the poet's fascination with the deep sea—which he associates with the depths within himself—is intermingled with terror: "I'll tell you one thing / nothing frightens me more / than the open sea / and the abyss / below" (2018, 16). Just as *Vistarverur* revolves around complex connections between inner and outer landscapes, this chapter focuses on the deep sea in recent Icelandic literature as a specific regional, ecological, historical, cultural, psychological, and imaginative space. In the spirit of cultural ecology (Zapf 2016), this approach assumes that, even as represented in literature, "marine life [i]s always

DOI: 10.4324/9781003647638-25

interconnected with particular environments, processes, and substances" (Alaimo 2016, 126). As Brandon Kaaz discusses in his chapter, "experiencing a certain area as space or place . . . is a subjective experience brought upon by knowledge and intimacy to one's environment" (Brandon Kaaz in this volume, xx-xx). The use of the term "space" in this analysis of the ocean deep in Icelandic literature is in line with Yi-Fu Tuan's (1987) distinction between place, which denotes security and stability, and space, denoting movement, freedom, and connectivity.

In earth science, our planet's major systems are the hydrosphere, cryosphere, lithosphere, biosphere, and atmosphere. But the area where life can be sustained—below, on, and above the Earth's surface—has also been referred to as "the Critical Zone" with emphasis on the interplay between the spheres. Taking into consideration both "the identity of a place" and our globe's heterogeneity with its "complex distribution of local and global phenomena", Arènes, Latour, and Gaillardet confront the problem of how to reconstruct "a view of the earth that is much more concrete, dynamic, complex, heterogeneous and reactive than what can be captured through the cartographic imaginary of points defined on a map by longitude and latitude" (2018, 123, 126). Highlighting the role of art and aesthetics in this regard (cf. Aït-Touati, Arènes, and Grégoire 2022), they propose, as an "exercise in *cosmography*" (Arènes, Latour, and Gaillardet 2018, 124), a visual representation in the form of diagrams. This chapter explores how such "cosmographic practices" also take place in literary imagery.

In recent Icelandic literature, the imagery of the deep sea and its creatures reflects how climate-related traumas are slowly sinking into the depths of the collective unconscious, bringing with them a heightened ecological sensibility, as can be seen in the micro-story "Skepnan" (2018; The Beast) by Þórdís Helgadóttir. Such imagery is used to explore complex interactions in the Earth's Critical Zone, as in *Urta* (2022; Cow Seal), a narrative poem by Gerður Kristný, which describes a relationship between a woman and a female seal and is here analysed through a hydroqueer lens, bringing together a hydrofeminist sense of fluidity and queer ecology's rejection of binary thinking. This point of view contributes to the increasing emphasis within the humanities on the more-than-human (Seymour 2020), as is demonstrated in an analysis of poems on the deep and the Greenland shark by Ingvarsson. These well-established Icelandic authors all have academic backgrounds: Helgadóttir has studied philosophy; Ingvarsson has a PhD in literary studies and is now an associate professor in cultural communication; and Gerður Kristný studied literature at the University of Iceland and still participates in academic events. They are therefore familiar with academic concepts and theories, and Ingvarsson sometimes overtly brings them into play while the allusions tend to be more indirect in Helgadóttir's works and Gerður Kristný stays decidedly clear of them. All these works are, however, good examples of how scientific, political, and academic theories relating to environmental issues are not isolated practices but rather part of a larger cultural discussion, and that they are in active dialogue with art and literature.

Monstrous seas

"It was on a beautiful Sunday in August that the beast rose from the depths."[2] The menace in these opening lines of a two-page story by Þórdís Helgadóttir (2018, 43) is one of many examples of nature being depicted as traumatic in Icelandic literature, representing how vulnerable people are to threats from external forces, and also to dangers from within (Aðalsteinsdóttir 2023a). Encounters with paranormal beings and monsters such as ghosts, trolls, and uncanny water creatures also abound in contemporary Icelandic fiction where people's interaction with their environment is brought into focus. They have obvious roots in folklore, where such confrontations have been interpreted as reflecting humans' relationship with nature (cf. Spaans and Williumsen in this volume; Sigurðardóttir et al. 2019). The sea, with its eternal transition and movement, is a place of particular ambivalence and contradictions in this respect; a source of life-giving food as well as a graveyard for countless fishermen, and symbolically a deep level from which traumas resurface (Aðalsteinsdóttir 2023b). A sea monster can turn out to be the sea itself and reveal some truths about the humans who encounter it.

In Helgadóttir's micro-story, the description of the beast rising from the depths brings to mind a tsunami that appears without warning on a beautiful day and destroys everything in its path. "The sea gave way", it says. "Then it cracked and came ashore."

> The beast's head filled the fjord. The skin was blue leather, covered in scratches and rough boils the size of villas. The wave passed over the rock wall. The distance between the eyes was such that it could contain a whole market square in a big city. A boulevard could have been placed between the eyes. The beast rose up, shadowing the sun, and the buildings along the fjord became a fine collar. All proportions in the world were turned upside down.
>
> The beast opened its jaws.
>
> (Helgadóttir 2018, 44)

The literary scholar E. Ann Kaplan believes that a permanent change in the relationship between humans and the natural world can be detected in twenty-first-century fiction: "No longer a passive object of the human gaze, nature in its dying is now active, a negative force, a violent presence, oftentimes an actor in its own right" (2016, 54). The short description above abruptly and dramatically closes a story which up to then has, at a very slow pace, been setting the scene for a family stroll on a beach. One could say the story's structure is marked by a clash of what is sometimes called "deep time", or geological time, which allows for cataclysmic events of a scale that challenges human perception, and an anthropocentric sense of temporality, which Amitav Ghosh associates with the modern novel's tendency towards individual perspective and compressed narrative time. The "realist" method of accentuating everyday regularity rather than the exceptional and unlikely is in fact a "concealment of the real", Ghosh claims (2016, 23–24, 32, 66). Although it has a parallel in science where the emphasis has shifted away from the geological history of singular events towards a history of a gradual, predictable evolution, such a worldview assumes that humanity can control its own destiny—and nature. It ignores the fault lines in evolutionary history that are caused by unpredictable catastrophes. Now, as Earth's climate has started to change, even our everyday lives are increasingly impacted by such events.

Motherhood is a theme in Helgadóttir's story, with a sinister emphasis on nature's constant threat to all life. Even before the beast appears, two female characters are alert. These are the mother of two human children and a duck with her one duckling. The mothers never leave the side of their offspring, aware that at any moment something could threaten them, even on this beautiful day (and, indeed, the duck has already lost one duckling to a predator). This parallel between ducks and humans protecting their offspring resonates with a scene from an older short story, "Under a Volcano" (1989), by the prominent Icelandic author and literary scholar Svava Jakobsdóttir. In that work, where an angry drake attacks humans approaching his family, the constant threat to life is present in the form of the volcano Hekla. The story ends, however, with a grandmother reflecting on the broader context where the cycle of life and death, in the form of parents and children, provides a sense of continuity. The abrupt ending of Helgadóttir's story, with the beast opening its jaws, deprives the reader of any such sense of continuity and perhaps conveys a changed perception of the natural world in times of climate change.

There is a long tradition of identifying mothers with Earth and women with nature in a social system where women and nature are positioned as the inferior opposites of men and culture (and the feminine is associated with monsters). The female body is often described in terms of symbiosis and flow, and Helga Kress is among the feminist scholars who have discussed water as a "feminine element and symbol" (1993, 31). Characterised by a lack of boundaries, flowing water is the opposite of a rigid patriarchal system of fixed laws and rules, and Kress interprets the element of water in the Sagas and other early medieval Icelandic literature as "dangerous to men", who in turn, and often with sexual undertones, try through shipbuilding and sailing to "defeat it" (1993, 31).

Some ecofeminists have sought empowerment from traditional associations between nature and the feminine, emphasising values that are often connected to women, such as softness, care, and cooperation. They stress the need to save the earth and its inhabitants from patriarchal ideology, where the constant struggle for domination is causing irreversible environmental damage. Such exaltation of the supposed connection between women and nature has also been criticised on the grounds that it supports the very essentialism that has been used to oppress women. However, as regards the ocean, Astrida Neimanis believes that the properties of water—fluidity and transmutation—are ill-suited for service to oppressive essentialism. Even if women's power and agency can, like water, be temporarily dammed by patriarchal representations and discourse, "[t]he desire of water to morph, shape-shift, and facilitate the new persistently overflows any attempt at capture. Is not 'woman' similarly uncontainable?" (Neimanis 2012, 85). Neimanis suggests that we adopt a way of thinking she calls "hydrofeminism". Water covers about 70 per cent of the Earth's surface and the range of the Critical Zone, often referred to in the geosciences as extending from treetops to bedrock, "could also be described in terms of fluids . . ., from the fluids in the atmosphere all the way down to the deepest aquifers" (Richter 2019). About 70 per cent of the human body is also water, but Neimanis points out that to "rethink embodiment as watery stirs up considerable trouble for dominant Western and humanist understandings of embodiment, where bodies are figured as discrete and coherent individual subjects, and as fundamentally autonomous" (2017, 2).

The perception of bodies as watery also allows for a queering of ecofeminism, a move which avoids essentialism, dismantles binary understandings of gender and sexuality, and regards sexuality and identities as fluid rather than stable. Queer studies have had their impact on ecology, drawing attention to how heterosexual norms have shaped human definitions of nature and society. In queer ecology, an anthropocentric and masculine worldview is rejected, and fluid gender and sexual orientation are seen as one aspect of continuous transformations in a world where boundaries—such as those between genders, species, or between the organic and the inorganic—are in flux rather than static (Henry 2023). The term "hydroqueer" therefore seems more appropriate for the growing awareness of how blurred the boundaries are between the environment and us—as individuals and as a species.

If we define ourselves as watery, we do not experience ourselves as a stable phenomenon but as part of a complex process. This means not only that environmental pollution, for example, the pollution of seas and rivers, results in the pollution of our own bodies but also that we must give up the traditional distinction between the human and the non-human: between us and other living things, between the organic and the inorganic, and even the distinction between human beings and monsters.

What kind of form?

The beach in Helgadóttir's story does not seem specifically Icelandic, and the deep is always hiding beneath the surface: the beast only becomes visible when it surfaces, and the duckling is described as disappearing when it dives. The result is an anthropocentric point of view where we can only imagine some "general" (physical and symbolical) deep and experience the danger it represents for landbound creatures. The sea creature that emerges can certainly be interpreted as the violent, negative force of nature that E. Ann Kaplan describes, but the contrast that is created when the sea monster bears down on a family on a beach trip—the difference between the safety of orderly society, on one hand, and the horror of uncontrollable natural forces on the other—is also deconstructed through subtle connections between them. Not only in the paragraph quoted earlier with its suggestions of the beast's face being able to contain a market square and boulevards but also in the quiet prelude to the catastrophe where parents with two young children are strolling along the beach. During the walk, one of the children picks up an item and places it in her pocket. On the beach, she has gathered "glittery stones, polished shards of glass and small pieces of colourful plastic" (Helgadóttir 2018, 43). The girl does not know when she grabs the latest treasure, whether it is a "conch or a bottle cap" (2018, 43). The difference between the man-made and the natural is blurred, and recent reports concerning hermit crabs come to mind. These crabs used to take up residence in abandoned conch shells but are now increasingly seeking shelter in the plastic waste left behind by humans.

The plastic on the beach is a reminder that when we disappear as a species, we will leave behind a layer of plastic and concrete as a stratigraphic marker of the Anthropocene. An era in which humans have become a geological force, affecting the planet's climate and thus, for example, sea levels and ocean currents. The terrifying force of nature threatening not only humanity but all life on earth turns out to be humanity itself and we are left to wonder which animal in nature the story's title refers to—duck, human, or sea monster—or even if all of us together represent "the beast".

Furthermore, the beach where the plastic items are discovered is the space where man meets the sea and its creatures. Such encounters are the theme of the narrative poem *Urta* by Gerður Kristný, where seals and humans meet in the sea, on land, or on the beach. There are obvious references to folklore where the boundaries between creatures of the sea and land sometimes become blurred, and women can, for example, take on the shape of a seal or return to human form, depending on the situation. Gerður Kristný uses the story of her foremother in Strandir, a midwife who had nineteen children but lost some of them to disease, her husband to the sea, and had to have a leg amputated. From this material, Gerður Kristný spins a description of people who survive on the margin between two worlds, the sea and dry land, and make a living both by farming and fishing: "Here we make use of / fields by furthest sea . . . We make use / of sea by furthest field" (2022, 42).[3]

The seals also live on these borders, often lingering in the shallows or resting on the shore. The existence of these two animal species, humans and seals, is depicted as intertwined, in both life and death. When the woman's son kills a seal pup, whose flesh will help the latest child in the woman's womb to survive, the seal cow "watches dark-eyed" (Gerður Kristný 2022, 27). Later, a series of misfortunes and deaths also strike the human family and underline the often-violent nature of the relationship between the human and the non-human. Images of water, in the form of liquid, ice, and vapour, are a recurring motif throughout the book, in a poetic enactment of how "the flow and flush of waters" connects us "to other worlds beyond our human selves" (Neimanis 2017, 2). In *Urta*, even the border between life and death becomes blurred. The sea-drowned husband wets the floor of his house when he steps ashore to check on his surviving wife, and a dead daughter, the victim of a winter with severe frost, seeks her mother's help when the seal cow has difficulty giving birth. Despite previous tragedies, a deep sympathy forms between these two females, the woman and the seal, both of whom have previously carried offspring "in the deep" (Gerður Kristný 2022, 89), in the waters within themselves—and lost an offspring to death.

In *Urta*, Gerður Kristný ties "together the tradition of old Norse meters and narrative poems, motifs from folklore and folk songs, references to the Sagas, and concise watery imagery" (Aðalsteinsdóttir 2026). Both *Urta*'s themes and form thus reflect a "watery logic" of interconnectedness where borders, whether it be between species or genre, are "always vulnerable to rupture and renegotiation" (Neimanis 2017, 2). The same can be said of the other works discussed in this chapter. Helgadóttir's microfiction oscillates between prose poems and short stories. The style of Ingvarsson's poems, where short lines and verses with concise imagery form longer bulks or chapters, is similar to that of Gerður Kristný, although the "narratives" one might detect are more fragmented and elusive. Instead, there is a complex, thematic flow between poems and chapters, illuminating how form and content are always entangled and how poetry, like all other art forms, can give "access to an embodied experience of our wateriness that might otherwise be too submerged, too subcutaneous, too repressed, or too large and distant (or even too obvious, mundane, and taken for granted), to readily sense" (Neimanis 2017, 55).

Loki's monstrous children

Changes and flow between life forms, natural phenomena, genders, and between nature and culture are common themes not only in folklore but also in myths. The imagery of the eschatological medieval poem *The Seeress's Prophecy* (*Völuspá*), with its descriptions of Earth sinking into the sea and a new world arising from it, seems particularly close at hand for Icelandic authors when it comes to describing terrifying geological events. In Ingvarsson's poem "Contemplation on Global Warming", from *Vistarverur*, an ocean-wet beginning (morning) and end (evening) are played against each other as in *Völuspá.* In the twenty-first-century poem, however, the speaker has difficulty distinguishing between the two, which may be due to the fact that these are not future events like the prophetess in *Völuspá* describes but the experience of someone caught in the midst of the chaos: "everything sinks / and the basement is flooded / it is terribly troublesome // the world perishes / and I don't know / if it's evening / or morning" (2018, 39). A declaration of "a new world / in the morning" (2018, 39) can be read both as a realisation that soon our world will be ruled by forces of nature more powerful than us, and perhaps by other organisms, and as a hope that humans can build some future on the ruins of our world.

The complex interaction of air, water, and land in the Critical Zone is a recurring theme in the three chapters of Ingvarsson's subsequent book of poetry, *Menn sem elska menn* (2021; Men who Love Men). In the first poem, the poet is pursuing literary research in a rainy Oslo, caught in the thick of the not-always-pleasant vaporosphere that runs through other spheres of the Critical Zone. In the next poem, Copenhagen is also "wet and cold", with "rotten ditches / foul canals" (Ingvarsson 2021, 25). From this city, the nineteenth-century-writer Tómas Sæmundsson had written declarations of love to the famous Romantic poet Jónas Hallgrímsson, who wandered the similarly wet and muddy streets of Reykjavík, and "speaks of a rendezvous / in the spirit / across seas / and lands // cosmic dimensions" (Ingvarsson 2021, 23). Ingvarsson's poems send the reader on precisely this kind of a journey, via myths and their complex cosmology where space is made for the convergence of various realms of existence. With a nod to the ecofeminism of the 1980s and 1990s and its importance for recent "questioning of the entrenched notion of the human, as well as the blurred boundaries between inorganic and organic matter" (Oppermann 2016, 273), he presents a climate-anxious reworking of Norse myths with a re-examination and queering of their gendered aspects (Aðalsteinsdóttir 2024). Ingvarsson thus participates in debunking "the objectification of the natural world, women, matter, bodily natures, and nonhuman species, and open[s] new eco-vistas into exploring the dynamic co-extensivity and permeability of human and nonhuman bodies and natures" (Oppermann 2013, 68). The title of the poem "The Sky from a Skull" refers to *Vafthrudnir's Sayings* (*Vafþrúðnismál*) in the Poetic Edda, in which the sky is said to be made from the skull of the male giant Ýmir. However, the twenty-first-century poet seeks shelter "in the skull of a woman" (Ingvarsson 2021, 60). Asking for female guidance, he espies cairns from a silenced women's culture in a landscape closely associated with the body. The reference to earlier ecofeminism is obvious, and a certain nostalgia can be detected for a world where "we were one mind" (Ingvarsson 2021, 72), but a nihilistic consolation is sought in the more posthuman thought that the particles which are now divided into different genetic chains can eventually merge again and become one mind. Mythology thus becomes a vehicle for overstepping not only gender divisions but all divisions between individuals, species, and matter.

The pre-modern worldview we experience through myth has been described as recognising that human beings are "unfixed and we are constantly liable to change" (van den Boom 2021), which is in line with a hydroqueer worldview. The myths do not, however, downplay the danger involved in overstepping boundaries. As in fairy tales and folklore, such practice tends to lead into the zone of monsters, the monstrous children of Loki, for example, who play a key role in Ingvarsson's poetry collection. In Norse mythology, Loki's boundaries are particularly fluid when it comes to both gender and species. For example, he gives birth to the eight-legged horse Sleipnir, after taking the form of a mare and enticing the stallion Svaðilfari. But his three children with the giantess Angurboða—the underworld queen Hel, the Midgard serpent Jörmungandur, and the wolf Fenrir—are so monstrous that the other gods perceive them as threats to the whole existence of the nine worlds. They therefore exile Hel to the underworld, cast the Midgard serpent into the depths of the ocean, and lay Fenrir in fetters. The problematics of this display of dominance and the repression of potentially hostile forces form the axis of the last part of *Menn sem elska menn*.

A critical zone

It is, however, another monstrous creature, the long-lived Greenland shark, that ties the complex themes of Haukur Ingvarsson's poetry collection together. When the shark is first introduced to the scene, the poet asks: "Have you looked into the black eyes / of the Greenland shark? / in them burns / the dull glow of radiocarbon" (2021, 27). These lines point to the fact that scientists have used radiocarbon dating of the eye-lens nuclei of Greenland sharks to determine these creatures' age, and in the process discovering individuals that had lived up to four hundred years. Bomb radiocarbon assays are one of the best techniques for determining the age of long-lived animals, because the "atmospheric testing of atomic bombs in the 1950s and 1960s resulted in a rapid and well-documented increase in atmospheric radiocarbon which was subsequently deposited in the calcified tissues of all surface-dwelling organisms" (Nielsen et al. 2017). The poems thus remind us that everything within Earth's ecosystems is connected. The oil from hunted shark bodies illuminated the city streets the Icelandic Romantic writers walked while thinking of each other and the oil's pungent fumes were inhaled and thus absorbed into their human bodies (Ingvarsson 2021, 43). In the same way, sharks (and all other living beings) carry the residues of human activity in their bodies. Furthermore, the poetry collection's structure, with myriads of interconnections between poems and their ecological, psychological, and mythical aspects, challenges the notion of any clear divide between culture and nature.

The Greenland shark can be seen as having a mythology of its own, tied up with its material existence as a part of a specific ecosystem. Primarily found in the mesopelagic and bathypelagic zones (deeper than 200 meters) of the North Atlantic and Arctic Oceans, it is an ancient species that is associated with a primitive and cruel nature as well as the masculine fisherman culture of its hunters (cf. Karls 2022). In *Menn sem elska menn*, a direct comparison is also made between the primitive shark-fishing of the poet's great grandfather and the story of how Thor, the god of thunder, caught the Midgard serpent on a fishing hook, arrogantly playing with destructive forces in a demonstration of power, just narrowly (and temporarily) saving himself and the world by severing the monster's head.[4] Descriptions of the shark's voyages in an "ocean without direction" and "from light to light" (Ingvarsson 2021, 27) can accordingly allude to the primordial void of Ginnungagap. But they also evoke associations with Christian myths, especially Milton's descriptions of how the fallen angel Satan voyages, in search of the realm of men, through "the hoary deep, a dark / Illimitable ocean, without bound, / Without dimension, where length, breadth, and height, / And time, and place, are lost" (2005, 70). The poet's question: "Did Loki so love his children that he did not sacrifice them?" (Ingvarsson 2021, 63), compares Loki to the Christian God, the father in heaven who "so loved the world that he gave his one and only [and, most significantly, human born] son" (John. 3:26). Loki, who has both mothered and fathered interspecies offspring, is depicted as a more loving parental figure, ready to risk Ragnarök, the end of the world, to save his children.

"I love men / who love men // who love men who love // men who love to love men / who love men / love", Ingvarsson concludes in the poem on the nineteenth-century-Romantics' correspondence (2021, 24). Pointing back to passionate declarations of love common in male friendships in the Romantic era, the love in *Menn sem elska menn* is, however, open not only in terms of gender but also in relation to the more-than-human. And here again, the Greenland shark provides a connection to myth. Like the Romantics, the shark is said to love men, albeit in a rather bloodthirsty fashion, similar to that of Milton's Satan, who admires the human creatures he "could love" and yet wants to destroy (2005, 115). Ingvarsson's description of the glow in the shark's eyes being "like a star deep in the sky" (2021, 27) reiterates not only the association with Satan as the fallen Lucifer (in Latin meaning "bearer of light" and a name for the morning star Venus) but also the theme of inversion, which is highlighted on the book's cover. An image of sky and ocean has been turned upside down, so the ocean is above and the sky is below. Similarly, Arènes, Latour, and Gaillardet's visual representation of the dynamic processes of the Critical Zone involve "the skin of the earth" being *reversed like a glove*. By placing the atmosphere in the centre and pushing the earth's core to the periphery, they aim to convey the "feeling of being inside and bound by revolving cycles" of moving matter, thus giving a more commonsensical "representation of the way we inhabit the earth" than the usual projection of a globe seen from outside (2018, 129–31). In Ingvarsson's poetic imagery, however, the usual world order is neither directly turned upside down nor inside out, but rather the distinction between the spheres is erased, primarily that between the ocean depths and space. They merge into the "directless ocean" that, with its interconnections of spheres, forms the Critical Zone of the poems.

Conclusion

Instead of projecting a holistic image, the word "zone" suggests something "irregular, fragmented, discontinuous", and the word "critical" implies that this zone is both important and fragile—that it is something that "could suddenly cross a tipping point" (Latour and Weibel 2020, 2), thereby providing a fitting view of life on Earth in the Anthropocene. In the twenty-first-century stories and poems discussed in this chapter, creatures of the sea are used to explore complex interactions in Earth's Critical Zone and to connect different spheres of existence. The mythical children of Loki, who in the poetry collection *Menn sem elska menn* herald the dissolution of humanity's anthropocentric, patriarchal worldview, are hybrid monsters that men and gods have feared for centuries. But the last line of the book seems to encourage rebellion and the liberation of monsters like the wolf Fenrir:

"Cut the fetter / set him / free" (Ingvarsson 2021, 71), presumably in the hope that this will make way for a different kind of world. However, swimming effortlessly between deep psychological, ecological, geological, and mythical waters, it is the Greenland shark that stands particularly for that which thrives in a critical zone; a zone marked by dissolution of the hierarchies, where man has been considered superior to woman, heterosexual love superior to homosexual love, and humans superior to other animals. In a radical rebellion against the worldview we are used to, the poetic imagery thus charts a hydroqueer cosmology that gives space for the mutable and relative nature of such distinctions and hierarchies.

Notes

1 Translations of quotes from this book come from an unpublished translation by Meg Matich.
2 All translations into English are mine, unless otherwise stated in the bibliography or in endnotes.
3 Translations of quotes from this book come from Rory McTurk's translation, forthcoming from Arc Publications as *Seal Mother*.
4 The novel *Havboka: eller Kunsten å fange en kjempehai fra en gummibåt på et stort hav gjennom fire årstider* (2015) by the Norwegian author Morten A. Strøksnes revolves around a hunt for a Greenland shark that also is connected to mythology.

Works cited

Aðalsteinsdóttir, Auður. 2023a. *Hamfarir í bókmenntum og listum*. Háskólaútgáfan.
Aðalsteinsdóttir, Auður. 2023b. "A Carnival of Sailing Homes and Shipwrecks: Ocean and Society in Hallgrímur Helgason's Historical Fiction." *Edda* 110 (1): 36–50. https://doi.org/10.18261/edda.110.1.4.
Aðalsteinsdóttir, Auður. 2024. "Orð úr undirdjúpum." *Són* 22: 15–39.
Aðalsteinsdóttir, Auður. 2027. "A Sea at Home: Watery Feminism as a Creative Practice in an Ecocentric World." In *Recent Environmental Approaches towards More-than-Human Humanities*, edited by Peggy Karpouzou and Nikoleta Zampaki. Routledge, forthcoming.
Aït-Touati, Frédérique, Alexandra Arènes, and Axelle Grégoire. 2022. *Terra Forma: Manuel de cartographies potentielles*. The MIT Press.
Alaimo, Stacy. 2016. "Oceanic Origins, Plastic Activism, and New Materialism at Sea." In *Exposed: Environmental Politics and Pleasures in Posthuman Times*. University of Minnesota Press.
Arènes, Alexandra, Bruno Latour, and Jerôme Gaillardet. 2018. "Giving Depth to the Surface: An Exercise in the Gaia-graphy of Critical Zones." *The Anthropocene Review* 5 (2): 120–35. https://doi.org/10.1177/2053019618782257.
Gerður, Kristný. 2022. *Urta*. Mál og menning.
Gerður, Kristný. 2025. *Seal Mother*. Translated by Rory McTurk. Arc Publications, forthcoming.
Ghosh, Amitav. 2016. *The Great Derangement: Climate Change and the Unthinkable*. Penguin Books.
Helgadóttir, Þórdís. 2018. "Skepnan." In *Keisaramörgæsir*. Bjartur-Veröld.
Henry, Claire. 2023. "Queer Posthumanism. Figures, Fluidity, and Fluids." In *Screening the Posthuman*, edited by Missy Molloy, Pansy Duncan, and Claire Henry. Oxford Academic.
Ingvarsson, Haukur. 2018. *Vistarverur*. Mál og menning.

Ingvarsson, Haukur. 2021. *Menn sem elska menn*. Mál og menning.
Jakobsdóttir, Svava. 1989. "Undir eldfjalli." In *Undir eldfjalli*. Forlagið.
Kaplan, E. Ann. 2016. *Climate Trauma: Foreseeing the Future in Dystopian Film and Fiction*. Rutgers University Press.
Karls, Unnur Steina K. 2022. "Af hákörlum, karlmennsku og kærleik: Um ljóðabókina *Menn sem elska menn* eftir Hauk Ingvarsson." *Hugrás*, November 29. https://hugras.is/2022/11/af-hakorlum-karlmennsku-og-kaerleik-um-ljodabokina-menn-sem-elska-menn-eftir-hauk-ingvarsson/.
Kress, Helga. 1993. *Máttugar meyjar: Íslensk fornbókmenntasaga*. Háskólaútgáfan.
Latour, Bruno, and Peter Weibel. 2020. "*Seven Objections Against Landing on Earth*." In *Critical Zones: The Science and Politics of Landing on Earth*, edited by Bruno Latour and Peter Weibel. MIT Press.
Milton, John. (1667) 2005. *Paradise Lost*. Oxford University Press.
Neimanis, Astrida. 2012. "Hydrofeminism: Or, On Becoming a Body of Water." In *Undutiful Daughters: Mobilizing Future Concepts, Bodies and Subjectivities in Feminist Thought and Practice*, edited by Henriette Gunkel, Chrysanthi Nigianni, and Fanny Söderback. Palgrave Macmillan.
Neimanis, Astrida. 2017. *Bodies of Water: Posthuman Feminist Phenomenology*. Bloomsbury Academic.
Nielsen, Julius, Rasmus B. Hedeholm, Jan Heinemeier, Peter G. Bushnell, Jørgen S. Christiansen, Jesper Olsen, Christopher Bronk Ramsey et al. 2017. "Eye Lens Radiocarbon Reveals Centuries of Longevity in the Greenland Shark (*Somniosus microcephalus*)." *Science* 353 (6300): 702–4. https://doi.org/10.1126/science.aaf1703.
Oppermann, Serpil. 2013. "Feminist Ecocriticism: The New Ecofeminist Settlement." *Feminismo/s*: 65–88.
Oppermann, Serpil. 2016. "From Material to Posthuman Ecocriticism: Hybridity, Stories, Natures." In *Handbook of Ecocriticism and Cultural Ecology*, edited by Hubert Zapf. De Gruyter.
Richter, Daniel, interviewee. 2019. In Eric Barstow. *The Education of Bruno Latour*. John Hope Franklin Humanities Institute at Duke University. Documentary film, 1 h., 10 min., 59 sec. https://www.youtube.com/watch?v=YKaVraLeYZ8.
Seymour, Nicole. 2020. "Queer Ecologies and Queer Environmentalisms." In *The Cambridge Companion to Queer Studies*. Cambridge University Press.
Sigurðardóttir, Ragnhildur, Anthony J. Newton, Megan T. Hicks, Andrew J. Dugmore, Viðar Hreinsson, A. E. J. Ogilvie, Árni Daníel Júlíusson, et al. 2019. "Trolls, Water, Time, and Community: Resource Management in the Mývatn District of Northeast Iceland." In *Global Perspectives on Long Term Community Resource Management*, edited by Ludomir R. Lozny and Thomas H. McGovern. Studies in Human Ecology and Adaptation. Springer.
Tuan, Yi-Fu. 1987. *Space and Place: The Perspective of Experience*. Arnold.
van den Boom, Frank. 2021. "New Bodies in an Ecological Crisis: The Unforeseeable Future in Ovid's Metamorphoses." *Leiden Elective Academic Periodical* (1): 155–82. https://graduatejournal-leap.universiteitleiden.nl/2021/06/frank-van-den-boom-new-bodies-in-an-ecological-crisis-the-unforeseeable-future-in-ovids-metamorphoses/.
Zapf, Hubert. 2016. *Literature as Cultural Ecology: Sustainable Texts*. Environmental Cultures Series, edited by Greg Garrard and Richard Kerridge. Bloomsbury Academic.

20

RISEN FROM THE FOAM

Hans Christian Andersen's fairy tales as aquatic texts

Klaus Müller-Wille

"Far out at sea"

"Far out at sea the water is as blue as the petals of the loveliest cornflower" (Andersen 2003a, 154).[1] Right from the first sentence of his fairy tale "The Little Mermaid" (1837; "Den lille Havfrue"), Andersen succeeds in portraying the sea as a fascinating foreign realm, which he associates with the blue flower of Romanticism and thus with the idea of a poetisation of the world. Indeed, the fairy tale goes on to be characterised by a whole series of opulent descriptions in which the underwater world is presented as a place of longing, where castles and gardens made of coral, shells, and amber appear between magnificent algae trees swaying in the current. Of course, Andersen was able to draw on a wide range of literary models for this fantasy, writings that dealt with seductive sirens, undines, mermaids, and mermen (cf. Kraß 2010 and Heller 2025). However, Andersen's preoccupation with aquatic landscapes is by no means limited to this text. Throughout his career, he repeatedly explored the relationship between humans and the sea, examining it from different perspectives not only in his fairy tales and stories but also in his poems and novels.

In this sense, it is not surprising that a whole series of studies has already been published that deal with the interrelationship between humans, water, the sea, and coastal landscapes in Andersen's writings. The imagination of the ocean in "The Little Mermaid" is often interpreted as a rather unspecific draft of a very human-like fantasy world, which allowed Andersen to indirectly address social and cultural issues of his time. Curiously, an article by Samuel Baker, which is included in an anthology on blue humanities, also employs a similar allegorical reading strategy, without paying much attention to how the sea is actually represented (Baker 2019). Closer to the idea of fluid imagination are essays in which

DOI: 10.4324/9781003647638-26

the evocation of a flowing environment in Andersen's fairy tales is linked to the theme of sexuality. Heinrich Detering links the depiction of amphibious existences in "The Little Mermaid" to Andersen's attempt to reflect indirectly on homoerotic desire and the escape from heterosexual normativity (2002, 173–219). This interpretative approach has influenced a whole series of current stage adaptations of the fairy tale, in which Andersen is literally claimed as the inventor of fluid identities.

Steve Mentz interprets the text quite differently in his history of oceanic imaginations (2020, 70–77). In his reading, the idea of oceanic desire is at the centre of the action. He connects this transgressive desire with the impossible and deeply ambivalent wish to merge with the water. In doing so, he operates with the idea that Andersen reverses yearning for the sea with the figure of the mermaid longing for land from the water in order to expose the deadly impulse of this pleasure principle even more clearly.

Frits Andersen's *Wonders: A Literary History of the Deep Sea* does not deal in detail with "The Little Mermaid".[2] He states that the text is "utterly devoid of wonder" in the sense of the depiction of a wonderful nature, but that the fairy tale hints at the "deep sea as the period's new zone of discovery and wonder" when the prince tells the mermaid about the sea and at the same moment reveals the "gaps in human knowledge" about this zone (2025, 88). He does not associate these reflections with reflections on the poetics of the fairy tale in which the wondrous often plays a major role. However, his observations indicate that different aesthetic categories may be more relevant for Andersen's aquatic poetology. In my chapter, I suggest that the metaphor of the flowing mirror (Barlby 1995), dynamic spatiality, forms in flux, and forces beyond human control are models for Andersen's attempt to redefine the fairy tale as a fluid genre.

Katie Ritson and Frederike Felcht offer interpretations of the sea in Andersen's texts that are based on more complex ecocritical considerations. However, both authors only refer to "The Little Mermaid" in passing. Ritson discusses Andersen's novel *The Two Baronesses* (1848; *De to Baronesser*) in detail (2019, 28–37). She focuses primarily on the central setting of the text—the coastal landscape of the North Frisian Wadden Sea, characterised by tides and constant change. In her view, this ecological context provides the narrative backdrop for events marked by transnational interconnections and a contingent plot. In her essay on Andersen's water worlds, Frederike Felcht (2025) deepens these interpretations by showing that the ecological perspective that characterises the novel's plot also gains poetological relevance. She embeds *The Two Baronesses* in a field of fairy tales, poems, and novels in which Andersen attempts to explore similar interrelationships between waterscapes and human actors.

My chapter follows on from these two inspiring studies. I too would like to focus on the question of the ecology of fluid relations between humans and waterscapes in Andersen's texts. In doing so, I will attempt to explore the poetological relevance of this issue in greater depth. I believe it is no coincidence that Andersen frequently draws on the fairy-tale genre to develop formal questions that can be linked to the representation of waterscapes as plot carriers. I would like to illustrate this by examining four fairy tales from three different phases of Andersen's creative work. However, the chapter begins with an analysis of two texts that Andersen published before "The Little Mermaid" and which serve here primarily to outline more precisely the formal innovations and poetological concepts associated with the treatment of the sea theme in his fairy tales.

Mirror, deep current, and abyss—the sea in "The Diving Bell" and *Agnete and the Merman*

As early as 1827, Andersen published a poem that he described as a "fairy tale from the bottom of the sea" (Andersen 2005a, 242). "The Diving Bell" ("Dykker-Klokken") appeared in the 98th issue of the journal *Kjøbenhavns Flyvende Post* and was later incorporated as the twelfth chapter in Andersen's experimental debut novel *A Journey on Foot* (1829; *Fodreisen*).

The poetic voice of the ballad recounts the experiences of diving to the bottom of the sea in a diving bell made of crystal. In doing so, the narrator first reflects on the unique perspective afforded by the water's refraction, where "the sun stood so red and large, but without the slightest ray" (Andersen 2005a, 243). Reflections on the observer's perspective also shape the following encounters with the sea creatures. To their amazement, the narrator discovers that the fish on the sea floor move around on two legs and lead an urban life similar to humans on earth. This observation goes hand in hand with the insight that water in the sea takes on the function of air. The fundamental similarity between the earth and the seabed ultimately leads the lyrical narrator to draw an analogy that eliminates the differences between fish and humans: "For when we look at things correctly, we are fish at heart; the whole air is like an ocean, and the earth is the bottom" (Andersen 2005a, 244). With this insight, the fantasy takes off. The two-legged fish on the sea floor indignantly draw the narrator's attention to all the human projections that are wrongly attributed to the sea creatures. They are particularly disturbed by the "cannibalistic" idea of "fish eating fish". Such phenomena are only known from the fringes of the upper sea, where (human-like) fish have been infected by human appetites, behind which lies the much more comprehensive cannibalism of (precisely these fish-like) humans:

> Fear not that we will eat you, no, no one here desires that;
> Only when we rise from the depths and seek the shores of the land,
> Your greedy minds infect us, and we eat our neighbours, ...
> For we are being eaten in a cannibalistic manner by you up there. (Andersen 2005a, 245)

At this point, at the latest, it becomes clear that the "fairy tale at the bottom of the sea" turns into a satire in which Andersen pokes fun at the anthropomorphising tendencies in the romantic nature poetry of his time and its penchant for unbroken analogies. The satire is underscored by the fact that Andersen, by pointing out the one-sided consumer behaviour of humans, clearly highlights the more complex interrelationships and imbalances between humans and sea creatures.

Given that Andersen dealt with depictions of the sea in such a bitingly ironic manner, it is interesting that he published the verse drama *Agnete and the Merman* (*Agnete og Havmanden*) in 1834. With this play, he not only ties in with a long ballad tradition in Northern Europe but also with models from Danish literature around 1800. Both Jens Baggesen and Adam Oehlenschläger deal with the subject of the married woman and mother Agnete, who is seduced by a merman and drawn into the sea, in their poems "Agnete from Holmegaard" (1808) and "Agnete" (1812). Oehlenschläger in particular draws on the idea of human-like life at the bottom of the sea that Andersen pokes fun at in "The Diving Bell". For example, he dresses the merman up as a kind of Neptune, who tries to impress Agnete with a whole series of theatrical attributes, including a scale armour, coral staff, turtle shell shield, snail-shaped helmet, and seaweed hair (Oehlenschläger 1861, 161–62). Accordingly, he also depicts his kingdom in the form of an operetta-like backdrop—for example, he names a hall whose walls are carved from the clearest crystal, a sleigh drawn by seals, and flowers that rise up in the water (1861, 162)—that anticipates a whole series of motifs from Andersen's later fairy tale "The Little Mermaid".

Baggesen's work, however, lacks such motifs. He seems to be primarily interested in the demonic powers of the merman who seduces Agnete with his singing. He depicts the sea floor in a completely different way, as a bottomless abyss:

> Agnete saw down to the Heaven
> In the depths of the sea —
> The wave was so silvery clear
> And the bottom was so blue!
> Oh! So blue! . . .
> And listen, beautiful Merman!
> I want to belong to you!
> If down into the abyss of the sea
> you will take me with you! (1845, 90)

In *Agnete and the Merman*, Andersen also plays with the motif of the abyssal sea, which unleashes a desire that ultimately cannot be fulfilled. At the same time, he modifies this motif by marking the sea as another space, one that is primarily characterised by the childlike experience of rocking, which is strongly influenced by the movements of the water. Thus, Agnete, who was almost literally born in the sea during a sudden storm tide, dreams of returning to her mother's womb when she sees the sea:

> By the sea view, when you see the shallow grounds
> Like green banks in the dark blue,

I don't know myself, I want to sleep there,
Where the big ships pass over me.
It is not a grave where the earth weighs heavy,
It is the cradle of the world where the child rocks. (Andersen 2005b, 382)

Agnete suffers above all from the idea of spending her life as a housewife and mother in the "quiet domesticity" ("stille Huuslighed"; Andersen 2005b, 386) of a "cramped home" ("snevre Hjem"; Andersen 2005b, 385). She is repelled by the ideal of a Biedermeier-style private idyll with a view of a fenced-in garden and neatly arranged flower beds. Instead, she longs for a space that has no boundaries and is characterised by continuous movement and dynamism. This movement, which results from direct interaction with the surrounding fluid matter, is linked to the idea of an essential form of life that cannot be realised on earth:

With flowers and with greenery, the land adorns itself.
A painted corpse, every summer, autumn, and spring;
No, life and soul belong to the mighty sea,
For it swells and crashes high against the shore.
While the earth must adorn itself to please,
The sea is all by itself, every day! (Andersen 2005b, 382)

In keeping with this idea, the merman does not seduce Agnete with a luxurious palace made of shells and precious materials, but with a seemingly endless castle made entirely of waves and currents:

I know a castle, its walls and roof
Are green waves, layer upon layer,
No sparrows build their nests there,
No ships are seen instead of small birds.
Its pillars are the rushing currents,
Where whales and dolphins swim!
It reaches from both poles
Down to the southern paradise. (Andersen 2005b, 383)

The view from the castle—with which the merman attempts to seduce Agnete—is also defined by moving images in the medium of the water that replace each other in a continuous stream of mirror images:

High above the water,
You gaze upon the land;
Above your head, with flowers and bushes,
It is reflected in the moonlit night!

> Countries with cities,
> Floating like clouds,
> Above you, across the calm sea. (Andersen 2005b, 400)

With these images of a completely fluid and seemingly boundless form of spatiality, the merman appeals to Agnete's oceanic desire. In this sense, it is not surprising that Andersen gives a voice not only to animals, plants, and nature spirits in the verse drama but also to the ocean waves themselves, which interact directly with Agnete.

Forms in flux—Sea foam and morphogenesis in "The Little Mermaid" (1837)

At first glance, the diegetic world of "The Little Mermaid" also revolves around a clear dichotomy between a terrestrial and an aquatic ecosystem. The emphasis with which the narrator attempts to draw attention to mirror-image similarities between the world above and below water is initially reminiscent of the highly anthropomorphised depiction of the sea in "The Diving Bell":

> Far out at sea the water is as blue as the petals of the loveliest cornflower and as clear as the purest glass, but it is very deep, deeper than any anchor cable can reach, many church towers would have to be placed on top of each other to stretch from the seabed to the surface. Down there the sea-folk live.
>
> Do not believe, though, that there is nothing but the bare, white sand on the seabed; no, the most marvellous trees and plants grow there that have such pliant stems and leaves that the slightest movement of the water causes them to move as if they were alive. All the fish, great and small, slip between their branches, just as birds up here do in the air. At the very deepest spot lies the sea-king's palace, . . . the roof is of mussel shells that open and close as the water passes. (Andersen 2003a, 154)

Perhaps the open or breathing shell roofs already provide an indication that the stable spatial order created by the juxtaposition of the upper and lower worlds will be undermined in the further course of the fairy tale. Deconstructive analyses of the fairy tale in particular have drawn attention to a long series of contradictions, aporias, and ambivalences that invalidate fundamental narrative patterns of order in the text. The wonderful metaphor of a "flowing mirror", with which Finn Barlby (1995) attempts to describe the dynamic character of the fairy tale, already suggests that these uncertainties also affect the basic spatial structure of the text.

This is primarily because the fixed location of the narrator—who in the quoted passage positions themself deictically with both feet firmly on the ground with the word "up here" —is suspended in the further course of the fairy tale, in which we are increasingly urged to adopt the view and perspective of the sea creatures, especially the view and perspective of the little mermaid (Dahlerup 1995, 21–22). Paradoxically, it is precisely the narrator's attempts to present this view as a strange, distorted perspective that leads to the dissolution of familiar ways of seeing. This is already hinted at in descriptions that tell how the little mermaid longed to leave the bottom of the sea and see what it was like "in our world" ("hos os"; Andersen 2003a, 155). Above all, however, the focus on the fundamental mediation of vision, which is achieved through air and water, prompts readers to reflect on the disturbing perception of the outside world from underwater: "they looked much larger than they do to our eyes" (Andersen 2003a, 156). The shifting perspectives ultimately lead to a destabilisation of the clear hierarchy between the human world and the marine world, which the narrator connects with the vertical distinction between "below" and "above".[3] In this sense, it is not surprising that readers are confronted with a loss of orientation in the introduction, in which "one" can no longer distinguish between above and below, "false" and "correct" perception (Egeberg 1995, 112):

> There lay a mysterious blue sheen over everything down there—it would be easier to believe one was high up in the air and could only see sky above and beneath one than down on the seabed. When the sea was calm, one could make out the sun, it seemed to be a purple flower with its entire light streaming out of the calyx.
>
> (Andersen 2003a, 155)

Here, the idea of a clear vertical order between sea and earth is replaced by the notion of a consistent mirroring, in which the seabed functions not as the "ground of the sea" but, confusingly, as the "sky of the sea" (according to this logic, the water surface would then have to function as the equivalent of the earth). To underscore this idea, the narrator resorts to the depiction of light sources on the seabed, which, interestingly, are also used at this very point to outline the idea of a dynamic space in constant motion:

> Outside the palace there was a large garden with bright-red and dark-blue trees, with fruit that shone like gold and flowers that blazed like fire in the constantly moving stems and leaves. The earth itself was the finest sand, but blue as a flare of sulphur.
>
> (Andersen 2003a, 154–55)

The playful approach to the basic topology of the text, in which vertical orders (seabed–water–earth–sky) stand alongside horizontal references (sea–earth) and various equivalence models (seabed–water = earth–air; seabed–water = sky–earth) is just the prelude to a wild game with different abstract spaces that can no longer be arranged into an order. In a central reply, the little mermaid's grandmother says that humans could emerge onto unknown heavenly shores just as sea creatures emerge from the sea onto the land of humans. It remains completely unclear how this topology, which is of course closely linked to the central religious concept of the afterlife of the soul, relates to the above-quoted descriptions of the narrative instance. It is also unclear how the air as the home of the air spirits can be integrated into the sequence of sea, earth, and sky. The semantics of the different states of aggregation (solid–liquid–gaseous), which are continuously referred to in the text, cannot be transferred one to one to any of the spatial arrangements presented. The same applies to the distinctive colour logic of blue, red, black, and white that the fairy tale operates with.

In other words, the text has an abstract spatial structure that shifts continuously, defining the marine habitat through constantly changing pairs of opposites, equivalencies, and hierarchies. In contrast to *Agnete and the Merman*, the dynamic spatiality here does not serve to outline the otherness of a fluid and constantly moving water world; rather, it forms the fundamental dynamic matrix of the fairy tale itself.

One consequence of the fluid spatial structure of the narrative is that the story cannot come to an end. Contrary to the witch's prediction that the little mermaid would turn into dead sea foam if the prince did not fall in love with her, she is transformed into an air spirit who must face new challenges. The heroine's ongoing morphogenesis—that is, the emergence of form—finds its counterpart in a dynamic spatial structure in which the diegetic world is constantly reconstituted on new foundations.

The principle of this dynamic formation, which constantly recreates itself in a continuous movement, is exhibited in two central scenes of the text. Andersen clearly attaches great importance to distinguishing the dynamic form principle of the fairy tale from static (petrified) form concepts in architecture, painting, and sculpture (it is no coincidence that the Pygmalion motif of animated stone plays a key role in the text). This is made clear by the description of the prince's castle, in the centre of which is a fountain that sets in motion a continuous, medially regulated light performance:

> [B]etween the columns that went round the entire building there were marble statues that looked as if they were alive. Through the clear glass in the tall windows one could glimpse the most magnificent halls that were hung with precious silk curtains and tapestries, and all the walls were adorned with large paintings that were a sheer joy to look at. In the middle of the largest hall there was a large plashing fountain, its jets shooting up towards the glass dome in the ceiling, through which the sun shone on the water and on the lovely plants growing in the large pool.
>
> (Andersen 2003a, 162)

The prince's castle finds a counterpart in the hall of the Sea King, in the middle of which a performance is staged, which attempts to combine songs, lighting effects, and dance figures into a unified whole. This performance is also created by a continuous stream of water:

> The walls and ceiling of the great dance hall were of thick but clear glass. . . . [A] blue-burning fire . . . lit up the entire hall and gleamed out through the walls so that the sea right outside looked quite illuminated; one could see all the innumerable fish, great and small, that swam towards the glass wall, on some of them the scales gleamed a purple-red, on others they seemed to be silver and gold. A broadly running stream ran through the middle of the hall, and on it mermen and mermaids danced to their own delightful singing.
>
> (Andersen 2003a, 164)

In addition to the reference to the two flowing sources of energy in the fairy tale's artistic centres, a key motif also points to the significance of the fluid concept of form that is discussed in the text. The fairy tale repeatedly points out that, in the worst case, the little mermaid will turn into "dead, salty sea foam" (Andersen 2003a, 173) or "deathly cold sea foam" (Andersen 2003a, 174). The dead sea foam automatically raises the question of "living sea foam", and this question is indeed answered in the second transformation scene in the text, in which the mermaid literally rises from the foam as an air spirit: "The little mermaid saw that she had a body like theirs [the daughters of the air], it rose more and more out of the foam" (Andersen 2003a, 174). Of course, Andersen alludes in this scene to the birth of Aphrodite, who—at least according to Hesiod's account—emerges from the foam created by Uranus' genitals being thrown into the sea, which is why she is nicknamed "Anadyomene" ("the rising one"). Andersen was well acquainted with this myth, as he reveals in a song of the waves in *Agnete and the Merman*, which revolves around the mythological connection between sea foam and the emerging "goddess of beauty and love" (Andersen 2005b, 401).

In both *Agnete and the Merman* and "The Little Mermaid", the motif of foam is repeatedly linked to the bodies of mermaids, who emerge briefly from the turbulent, foam-crested waves before disappearing back into the sea. As early as the aesthetic debates of the early modern period, this phenomenon was associated with the aesthetic principle of a form emerging from the movement of formless matter itself, no longer regulated by a predetermined idea, but caused by turbulence, folds, eddies, and the like (Fricke 2010). By the Romantic period at the latest, this self-referential, poetological reading of the myth had become firmly established in the visual arts and literature (Serres 1995; Groth 2015; Böhme 2025).

What is important is that the fluid conceptualisation of the text is revealed not only through the image of the fountain but also through the reference to the far-reaching aesthetic implications of the myth of Aphrodite. The ongoing metamorphosis of the fundamental spatial patterns of order on which the diegesis rests is tied back to the central setting of the fairy tale. In this way, the strange form of the narrative is semanticised as an aquatic, flowing, or oceanic quality of narration. This fluid quality, in turn, seems to be very closely linked to Andersen's specific use of the fairy-tale genre. This is defined not by traditional narrative patterns, but by ever-new forms obtained through complex self-referential processes. In this sense, it could be argued that Andersen deliberately stages his novel conception of the fairy tale in this central poetological text in the form of a birth from the sea.

Humans from the perspective of fish—the sea and the changing genre of fairy tales

Given the far-reaching implications of the exploration of the sea in "The Little Mermaid", it is not surprising that Andersen published a whole series of texts after the publication of this fairy tale in which the motifs of water and the sea are used for far-reaching poetological reflections. Felcht (2025) refers to a large number of these texts. At the end of this chapter, I will focus only on those examples in which Andersen's reference to the sea is used to reflect on a new conception of the genre of his fairy tales and stories.

At first glance, the depiction of the sea in "A Story from the Sand Dunes" (1860; "En Historie fra Klitterne") is far more realistic than the fantastic representation in "The Little Mermaid". In contrast to the fairy tale, the story can be located historically and geographically and understood as a tale about the instability that characterises coastal environments (cf. Ritson 2019, 27–37). It takes place at the tip of Denmark's northwestern coast (near Skagen) during the reign of Christian VII (1766–1808). The wild coastal areas, the hinterland with its dunes and heathland, and the Kattegat are described in great detail. In fact, the first drafts of the text date back to a trip to Jutland in 1859, which Andersen initially documented in diaries and manuscripts and then in shorter travelogues, which he published under the title "Skagen" in the ninth volume of the *Folkekalender for Danmark* (1860).

Nevertheless, it would be too simplistic to describe the text as merely a realistic local narrative. Andersen tries too hard to emphasise the wonderful features of the North Jutland landscape, which is explicitly referred to in the text as the "land of fairy tales" ("Eventyrernes Land"; Andersen 2003b, 213). On the one hand, the fairy-tale quality of northern Denmark is created by long descriptions of the interstice of the coastline, which seems to belong neither to the sea nor to the land. On the other hand, Andersen draws attention to local legends, fables, and ballad traditions, which he gleaned from the collections and publications of oral narrative genres available at the time. Twice he draws on fairy-tale material that is at least remotely reminiscent of his earlier depictions of the sea. For example, he tells of the funeral of a man washed ashore, which is followed by a series of severe storms and floods. Only when the grave is opened do people realise that they have buried a merman whom the sea is trying to take back. In the second fairy tale, Andersen again lets fish speak. An eel mother warns her daughters not to stray too far from home, as they risk being speared by eel spears. Of course, the daughters do not heed their mother's words, and so only three of her eight children return from their trip. The mother still believes that her children will return, even after she learns that her daughters have been skinned, cut up, fried, and eaten. Only when she learns that alcohol was also involved does she burst into tears, knowing that "brandy buries the eel" ("Brændevinet begraver Aalen"; Andersen 2003b, 312).

At first glance, both embedded stories contribute to shaping the landscape literarily by linking natural phenomena with narratives. However, the tragic events of "A Story from the Sand-Dunes" seen to defy this logic. They revolve more around the drastic experience of contingent strokes of fate that can no longer be pieced together into a meaningful narrative. The story tells of the fate of a rich Spanish child who loses all his relatives in a shipwreck off the Danish coast and grows up in poverty in North Jutland. The hero is suspected of murder and spends several years in prison, even though he is innocent. With a new love, his life seems to take a turn for the better, but another shipwreck leads to her death as well. At the end of the story, the hero descends into madness and wanders through the dunes. During a hurricane, he is buried in a church, which now serves as his tomb. Almost all of the strokes of fate are triggered by forces of nature, to which the hero must surrender. The readers wait the entire time for a reunion with the hero's Spanish relatives, but the story does not come to such a redeeming conclusion, just as the fairy tale of the eels didn't. In the process, the dramatic events of the eel story are consistently reversed: It is not a drastic tale of nature that falls victim to human appetites, but an account of humans who fall prey to the unpredictable and unaccountable agencies of nature. For it is particularly the long scenes describing how the hero tries in vain to save his beloved in a stormy sea that make it clear that in this text, sand, wind, and sea have greater power to influence the plot than the human characters. Thus, the formal peculiarity of the text, which manifests itself in the loosely assembled conglomeration of fragmentary ballads, anecdotes, and fables inserted into the text as internal narratives, is ultimately traced back to an articulation of the power of the sea (for a corresponding reflection about the interplay between literary form, aquatic agency, and the visibility of the non-human, see Kluwick 2020).

Conclusion

Recent Andersen research has repeatedly pointed out that the late Andersen attempted to reorient himself by using the fairy-tale genre to poetically do justice to the technical achievements of his century. He formulated this modernist programme in a text with the telling title "The Muse of the New Century" (1861; "Det nye Aarhundredes Musa"). In this context, too, encounters with the sea play a decisive role. For example, he has the heroine of his urban adventure "The Dryad" (1868; "Dryaden") visit the extensive Paris World's Fair of 1867, where she admires an aquarium installation (Andersen 2025, 87–104). Readers of the fairy tale "The Great Sea Serpent" (1872; "Den store Søslange") are also amazed when they are taken back to the bottom of the sea to encounter the recently laid telegraph cables between Europe and America (Andersen 2025, 41–61).

Referring back to Andersen's own statements, it seems appropriate to describe the texts as fairy tales of technology (cf. Felcht 2013, 48–50, 74–75, 166–69, 218–21). Following Thomsen's and Felcht's readings of "The Great Sea Serpent", what strikes me as more interesting is that Andersen resorts to the fiction of looking back through the eyes of fish to question the human perspective on simplistic stories of progress (Thomsen 2019, 202–7; Felcht 2025, 244–45). Particularly drastic are the observations of the fish, which were carted in barrels on railways to the Paris exhibition halls to critically observe the "human mire" ("Menneskemudderet"; Andersen 2003c, 212) parading past their windows. In the process, an observer with "paper, ink and pen, writing everything up, writing everything down" (Andersen 2003c, 213) also catches their eye. They immediately identify him as "a kind of octopus among humans" (Andersen 2003c, 213). Perhaps this would indeed be a good description for the marine poet Andersen.

Notes

1 All translations of Andersen's texts are by John Irons (Andersen 2022). All other translations are auxiliary translations created by me for this chapter.
2 I would like to thank Frederike Felcht for pointing out Frits Andersen's brief but important discussion of *The Little Mermaid.*
3 Alexia Panagiotidis will explore the topological paradoxes of the fairy tale in her doctoral thesis on Andersen and ancient myths.

Cited works

Andersen, Frits. 2025. *Wonders: A Literary History of the Deep Sea.* Aarhus University Press.
Andersen, Hans Christian. 2003a. "Den lille Havfruen." In *Eventyr og historier*, Vol. 1 of *H.C. Andersens samlede værker*, edited by Klaus P. Mortensen, 154–75. Gyldendal.
Andersen, Hans Christian. 2003b. "En Historie fra Klitterne." In *Eventyr og historier*, Vol. 2 of *H.C. Andersens samlede værker*, edited by Klaus P. Mortensen, 303–33. Gyldendal.
Andersen, Hans Christian. 2003c. "Dryaden." In *Eventyr og historier*, Vol. 3 of *H.C. Andersens samlede værker*, edited by Klaus P. Mortensen, 195–215. Gyldendal.
Andersen, Hans Christian. 2005a. "Dykker-Klokken." In *Digte og blandinger*, Vol. 9 of *H.C. Andersens samlede værker*, edited by Klaus P. Mortensen, 243–47. Gyldendal.
Andersen, Hans Christian. 2005b. *Agnete og havmanden.* In *Skuespil*, Vol. 10 of *H.C. Andersens samlede værker*, edited by Klaus P. Mortensen, 363–453. Gyldendal.
Andersen, Hans Christian. 2022. *Fairy Tales and Stories of Hans Christian Andersen.* Translated by John Irons. The Hans Christian Andersen Centre. Accessed August 6, 2024. https://hcams.andersen.sdu.dk/exist/apps/andersen-irons/index.html.
Baggesen, Jens. 1845. "Agnete fra Holmgaard." In *Jens Baggesens danske Værker: Anden udgave. Andet Bind. Lyriske Digte*, 188–99. Reitzel.
Baker, Samuel. 2019. "*The Forsaken Merman, The Little Mermaid,* and Early Modernism: Undersea Imagery and Dissolution of Culture." In *The Aesthetics of the Undersea*, edited by Margaret Cohen and Killian Quigley, 97–110. Routledge.

Barlby, Finn. 1995. "Det flydende spejl." In *Det flydende spejl: Analyser af H.C. Andersens "Den lille Havfrue"*, edited by Finn Barlby, 7–16. Dråben.
Böhme, Hartmut. 2025. "Wasser als morphologische Kraft." In *Wasserlandschaften. Ökologien des Fluiden um 1800*, edited by Roland Borgards, Frederike Middelhoff, and Barbara Thums, 21–46. Metzler.
Dahlerup, Pil. 1995. "*Den lillle Havfrue* dekonstrueret." In *Det flydende spejl: Analyser af H.C. Andersens "Den lille Havfrue"*, edited by Finn Barlby, 17–27. Dråben.
Detering, Heinrich. 2002. *Das offene Geheimnis: Zur literarischen Produktivität eines Tabus von Winckelmann bis zu Thomas Mann*. Studienausgabe. Wallstein.
Egeberg, Ole. 1995. "Gentagelse og gengivelse." In *Det flydende spejl. Analyser af H.C. Andersens "Den lille Havfrue"*, edited by Finn Barlby, 108–19. Dråben.
Felcht, Frederike. 2013. *Grenzüberschreitende Geschichten: H.C. Andersens Texte aus globaler Perspektive*. A. Francke.
Felcht, Frederike. 2025. "Von Wassertropfen bis Wattenmeer: Hans Christian Andersen." In *Wasserlandschaften*: *Ökologien des Fluiden um 1800*, edited by Roland Borgards, Frederike Middelhoff, and Barbara Thums, 243–70. Metzler.
Fricke, Beate. 2010. "Schaumgeburten: Zur Topologie der creatio ex nihilio bei Albrecht Dürer und ihrer Vorgeschichte." In *Das Meer, der Tausch und die Grenzen der Repräsentation*, edited by Hannah Bader and Gerhard Wolf, 41–66. Diaphanes.
Groth, Sebastian. 2015. "Venus Anadyomene: The Birth of Art." In *Venus as a Muse: From Lucretius to Michel Serres*, edited by Hanjo Berresem, Günter Blamberger, and Sebastian Groth, 15–40. Brill.
Heller, Jakob Christoph. 2025. "Figurationen des Amphibischen als potologische Reflexionsinstanzen und kulturelle Archivare bei Ludwig Tieck (*Das Donauweib, Der Wassermensch*)." In *Wasserlandschaften*: *Ökologien des Fluiden um 1800*, edited by Roland Borgards, Frederike Middelhoff, and Barbara Thums, 169–90. Metzler.
Kluwick, Ursula. 2020. "The Global Deluge: Floods, Diluvian Imagery, and Aquatic Language in Amitav Ghosh's *The Hungry Tide* and *Gun Island*." *Green Letters* 24 (1): 64–78.
Kraß, Andreas. 2010. *Meerjungfrauen: Geschichten einer unmöglichen Liebe*. Fischer.
Mentz, Steve. 2020. *Ocean*. Bloomsbury.
Oehlenschläger, Adam. 1861. "Agnete." In *Oehlenschlägers Poetiske Skrifter: 24. Deel. Digte*, edited by F. L. Liebenberg, 161–64. Reitzel.
Ritson, Katie. 2019. *The Shifting Sands of the North Sea Lowlands: Literary and Historical Imaginations*. Routledge.
Serres, Michel. 1995. *Genesis*. University of Michigan Press.
Thomsen, Torsten Bøgh. 2019. *Skyggepunkter: Menneske, natur og materialitet i H.C. Andersens forfatterskab*. Spring.

INDEX

Andersen, Hans Christian 3, 57, 74, 78–9, 115, 162, 269–81
anthropocentrism 30, 54, 63, 68, 232, 234, 240, 259, 261, 266
archipelago 8, 34, 35, 37, 45–56, 126–7, 134–5, 203–15, 243–4
Arctic Ocean 5, 11, 133, 150, 153, 165–75, 257, 265
Arrebo, Anders 23, 187, 188
Atlantic 2, 3, 5, 7, 11, 40, 45, 89, 139–48, 153, 157, 179–90, 194, 197–8, 203, 212–13, 216–26, 235, 265

Baggesen, Jens 272
Baltic Sea 3, 5, 19–31, 36–7, 40, 48, 54, 57–69, 244
Blicher, Steen Steensen 73–83
Blixen, Karen (Isak Dinesen) 232–42

Calvino, Italo 245, 246
Carson, Rachel 128
Chinese Sea *see* South China Sea
Claussen, Sophus 57
Claussøn Friis, Peder. 22, 180, 183, 184, 189
climate change 5, 66–9, 99, 128–9, 248, 257–68
colonialism 2, 3, 5, 6–10, 11, 12, 62–3, 67, 165–76, 206–7, 210, 216–27, 235
colour (of the sea) 1–2, 10–12, 38, 78, 80, 129–30, 140, 194, 195, 197–8, 224, 231, 236–9, 251, 269–76.
colour (skin colour) 220
contact zone 62–3, 69
Critical Zone 257–68

Dahlstrøm Johnsgaard, Magnus 3
Dante 147, 150
Dass, Petter 131, 180, 187, 189–90
deep sea 51, 105–9, 110, 123, 129, 132–4, 224, 237, 244 257–68, 270, 271–4
deep time 50, 52, 54, 74, 77, 78, 259
deluge *see* flood
Dickinson, Emily 45–8
Dinesen, Isak *see* Blixen, Karen
diver, diving 239–41, 271–4
Drachmann, Holger 57–69, 74
Duun, Olav 3, 126–36

fishing, fishermen, fishwork 3–4, 76, 89, 123, 126–7, 129–31, 132–4, 152–64, 180, 188, 189, 190, 235, 239–40, 259, 262, 265, 271, 279
fjords 6, 89, 103, 107, 113, 116, 117, 118–21, 155, 156, 161, 173, 224, 259
flood 57–70, 73, 77, 79, 126–36, 235, 239–40, 278

gender (queerness, sexuality) 12, 79, 117, 123–4, 192–202, 172, 220, 225, 232, 258, 260–1, 263–7
Gerður Kristný 3, 258, 262, 263
Goldschmidt, Meïr Aaron 74
Greenland Sea 5, 156, 157, 159, 160, 257

Gulf of Bothnia 25
Gulf of Finland 1, 25, 33–42
Gulf of Florida 193, 194, 198–9
Gulf of Riga 25
Gundersen, Gunnar Ball 86, 91–4

Hallgrímsson, Jónas 264
Halse, Sven 57
Hamsun, Knut 127
Hansen Resen, Peder 22
Hartvedt, Gunnar Hagen 86, 94, 95
Heinesen, Jens Pauli 212
Heinesen, William 3, 203, 205, 212–13
Helgadóttir, Þórdís 258, 259–61, 262, 263
Høeg, Peter 165–75
Holberg, Ludvig 185
Hoydal, Gunnar 212
hydroqueer 258, 261, 264, 267

Ibsen, Henrik 113–25, 249, 251
Icelandic sagas 141–2, 152–64
Iceland Sea 5, 257
Idriss, Marjam 216–27
Indian Ocean 235
Ingvarsson, Haukur 257, 258, 263–7

Jacobsen, Jørgen-Frantz 203–13
Jakobsdóttir, Svava 260
Jansson, Tove 3, 33–42
Jensen, Carl-Jóhan 212, 213
Jensen, Peder Frederik 57, 58
Jessen, Erik Johan 188
Johnsgaard, Magnus 3, 86, 87, 90, 94, 95

Leem, Knud 188
London, Jack 246, 250, 251

Macfarlane, Robert 133, 134
MacKenzie, George Stuart 145, 148
Magnússon, Jón 184, 190
Månsson, Johan 3, 19–31
Maurer, Konrad 142, 144, 146, 149
Mediterranean Sea 2, 4, 45, 46, 201, 235
Melville, Herman 49
mermaids and other human-fish hybrids 115–17, 118, 120, 132, 269–79
Milton, John 180, 187, 265–6
Modern Breakthrough (*moderne gennembrud*) 57, 69
more-than-wet ontology 9, 45, 61
Munk, Jens 3, 30
Münster, Sebastian 141, 180, 182, 183

Nors, Dorthe 73, 75, 81–3
North Sea 3, 4, 5, 73–84, 86–96, 97–111, 128, 133, 143, 145, 188, 194, 201, 213, 235
Norwegian Sea 3, 5, 105, 116, 122, 123, 126–36, 145, 189, 257

Oehlenschläger, Adam 272
oil (petroleum) 3, 5, 11, 33–42, 86–96, 97–111, 127–8, 133–4, 167–9, 217–19, 223–6
Olaus Magnus 140, 143, 145, 179–90
Olsen, Einar Grønlie 86, 94, 95

Pacific Ocean 2, 219, 235
Pellinge Archipelago 37, 42
petroleum, petroculture *see* oil
Pfeiffer, Ida 142, 143, 144, 146, 147, 149
Poestion, Josef C. 142, 143, 144, 146, 149
Pontopiddan, Henrik 74

queer, queerness *see* gender

Richardt, Christian 57
Rimbereid, Øyvind 97–111
Rouquette, Louis-Frédéric 143, 144, 147, 149

sailing (sea travel, shipwrecks) 19–31, 49, 54, 59–60, 74, 91–4, 117, 121, 130, 139–48, 152–64, 179, 185, 192–202, 210, 216–18, 226, 232–41, 279
Schønnebøl, Erik Hansen 180, 182, 183, 184, 187, 189
sea monsters *see* troll whales
sea travel *see* sailing
Seeberg, Peter 73, 75, 79–81, 82, 83
Shakespeare, William 233
shipwreck *see* sailing
Skram, Amalie 3, 11, 192–202
Södergran, Edith 1–2
Sølberg, Sara 134
South China Sea 235, 237, 239
Spielhagen, Friedrich 57
Steenbuch, Hans *see* Jessen, Erik Johan
Stockholm Archipelago 24, 25, 48, 243
Storm, Theodor 74
storm surge *see* flood
Strindberg, August 3, 243–54
Strøksnes, Morten 134, 267

Tranströmer, Tomas 3, 45–55
troll whales and other magical sea creatures 140, 144, 152, 157–9, 160, 179–90, 259–67, 280
Wadden Sea 5, 11, 73–84, 270
waves 12, 13, 34, 48, 54, 67–8, 131, 144, 157, 159, 161, 234, 243–54, 272–4, 277